Advances in Networks, Security and Communications: Reviews

Volume 1

Sergey Y. Yurish
Editor

Advances in Networks, Security and Communications: Reviews

Volume 1

International Frequency Sensor Association Publishing

Sergey Y. Yurish
Editor

Advances in Networks, Security and Communications: Reviews
Volume 1

Published by International Frequency Sensor Association (IFSA) Publishing, S. L., 2017
E-mail (for print book orders and customer service enquires): ifsa.books@sensorsportal.com

Visit our Home Page on http://www.sensorsportal.com

ISBN: 978-84-697-8994-0
e-ISBN: 978-84-697-8993-3
BN-20171229-XX
BIC: TJK

Acknowledgments

As Editor I would like to express my undying gratitude to all authors, editorial staff, reviewers and others who actively participated in this book. We want also to express our gratitude to all their families, friends and colleagues for their help and understanding.

Contents

Contents 7
Contributors 15
Preface 19

Networks **21**

1. WSN Routing Improvement Techniques **23**
  1.1. Introduction 23
  1.2. Architecture of a Wireless Sensor Node 24
    1.2.1. The Sensing Unit 24
    1.2.2. The Processing Unit 25
    1.2.3. The Wireless Transceiver Unit 25
    1.2.4. The Power Unit 25
  1.3. The Offer of Sensor Manufacturers 26
    1.3.1. The Tmote Sky of Moteiv 26
    1.3.2. Sun SPOT from Sun Microsystems 27
    1.3.3. The WiEye of EasySen 28
    1.3.4. The Micaz Mote of Crossbow 28
    1.3.5. Jennic Sensors 29
    1.3.6. Other Commercial Applications 30
  1.4. Wireless Sensor Network Architecture 32
    1.4.1. Wireless Sensor Network Architecture 32
    1.4.2. Energy Consumption Factors in WSNs 33
      1.4.2.1. Collisions..*33*
      1.4.2.2. Idle Listening..*33*
      1.4.2.3. Overhearing..*33*
      1.4.2.4. Overmitting...*34*
      1.4.2.5. Control Packets Overhead...*34*
      1.4.2.6. Network Protocols..*34*
      1.4.2.7. Attacks and Security Issues..*34*
      1.4.2.8. Cryptography and Security Solutions.......................................*34*
  1.5. Application Domains of Sensor Networks 35
    1.5.1. Military Applications 35
    1.5.2. Security and Safety Related Applications 35
    1.5.3. Medical Applications 35
    1.5.4. Environmental Applications 36
    1.5.5. Commercial Applications 36
    1.5.6. Cold Chain Monitoring System 36
      1.5.6.1. Issues and Challenges...*36*
      1.5.6.2. Needs and Operating Principles...*38*
  1.6. Routing Mechanisms for WSN Cold Chain Monitoring System 40
    1.6.1. Introduction 40
    1.6.2. Backgrounds and Related Works 42
    1.6.3. Routes Selection Criteria 43
      1.6.3.1. Remaining Energy Level...*43*
      1.6.3.2. Sensor Proximity with Respect to the Base Station (Proximity-BS)............*43*
      1.6.3.3. Degree of Connectivity...*45*
      1.6.3.4. LQI: Link Quality Indicator..*45*

1.6.3.5. Composite or Hybrid Metric 46
1.6.4. Routing Mechanisms 46
 1.6.4.1. Simple Routing 46
 1.6.4.2. Round Robin Routing 47
 1.6.4.3. Weighted Round-Robin Routing (W2R routing) 48
1.7. L2RP: The Link Reliability Based Routing Protocol 50
1.8. Routing Protocol Performance Criteria 51
 1.8.1. Average Ratio of the Remaining Energy 51
 1.8.2. Average Path Lengths 52
 1.8.3. LIF: Load Imbalance Factor 52
 1.8.4. Network Lifetime 52
 1.8.5. Average Percentage of Lost Packets 52
1.9. L2RP Routing Protocol Performance Evaluation Model 53
 1.9.1. Energy Consumption Model 53
 1.9.2. Network Deployment and Performance Evaluation Parameters 53
 1.9.3. LQI Model for Performance Evaluation Purposes 55
1.10. L2RP Routing Protocol Performance Evaluation 56
 1.10.1. Average Path Length 56
 1.10.2. LIF: Load Imbalance Factor 59
 1.10.3. Average Percentage of Packet Losses 62
 1.10.4. Composite or Hybrid Metric 63
 1.10.5. Average Network Lifetime 65
 1.10.6. Average Ratio of the Remaining Energy 67
 1.10.7. Impacts of Increasing the Number of Achtophorous Nodes 68
 1.10.8. Impacts of the Unreliability of Wireless Links 69
1.11. Energy Over-Consumption Induced by Securing Routing Operations 72
 1.11.1. Introduction and Backgrounds 72
 1.11.2. Energy Over-Consumption Induced by Adding an Integrity Key to L2RP 73
1.12. Conclusions 81
Acknowledgements 83
References 83

2. Connectivity Recovery and Augmentation in Wireless Ad Hoc Networks 89
2.1. Introduction 89
2.2. Articulation Nodes and Bridges 91
 2.2.1. Definitions and Terms 91
 2.2.2. Location of Articulation and Bridges 92
 2.2.3. Locating Articulations and Bridges in Wireless Ad Hoc Networks 93
2.3. Cooperative Communication 102
2.4. Connectivity Recovery 104
2.5. Connectivity Augmentation 110
2.6. Conclusion 112
References 113

3. QoS Routing in Ad Hoc Network 115
3.1. Introduction 115
3.2. Routing in Ad Hoc Network 115
3.3. Usual Metrics 117
 3.3.1. Hop-Count Metric 117
 3.3.2. Delay-Based Metrics 118
 3.3.3. ETX Metric 118
 3.3.4. BER-Based Metric 119
 3.3.5. Retransmission-Based Metric 121
3.4. Taking Into Account the Quality of Links in the Choice of Route 122

3.5. Conclusion..124
Acknowledgements 125
References 125

4. A Knowledge-Based Modeling and Simulation Approach for the Management of Sensor Networks 129
    4.1. Introduction 129
    4.2. Related Work..130
        4.2.1. Modelling and Simulation Tools for Sensor Networks 130
        4.2.2. Environment Representation 131
        4.2.3. Spatial Behaviours and Knowledge Management 132
    4.3. Generation of Informed Virtual Geographic Environments.....................133
        4.3.1. GIS Input Data Selection...133
        4.3.2. Spatial Decomposition 134
        4.3.3. Map Unification 134
        4.3.4. Informed Topologic Graph 134
    4.4. From Semantic Information to Environment Knowledge...........................135
        4.4.1. Environment Knowledge 136
        4.4.2. Inference Engine 137
        4.4.3. Mapping Knowledge Using Environment Knowledge and Inference Engine 138
    4.5. From Environment Knowledge to Spatial Behaviors 139
        4.5.1. Agent Archetypes...139
        4.5.2. Action Archetypes..141
    4.6. A Knowledge-Based Approach 141
        4.6.1. Agent-Based Simulation Using Informed Virtual Geographic Environments....142
        4.6.2. Spatio-Temporal Knowledge...142
            4.6.2.1. Representation Formalism..142
            4.6.2.2. Knowledge Categories..143
    4.7. Case Study: Simulation of a Sensor Network for Weather Monitoring.......146
    4.8. Discussion 148
    4.9. Conclusion and Future Perspectives 149
    Acknowledgements 149
    References 150

Security 153

5. Reference Monitor-based Security Framework for Trust in Mobile Agent Computing 155
    5.1. Introduction 155
    5.2. Background Study 156
    5.3. Reference Monitor Based Security Framework..159
    5.4. Integration and Evaluation of RM-Agent 161
        5.4.1. Distribution and Integration of RM-Agent into an AP 161
        5.4.2. Verifying Authenticity and Integrity of RM-Agent...........................162
            5.4.2.1. Authentication and Integrity Checking Protocol (AICP) 164
    5.5. Experiments and Results 167
    5.6. Conclusions and Future Work 167
    References 168

6. Risk Assessment Considering Configuration of Hybrid Cloud Computing 171
    6.1. Introduction 171
    6.2. Overview of Hybrid Cloud Computing Configuration 172
        6.2.1. Reference Model of Cloud Computing...172

6.2.2. Hybrid Cloud Computing 172
6.2.3. Related Work.. 174
6.3. Risk Identification and Analysis: Qualitative Risk Analysis of Hybrid Cloud Computing
Configuration 175
6.3.1. Risk Identification: Extraction of Risk Factors in Hybrid Cloud Computing
Configuration.. 175
6.3.2. Risk Analysis: Qualitative Analysis of Risk Factors in Hybrid Cloud Computing
Configuration.. 175
6.3.2.1. Risk Analysis Method... 175
6.3.2.2. Risk Analysis Result from Qualitative Viewpoint......................... 176
6.3.2.3. Summary of Risk Analysis Results... 176
6.4. Risk Evaluation: Quantitative Risk Evaluation of Hybrid Cloud Computing Configuration......... 177
6.4.1. Risk Formula 177
6.4.1.1. Approximation of Asset Value... 179
6.4.1.2. Approximation of Threat Value.. 180
6.4.1.3. Approximation of Value of Vulnerability 180
6.4.2. Calculation of Risk Value.. 181
6.4.3. Discussion 181
6.5. Conclusion and Future Work 184
Acknowledgements 184
References 184

**7. The Mathematical Modeling of Road Transport in Context of Critical
Infrastructure Protection 187**
7.1. Introduction 187
7.2. Critical Infrastructure 188
7.2.1. The Cross-Cutting Criteria 188
7.2.2. The Sectoral Criteria.. 188
7.3. Critical Road Infrastructure 189
7.4. Mathematical Models of Road Transport 190
7.4.1. Macroscopic Model 191
7.4.2. Macroscopic Model 191
7.5. The Use of Dynamic Modeling of the Impacts of Road Transport 194
7.6. Conclusion 197
Acknowledgements 198
References 198

8. The Assessment of the Soft Targets 201
8.1. Introduction 201
8.2. Methodology of the Solving 203
8.2.1. Deming's Circle Methodology 203
8.2.2. Processes Management... 204
8.2.3. Crisis Escalation 204
8.2.4. Failure Mode and Effect Analysis 206
8.3. The Proposed Static Assessment.. 207
8.4. The Proposed Dynamic Assesment.. 209
8.5. The Proposal of the Preventive Actions 210
8.6. Conclusions 211
Acknowledgements 212
References 213

Communications **215**

9. Performance Analysis of Quadrature Amplitude Modulation Schemes in Amplify-and-Forward Relay Networks over Rayleigh Fading Channels **217**
9.1. Introduction 217
9.2. System Model 219
9.3. ASER Performance Analysis 221
    9.3.1. ASER of General Order RQAM for MRC Scheme 221
    9.3.2. ASER of General Order XQAM for MRC Scheme 223
    9.3.3. ASER of General Order XQAM for BRS Scheme 224
9.4. Numerical and Simulation Results 224
9.5. Conclusion 229
Acknowledgements 230
References 230

10. High-Gain Low-Cost Microstrip Antennas and Arrays Based on FR4 Epoxy **233**
10.1. Introduction 233
10.2. The Studied Methods 234
10.3. Parameter Study and Performance Comparison 235
10.4. Design Examples 239
10.5. Conclusions 242
References 242

11. Interference Management and System Optimisation for Femtocells Technology in LTE and Future 4G/5G Networks **247**
11.1. Introduction 247
    11.1.1. Motivation Towards Small Cells 248
    11.1.2. Challenges of Vehicular Environment 250
11.2. Related Work 251
    11.2.1. LTE Vehicular UEs Penetration Loss 252
    11.2.2. Vehicular UEs Performance 252
    11.2.3. Small Cells Interference 256
11.3. Mobile Femtocell Technology 259
11.4. Vehicular UEs Performance Analysis in LTE Networks 261
    11.4.1. System Model 261
    11.4.2. Results and Discussion 271
11.5. Interference Management for Co-Channel and Co-Tire Femtocells Technology 279
    11.5.1. Coverage Optimisation 280
    11.5.2. Transmission Power Control 283
    11.5.3. The Proposed Interference Management Scheme 284
    11.5.4. The Proposed Scheme (System Performance Analysis) 288
    11.5.5. Results and Discussion 291
11.6. Conclusions and Further Work 302
    11.6.1. Conclusion 302
        11.6.1.1. Performance Evaluation *302*
        11.6.1.2. Interference Management Evaluation *303*
    11.6.2. Future Work 304
References 305

12. The Buffer Delay Correction Algorithm for VoIP Communication **311**
12.1. Background 311
12.2. Theatrical Aspects of De-Jitter Buffer 314

11

12.3. Buffer Delay Correction Algorithm (BDCA) 317
12.4. Implementation and Results 319
12.5. Conclusion 321
References 323

13. Opportunistic Max²-Degree Network Coding for Wireless Data Broadcasting 325
13.1. Introduction 325
    13.1.1. Motivation and Related Work 325
        13.1.1.1. WDNC Based WBC 326
        13.1.1.2. IDNC Based WBC 326
    13.1.2. Contributions 326
13.2. System Model and Parameters 327
    13.2.1. Mathematical Description 327
        13.2.1.1. Signal Transmission 327
        13.2.1.2. Signal Reception 328
    13.2.2. Assumptions 328
13.3. The Proposed OM²DNC Based WBC Protocol 329
    13.3.1. OM²DNC Strategy 331
    13.3.2. Joint Network-RSC Decoding 333
13.4. Performance Analysis 334
13.5. Simulation Results 335
13.6. Conclusions 340
References 340

14. Analytical Model for Vehicular Mobility: A Microscopic Approach 343
14.1. Introduction 343
14.2. Communication Patterns 344
14.3. Information Dissemination 344
14.4. Vehicular Mobility Models 345
14.5. Network Architecture 345
14.6. Microscopic Analysis 347
    14.6.1. Joint Poisson Spatial Distribution 347
        14.6.1.1. Two-Lane Highway 347
        14.6.1.2. K-Lane Highway 352
    14.6.2. Conditional Probability of Number of Vehicles 354
        14.6.2.1. Two-Lane Highway 354
        14.6.2.2. Three-Lane Highway 354
    14.6.3. Conditional Expected Number of Vehicles 355
        14.6.3.1. Two-Lane Highway 355
        14.6.3.2. Three-Lane Highway 356
        14.6.3.3. K-Lane Highway 358
    14.6.4. Tail Probability of Number of Vehicles 358
        14.6.4.1. Two-Lane Highway 358
        14.6.4.2. Three-Lane Highway 359
14.7. Conclusions and Future Work 359
References 360

15. USRP-based Implementations of Various Scenarios for Spectrum Sensing 363
15.1. Introduction 363
15.2. Theoretical Aspects 364
    15.2.1. Energy Detection 364
        15.2.1.1. System Model 364

15.2.1.2. Conventional Energy Detector...*365*

15.2.1.3. Energy Detection with Uncertainty..*368*

15.2.2. Eigenvalue Based Detection...370

15.2.2.1. Maximum-Minimum Eigenvalue (MME) Detection Method.....................*371*

15.2.2.2. Energy with Minimum Eigenvalue (EME) Detection Method..................*372*

15.2.2.3. Cooperative Spectrum Sensing *372*

15.2.2.4. Soft Data Fusion...*373*

15.2.2.5. Hard Decision Fusion..*374*

15.3. Experimental Setup 375

15.3.1. GNU Radio...375

15.3.2. GNU Radio Blocks...375

15.3.3. Results..376

15.3.3.1. Simulation Results..*376*

15.3.3.2. Experimental Results for Single Energy Detector..................................*378*

15.3.3.3. Eigenvalue Based Spectrum Sensing...*380*

15.3.3.4. Cooperative Spectrum Sensing Results *381*

15.4. Conclusions 387

Acknowledgements 387

References 387

Index **389**

Contributors

Ulisses Rodrigues Afonseca
Instituto Federal de Goiás - Campus Luziânia, Luziânia-GO, Brazil,

Mahdi Aiash
Faculty of Science & Technology, Middlesex University, London, United Kingdom

Sandhya Armoogum
Dept. Industrial Systems & Engineering, University of Technology, Mauritius

Vimal Bhatia
Discipline of Electrical Engineering, Indian Institute of Technology Indore, India,

Jacir Luiz Bordim
Department of Computer Science, University of Brasília - UnB, Brasília-DF, Brazil,

Gean D. Breda
Faculty of Electrical and Computing Engineering, University of Campinas, Brazil

Robert H. Colson
Faculty of Science & Technology, Middlesex University, London, United Kingdom

Cherif Diallo
Laboratoire Algèbre, Cryptographie, Codes et Applications (ACCA),
UFR Sciences Appliquées et de Technologies (UFR SAT), Université Gaston Berger,
Saint-Louis, Sénégal

Dharmendra Dixit
Department of Electronics and Communication Engineering, The LNM IIT Jaipur, India,

Lucia Duricova
Tomas Bata University in Zlin, Faculty of applied informatics, Zlín, Czech Republic

Babak Honarbakhsh
Electrical Engineering Department, Shahid Beheshti University, Tehran, Iran

Robinson S. V. Hoto
Mathematics Department, University of Londrina, Brazil

Martin Hromada
Tomas Bata University in Zlin, Faculty of applied informatics, Zlín, Czech Republic

Motoi Iwashita
Chiba Institute of Technology, Japan

Atsushi Kanai
   Hosei University, Japan

Tsutomu Konosu
   Chiba Institute of Technology, Japan

Nagendra Kumar
   Discipline of Electrical Engineering, Indian Institute of Technology Indore, India,

Aboubaker Lasebae
   Faculty of Science & Technology, Middlesex University, London, United Kingdom

Jonathan Loo
   Faculty of Science & Technology, Middlesex University, London, United Kingdom

Galia Marinova
   Faculty of Telecommunications, Technical University, Sofia, Bulgaria

Shinsuke Matsui
   Chiba Institute of Technology, Japan

Mehdi Mekni
   Department of Computer Science and Information Technology, St. Cloud State University, St. Cloud, MN, USA

Leonardo S. Mendes
   Faculty of Electrical and Computing Engineering, University of Campinas, Brazil

Nawaz Mohamudally
   Dept. Industrial Systems & Engineering, University of Technology, Mauritius

Jan Mrazek
   Tomas Bata University in Zlin, Faculty of applied informatics, Zlín, Czech Republic

Amor Nafkha
   Centrale-Supelec SCEE, Avenue de la Boulaie 35576, Rennes, France

Nimal Nissanke
   Dept. Industrial Systems & Engineering, University of Technology, Mauritius

Ali Raheem
   Faculty of Science & Technology, Middlesex University, London, United Kingdom

Rand Raheem
   Faculty of Science & Technology, Middlesex University, London, United Kingdom

Amir Masoud Rahmani
   Department of Computer Engineering, Shahr-e-Qods branch, Islamic Azad University, Tehran, Iran

Camila Feitosa Rêgo
Department of Computer Science, University of Brasília - UnB, Brasília-DF, Brazil,

Fábio Sakuray
Computer Science Department, University of Londrina, Brazil

Hiroyuki Sato
The University of Tokyo, Japan

Yoshiaki Seki
Tokyo City University, Japan

Shigeaki Tanimoto
Chiba Institute of Technology, Japan

Zdravka Tchobanova
Faculty of Telecommunications, Technical University, Sofia, Bulgaria,

Jian Wang
Army Engineering University of PLA, China

Wei Xie
Army Engineering University of PLA, China

Kui Xu
Army Engineering University of PLA, China

Tiguiane Yélémou
Polytechnic University of Bobo-Dioulasso, BP 1091, Bobo-Dioulasso, Burkina Faso,

Mani Zarei
Department of Computer Engineering, Shahr-e-Qods branch, Islamic Azad University, Tehran, Iran

Dongmei Zhang Army Engineering University of PLA, China

Preface

Networks are everywhere, from the Internet to sensor networks. They are used widely for various type of communications. Their safe existence and functioning need deep consideration of different security aspects. It was the reason to combine these three topics such as Networks, Communication and Security in one book volume.

Every research and development must be started from a state-of-the-art review, and networks, security and communication are not an exception. The review is one of the most labor- and time-consuming parts of research. It is strongly necessary to take into account and reflect in the review the current stage of development, including existing technologies, security aspects and hardware. Many PhD students and researchers working in the same area are doing the same type of work. A researcher must find appropriate references, to read it and make a critical analysis to determine what was done well before and what was not solved till now, and determine and formulate his future scientific aim and objectives. The professionally made state-of-the art must take into account not only open access books, articles and conference papers available online for free, but also traditional monographs and journal articles, which are available only off-line and in print (paper) format.

After very successful publication of our popular Book Series on *"Advances"* from different technical areas in 2012-2017, it was decided to start publication of new Book Series on *'Advances in Networks, Security and Communications: Reviews'*.

It is my great pleasure to present the 1st volume of Book Series titled *'Advances in Networks, Security and Communications: Reviews'* started by the IFSA Publishing in 2017. The volume is published as an Open Access Book in order to significantly increase the reach and impact of this Book Series, which also published in two formats: electronic (pdf) and print (paperback).

The 1st volume of new Book Series contains fifteen chapters submitted by 42 contributors from academia and industry from 13 countries: Brazil, Bulgatia, Burkina Faso, China, Czech Republic, France, India, Iran, Japan, Mauritius, Senegal, UK, and USA. The book is devided into three main parts: Networks (four chapters), Security (four chapters) and Communication (seven chapters). *'Advances in Networks, Security and Communications: Reviews'* provides focused coverage of these three main technologies. It explores practical solutions to a wide range of network, communication and security issues. Chapters are written by experts in the field and address the immediate and long-term challenges in the authors' respective areas of expertise. Coverage includes wireless sensor network routing improvement; connectivity recovery, augmentation and routing in wireless Ad Hoc networks; advanced modeling and simulation approach for the sensor networks management; security aspects for mobile agent and cloud computing; various communication aspects (algorithms, coding and hardware design) and others.

Every chapter of this book is independent and self-contained. Fifteen chapters of the book have the similar structure: introduction, state-of-the art, description of advances and list of well selected references, including books, journal articles, conference proceedings and web sites.

This book ensures that our readers will stay at the cutting edge of the field and get the right and effective start point and road map for the further researches and developments. By this way, they will be able to save more time for productive research activity and eliminate routine work.

This book will be a valuable tool in both learning how to design various networks, as well as a reference as an advance in readers' research careers.

Dr. Sergey Y. Yurish

Editor
IFSA Publishing   *Barcelona, Spain*

Networks

Chapter 1
WSN Routing Improvement Techniques

Dr. Chérif Diallo

1.1. Introduction

The topics in this chapter detail the structure of sensor devices which are utilized to form wireless sensor networks (WSNs) intended to be used for many applications in various fields. We will be particularly interested in improving the efficiency of routing mechanisms, as well as the impact of securing the routing process.

In a wireless sensor network (WSN), energy efficiency of routing protocols is of primary importance. Embedded with local load balancing mechanisms, the studied L2RP protocol is a Link Reliability based Routing Protocol which aims to help source nodes to exploit the potential capabilities of their respective neighbours. As it is a reliability-oriented protocol, L2RP discards unreliable links to avoid the substantial energy cost of packet losses. We will detail some performance evaluation results to show major efficiency benefits that stem from load balancing which helps in lengthening the network lifetime while minimizing packet losses. In WSN, the choice of a routing protocol and its key parameters depends on the nature of the application and on its primary mission. Lot of works addressed routing issues with more or less effectiveness, some of which pointed out the use of the link quality indicator (LQI) as a route selection criterion (metric). In a previous work, following an experimental study, we have shown, under some conditions, the inefficiency of the LQI based routing.

In this chapter, we propose through L2RP a simple way to improve reliability and efficiency of the LQI based routing in WSN. We also give a comparative study of several metrics including new definitions of LQI based metrics. The performance evaluation results show that our adaptation of the LQI metric is among the best route selection criteria regardless of the performance criterion under consideration [1-3]. Finally, as security has become critical for these networks too, we address in the last part the problem of the additional energy cost induced by solutions used to secure the routing process [4].

Dr. Chérif Diallo
Laboratoire Algèbre, Cryptographie, Codes et Applications (ACCA),
UFR Sciences Appliquées et de Technologies (UFR SAT), Université Gaston Berger, Saint-Louis, Sénégal

1.2. Architecture of a Wireless Sensor Node

A sensor node Fig. 1.1 is mainly composed by four basic units: the sensing unit, the processing unit, the wireless transceiver unit, and the power unit. It can also contain, in accordance with its domains of application, supplementary modules such as a Global Positioning System (GPS) or an energy generator system (solar cell or photovoltaic). There are also sensors, more voluminous, equipped with a mobility system responsible for moving the node in case of need.

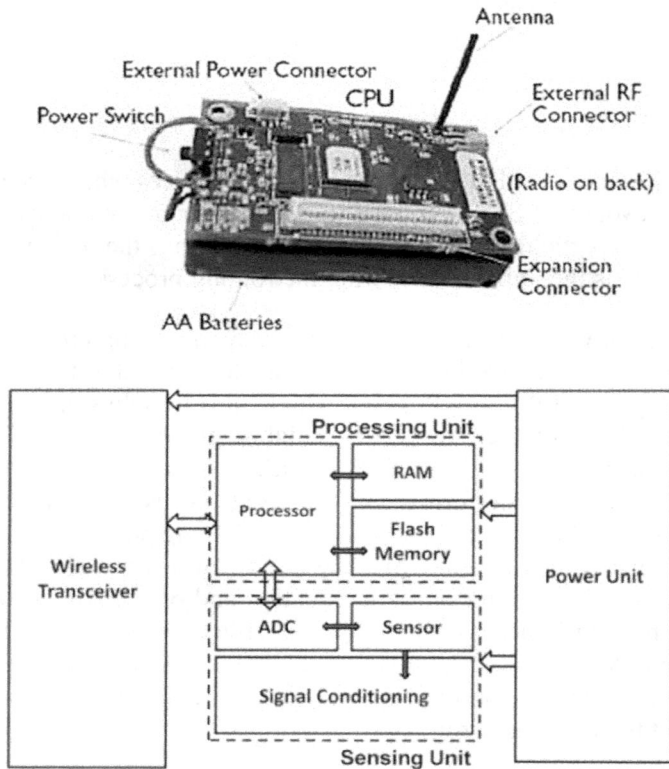

Fig. 1.1. A sensor node and its main units.

1.2.1. The Sensing Unit

The sensing unit is a device transforming the state of an observed physical quantity into a human known quantity such as electrical voltage, luminous intensity or temperature. It is distinguished from the measuring instrument by the fact that it is merely a simple interface between a physical process and manipulable information. The sensing unit is generally composed by two subunits: The sensor itself and the analog-digital converter (ADC) which is responsible for converting the original signal based on the observed phenomenon. ADC transforms these signals in digital signals understandable by the processing unit.

The sensor is characterized according to several criteria. The most common are: the observed physical quantity (light, temperature, noise, humidity, etc.), its transmission range, its sensibility and its functioning temperature range. To use a sensor probe in the best condition, it is often useful to practice a calibration in order to know the uncertainly of measures provided.

1.2.2. The Processing Unit

It includes a processor generally associated to a small storage unit. It works with an operating system specially designed for the sensor (for example: TinyOS). It executes the communication protocols which make it possible to send data to the other nodes of the network, to receive data, and to participate in many network operations such as auto-configuration, auto-organization, clustering process, routing computation, data aggregation and security mechanisms.

1.2.3. The Wireless Transceiver Unit

It carries out all the transmissions and receptions of data on a wireless media. It may be of the optical type (like in the smart dust node), or of radio-frequency type:

- The optical type communications can prevent from electrical interferences. Nevertheless, it's not possible to establish links through the obstacles. Such sensor devices have the disadvantage to requiring a permanent sight of view between the communicating entities.

- The radio-frequency unit transmission types include modulation circuits, demodulation, filtering and multiplexing. This implies an increase in the complexity and cost of the sensor manufacturing.

Designing radio-frequency transmission units to enable large communication range with low energy consumption capabilities is an exciting challenge because powerful signals often consume more energy. An alternative could consist generally to use long antennas, but it is not possible in the domain of wireless sensors because of the small size of the devices.

1.2.4. The Power Unit

A sensor device is provided with an energy source (generally a battery). Considering its small size, this energetic resource is limited and generally not replaceable. What often makes the energy the most precious resource of a sensor network, because it directly affects sensor nodes lifetime and so the entire lifetime of the network. Consequently, the energy control unit constitutes an essential system. It must distribute the available energy to the other modules, in an optimal way, for example by reducing the useless spending by putting components on (deep) sleep mode. This unit can also handle the energy loading systems from the environment via photovoltaic cells for instance.

1.3. The Offer of Sensor Manufacturers

The double protocol stack Bluetooth / ZigBee are among the best standards to be exploited in wireless sensor networks. Bluetooth is a specification of telecommunication industries. It uses a short-range radio technology in order to simplify the connexions between the electronic devices. Initiated in 1994 by Ericsson, The Bluetooth technique has been standardized under the standard IEEE 802.4.15 and it is destined to the creation and the maintain of Personal Area Network (PAN). Such a network is used for the transfer of low data bandwidth and low transmission range between compatible devices. Unfortunately, the big drawback of this technique is its high energy consumption and therefore can not be used by sensors that are powered by non-replaceable battery and that ideally should work without outage for several years.

The standard ZigBee [5] combined with the IEEE 802.15.4 offers features that respond even better to sensor networks needs in terms of energy network saving. ZigBee offers also lower data rates, but it also consumes significantly less than Bluetooth. A small rate of data is not handicapping for a sensor network where the transmission frequencies are not important. Nevertheless, the current tendency of the manufacturers is to use proprietary techniques which have the advantage of being specifically optimized for a specific use, but which have the big drawback of not being compatible between them.

From a hardware point of view, some new techniques will considerably influence the future of the wireless sensor networks. The UWB (Ultra Wide Band) technique is a good example for that. This transmission technique will allow reaching extremely low consumption levels thanks to its simplicity at materiel level.

Moreover, the attenuation of the generated signal by obstacles is lower compared to conventional narrow-band radio systems.

From a software point of view, the field of wireless sensors is therefore moving towards a great rise and many new products continually flooding the sensors market every year. All the more, open-source techniques are now largely used, such as TinyOS [6]. This last one was developed at the University of Berkeley. TinyOS [6] is an open-source operating system designed for wireless sensors and which is used by more than 500 Universities and academic research centers in the world. The realization of programs on the platform is done exclusively in NesC (dialect of the C programming language). The main particularity of this OS is its extremely small size in terms of memory (some kilobytes).

In the following, we review some main devices provided by few sensor manufacturers.

1.3.1. The Tmote Sky of Moteiv

Specifically designed for promoting a rapid deployment of prototypes, the Tmote Sky sensors of Moteiv [7] are adapted to many industrial applications in the fields of light, humidity and temperature control (Table 1.1). These sensors are often equipped with an integrated antenna covering 50 m indoor transmission range and 125 m outdoors. They are also accompanied by a USB interface to enable interoperability with other devices and peripherals.

Designed with the native support of TinyOS [6], the Tmote Sky sensors of Moteiv offer the advantage of developing innovative technologies related to wireless protocols while also benefiting from the open source software capabilities (Fig. 1.2).

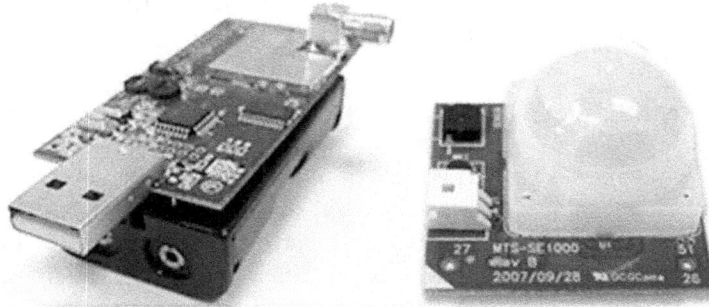

Fig. 1.2. Tmote Sky sensor.

1.3.2. Sun SPOT from Sun Microsystems

Sun SPOT [8] (Sun Small Programmable Object Technology) is a Mote based wireless sensor network developed by Sun Microsystems. This device is based on the IEEE 802.15.4 standard. Contrarily to the others available Mote systems, Sun-SPOT is make on the Java Squawk virtual machine.

The first limited series of Sun-SPOT development kits was launched since April 2007. These first introductive kits have two sensors, one base station, the software development tools and a USB wire. The software is compatible with Windows XP, Mac OS X 10.4, and most common Linux distributions. The SunSPOT sensor node is shown in Fig. 1.3.

Fig. 1.3. SunSPOT sensor node.

Another SunSPOT project is the hardware and software platform developed by Oracle Labs [8]. This project was created to encourage the development of new applications and devices. It is designed from the ground up to allow programmers who never before worked with embedded devices to think beyond the keyboard, mouse and screen and write programs that interact with each other. A Java programmer can use standard Java development tools such as NetBeans to write specific codes [9].

The whole project (hardware, operating environment, java virtual machine, drivers and applications) is available as open source materials.

1.3.3. The WiEye of EasySen

Developed by EasySen, WiEye [10] is a set of modular cards for wireless sensor network applications that meet the needs of professional surveillance and security. The device is based on the IEEE 802.15.4 standard and is particularly adapted to the detection of presence and movement of individuals and/or vehicles. Composed by several sensing units Fig. 1.4, the WiEye components are designed as complementary modules to be connected to the other sensor devices such as Tmote Sky for instance Fig. 1.4.

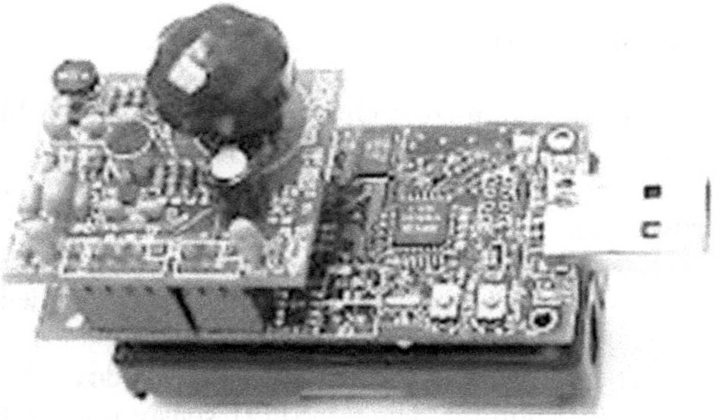

Fig. 1.4. Illustration of a WiEye sensor node.

1.3.4. The Micaz Mote of Crossbow

The Crossbow Company proposes a hardware system [11] including the process unit (processor and storage), the energy control unit and the transmission unit, which insure data sending to the base station. The Micaz Mote of Crossbow Fig. 1.5 is equipped with a connector allowing the integration of different sensing units.

The MICAz is a 2.4 GHz Mote module used for enabling low-power wireless sensor networks and it is intended for Indoor Building Monitoring and for Security purposes. It enables acoustic, video, vibration and other high speed sensor data applications (Table 1.1). Such devices can also operate in large scale wireless sensor networks including more than one thousand (1000+) nodes [12].

Fig. 1.5. Micaz mote sensors provided by Crossbow.

1.3.5. Jennic Sensors

Equipped with a 32-bit RISC processor, Jennic [13] solutions are designed for both home automation and industrial applications (Fig. 1.6). Characterized by their very low power consumption (Table 1.1), which ensures a long battery lifetime, they specifically target the areas of location, telemetry, remote control, toys and gaming peripherals.

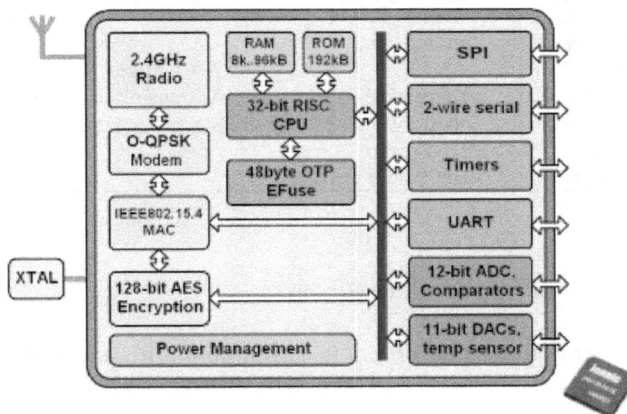

Fig. 1.6. Sensor development kit provided by Jennic.

Based on the well-established IEEE 802.15.4 standard and a 2.45GHz low power radio technology, it adds the self-healing JenNet tree networking stack, the IETF 6LoWPAN IP layer and "JIP", a powerful and flexible application layer enabling interoperability between devices. It is initially targeted at Lighting and Home Automation systems where it is rapidly becoming established as the system of choice, with provision for many different types of devices, all connecting to the same wireless network.

With this solution, every device in the home can have its own IPv6 address, turning the home into a simple extension of the internet and enabling new and exciting innovations to be created in home controls and energy management (IoT, Internet of Things).

The JenNet-IP-EK040 evaluation kit provides all the components needed to allow developers to create these new applications: domestic lighting control, home automation (heating and ventilation control, energy management, access control, blind and window control), building automation (HVAC, access control, security, fire detection and alarm), commercial and outdoor lighting, security systems, etc. [14, 15].

Table 1.1. Main features of some wireless sensors.

	Tmote SKY	Sun SPOT	Crossbow	Jennic
Processor	8 MHz Texas Instruments MSP430	180 MHz 32 bit ARM920T core	MICAz ZigBee Series (MPR2400)	Low power 32bit RISC CPU, 4 to 32MHz clock speed
Data capacity	10 Kb RAM, 48 Kb Flash	512 Kb RAM 4 Mb Flash	4 Kb RAM 512 Mb Flash	128 Kb RAM 128 Kb Flash
Wireless radio	2.4 GHz IEEE 802.15.4 Radio 250 kbps, High Data Rate Radio	2.4 GHz IEEE 802.15.4 Radio	2.4 GHz IEEE 802.15.4 Radio 250 kbps, High Data Rate Radio Outdoor transmission range 150 m	2.4 GHz IEEE 802.15.4 Radio Transmission power 2.5dbm
Sensing feature	Temperature, Humidity, Light	Accelerometer, Temperature, Light	Temperature, Humidity, Light, pressure, acoustic activity, video, seismic movements, magnetic	Temperature, Humidity, Light
Energy	2 AA batteries, 21-23 mA (reception) 19-21 mA (transmission) 5-21 μA (deep sleep mode)	750 mAh Lithium-ion, 3.7 V rechargeable battery 30 μA (deep sleep mode)	2 AA batteries 25 mA (transmission) 15 μA (deep sleep mode)	2 - 3.7 V battery 17.5 mA (reception) 15.0 mA (transmission) 1.25 μA (deep sleep mode)

1.3.6. Other Commercial Applications

There are several other commercial products available on the sensor market:

- ShockFish SA, now Spotme, offers the TinyNode devices Fig. 1.7. Composed of sensors and modular cards for industrial applications [16], this offer is accompanied by a range of products intended to academic research and is also compatible with TinyOS [6].

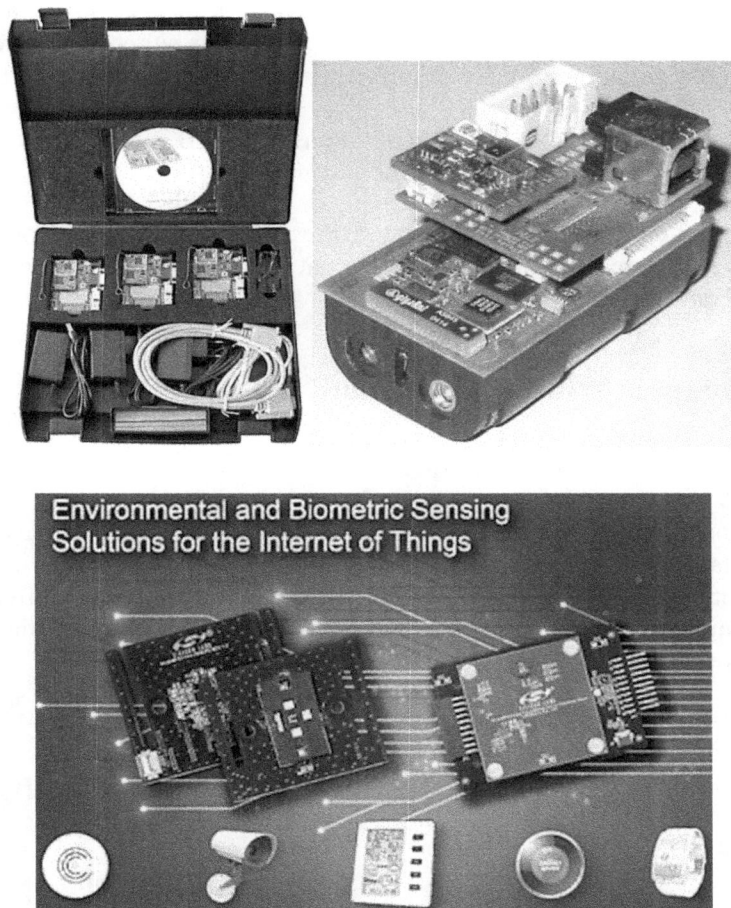

Fig. 1.7. ShockFish, BTnode and Silicon Labs products.

- The T-Node products, G-Node and L-Node [17] from SOWNet Technologies are few responses to the several specific needs of academic research and sensor network application developers. This Dutch company offers several TinyOS compatible development kits, allowing easy integration with open source products. It also offers a monitoring application for museums and art objects, which uses remotely managed sensors to provide discrete (invisible) and individual monitoring of each art piece in a museum.

- The Tinynode 584 is an ultra-low power OEM module that provides a simple and reliable way to add wireless communication to sensors, actuators and controllers. The Tinynode evaluation kit contains everything you need to get started. The kit ships with a full TinyOS environment along with demo applications and protocol software [18].

- The BTnode rev3 provides a distributed environment for prototyping Ad Hoc networks and supports different sensor add-ons using the IO provided on the extension connector J1 and J2. There are different UART, SPI, I2C, GPIO, ADC available. The J2 is suitable for prototyping, as you don't need to make a PCB or solder anything. Just connect you sensor(s) to appropriate IO lines. The J1 is suited if you need more IO lines and/or are going to produce sensor boards that can just be plugged in [19].

- Silicon Labs [20] and Microchip [21] solutions are ZigBee-compatible offerings for both industrial applications and academic research.

As we can see, given the enthusiasm of researchers for wireless sensors, the number of manufacturers has increased enormously in recent years to provide concrete answers to both industrial and academic needs.

1.4. Wireless Sensor Network Architecture

1.4.1. Wireless Sensor Network Architecture

A Wireless Sensor Network (WSN) Fig. 1.8 consists of a set of nodes deployed in a given environment (target) for accurate application (temperature, pressure, control, monitoring, intrusion detection, air humidity, agriculture, home automation, medical field, etc.). The sensors communicate with each other to relay information to a Sink node that communicates with the user interface.

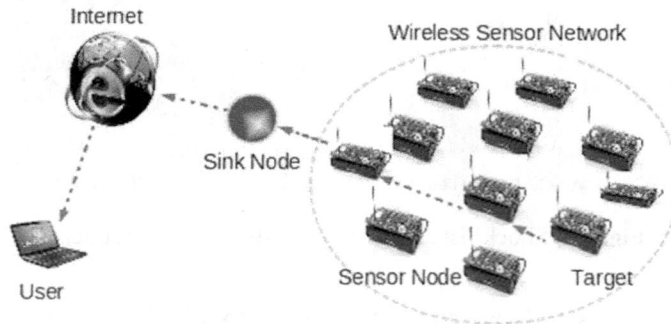

Fig. 1.8. Illustration of a wireless sensor network [22].

A WSN is therefore composed of a number of autonomous sensors capable of collecting and collaborating together to transmit environmental data. The position of these sensors is not necessarily predetermined. They may be randomly dispersed in a geographic area, known as coverage area or target area.

The data collected by the sensors are routed via multi-hops routing to a node considered as a collection point, called a sink node. From the processing center, the user connects to the Sink using Internet infrastructure or a satellite communication system or other telecommunications systems. The user can, via the Sink, send requests to the other sensors of the network specifying the type of data required and collect and then analyze the events recorded by the sensors in the coverage area.

Wireless sensor networks are designed to collect information related to some events in hostile environments where human presence is not often possible. That is why we consider that once they are deployed, the sensors are autonomous. Their lifetime tightly rely on the manner of their batteries are utilized during network operations. Then, the most critical feature in sensor networks is the hardship of its energy resources because each sensor node has initial low resources in terms of energy and it's not often possible to replace batteries during network operations. That is why it is important to take into account the energy consumption factors when designing wireless sensor networks.

1.4.2. Energy Consumption Factors in WSNs

1.4.2.1. Collisions

Collisions are the primary source of energy loss. When two frames are issued simultaneously and then collide, those become unusable and must be retransmitted. In the same manner, when a node simultaneously receives two packets which have not necessarily been sent at the same time or which have been transmitted by two nodes out of reach one from the other. These two cases require retransmission of frames. MAC protocols, like CSMA, are trying in their own way to avoid collisions in order to backup network performance by reducing packet retransmission.

1.4.2.2. Idle Listening

"Idle listening" or "empty Listening" occurs when listening to the radio channel to receive contingently any data. Its energy cost is particularly high. Many MAC protocols always listen to the channel to the active state, assuming the entire device would be turned off by the user if there is no data to send. The exact cost of the idle listening depends on the hardware radio and on the operation mode. Turning off the radio channel is a solution; nevertheless the cost of the transition between different modes (sleep/deep/active) also consumes energy. Then the frequency of transition between modes must be reasonable when designing protocols.

1.4.2.3. Overhearing

Another common cause of excessive energy consumption is "overhearing". This occurs when a node hears packets that are not destined for it. The cost of "overhearing" can be an aggravating factor of energy losses especially when traffic or node density is notably high during network operations.

1.4.2.4. Overmitting

Overmitting occurs when a source node sends data to a destination node which is not ready to receive. Moreover, if the transmitting node is waiting for an acknowledgment (ACK) from the destination node, it then retransmits the same packet several times because it remains in transmission mode (pending receipt of the eventual ACK frame). This could lead to high energy losses.

1.4.2.5. Control Packets Overhead

Some media access control techniques use specific packets (RTS, CTS) for collision avoidance. Running a network protocol still needs few control packets and necessary headers to encapsulate application data. Sending, receiving, and listening to the control packets consume more energy. As control packets do not carry user data directly, those also reduce the overall effective throughput of the network.

1.4.2.6. Network Protocols

The most protocols designed for network auto-configuration and/or organization, such as routing and clustering algorithms, are also energy consumption factors. Indeed, routing and clustering schemes often involve various message exchanges between sensors before a node processes its routing table or data aggregation operations.

1.4.2.7. Attacks and Security Issues

Many attacks are carried out against the sensor networks in order to exhaust the batteries of the nodes and then to reduce their lifetimes, and consequently that of the network. For instance, with a very high power transmitter node, an attacker can perform Hello Flood attack by broadcasting Hello messages on a large range transmission to some sensors of the network. Then sensors will try to respond to the attacker node which is actually out of range. In such a case, an attacker uses this mechanism to cause energy exhaustion in the hacked network.

1.4.2.8. Cryptography and Security Solutions

Security mechanisms, including cryptography, are aggravating factors of energy consumption in sensor networks. Many works are concerned with the establishment of cryptographic keys without worrying about key lengths. This is the strong point of cryptographic algorithms for wired networks. However, this becomes a major disadvantage in wireless sensor networks where the safeguarding of energy is a sine qua non condition for longer node network lifetime. Moreover, adding integrity key to secure routing protocol exchanges should increase packet size and therefore energy expenditure. In the Section 1.11, we will try to estimate how, that's to say in what proportion, adding an integrity key will cause energy exhaustion.

1.5. Application Domains of Sensor Networks

The fast evolution of wireless communication media in recent years, the miniaturization, and the low cost of sensor components have enable emergence of large innovative applications of sensor devices (thermal, humidity, optical, vibration, etc.). Sensor networks could also revolutionize the way we understand and build complex physical systems, especially in the military, environmental, domestic, health, security, and so on.

1.5.1. Military Applications

The military domain is, as in the most cases of several technologies, a precursor for the development of sensor network applications. The rapid deployment, the reduced cost, the self-organization and fault tolerance of sensor networks are characteristics that make this type of network an appreciable tool in such a field. A WSN deployed in a strategic sector which is often difficult to access, allows for example to monitor all movements (friends or enemies), or to analyze the ground (detection of chemical agents, biological or radiation) before sending troops.

1.5.2. Security and Safety Related Applications

Sensor networks enable real-time monitoring of aircraft, ship, automotive or metro structures, as well as electrical and gas supply and distribution networks. A WSN designed for movement or noise detections can constitute an intrusion detection system in an industrial site or a critical sector. With WSN, it would not be so simple to defeat such a system, since there is no critical point. Prevention of accidents by monitoring roads or railways is one of the envisaged applications of wireless sensor networks.

The evolution of seismic movements, alterations in the structure of a building, a road, a wharf, a railway, a bridge or a hydroelectric dam (following an earthquake or aging) could be detected by sensors. Some sensors that activate only periodically can work for years or even decades.

Another potential area of sensor application use is safety. Experts agree that terrorist threats in the future will be exerted in particular through maritime transport. The cargo containers could thus be monitored and located with sensor devices. The presence of explosives, conventional or nuclear, or even human beings in these containers could be also detected. Consequently, it becomes possible to avoid potential terrorist attacks on cargo ships (90 % of the world's trade passes through the ocean and 200 million containers pass through each year).

1.5.3. Medical Applications

Monitoring the vital functions of a living organism can be facilitated by small sensors swallowed or implanted under the skin. Capsules in the form of sensors or small cameras that can be swallowed already exist and allow, without using surgery, to transmit images from the inside of a human body. A recent study presents sensors that can function in the human body for the treatment of certain diseases. The monitoring of blood glucose, the monitoring of vital organs or the early detection of certain cancers are intended biomedical applications. Sensor networks would also allow for real-time monitoring of motherhood (detection of babies) or patients.

1.5.4. Environmental Applications

The detection of movements of birds, small animals or insects is one of the possible environmental applications of wireless sensor networks. Thus, it becomes easier to observe the bio-diversity, without disturbing vulnerable animal species, and to propose more effective solutions for the conservation of the fauna and the flora. Thanks to sensor networks, it is also possible to detect environmental problems such as pollution, forest fires and the possibility of assessing the rate of pesticides in water. On industrial sites, such as nuclear or oil tankers, sensors can be deployed in a network to detect leaks of toxic products (gas, chemicals, radioactive elements, petroleum, etc.) in real time in order to alert the users and to enable more rapidly effective intervention.

1.5.5. Commercial Applications

In the field of trading, sensor networks could be used to improve the storage and the delivery process (in particular, as a cold chain monitoring system). Thus, the created network may be used to determine the position, condition and direction of a package, pallet or cargo. Manufacturers, via sensor networks, could follow the production process from the raw material to the delivered product.

In buildings, the home automation system of heating and air conditioning, lighting or water distribution could optimize its efficiency thanks to small sensors. Used on a large scale, such applications would help in reducing global energy demand and indirectly emissions of greenhouse effect gas. For USA, this economy is estimated at around $55 billion per year, with a decrease of $35 million tons of carbon emissions [23].

1.5.6. Cold Chain Monitoring System

1.5.6.1. Issues and Challenges

The cold chain monitoring domain and, more generally, the monitoring of the storage temperature of foodstuffs is becoming increasingly important in the economies of industrialized countries. It concerns frozen products, fishery products, fruit and vegetables but also, and increasingly, pharmaceutical products, vaccines and other biotechnology products. At the same time, the health problems caused by non-compliance with storage conditions are increasing. According to a report published in 1999, the non-respect of the compliance conditions of the cold chain would cause in the United States: 76 million poisonings; 325,000 hospitalizations; and 5,000 deaths per year. That is why the European Community has responded to this social problem by introducing new instructions on the transport of frozen products (Directive 92/1/EEC) requiring the automatic recording of air temperature in the various stages of the transport (vehicle and warehouse), using fixe devices. However, these devices provide only partial response to the problem because the ambient storage temperature is not necessarily the temperature of the monitored product, and they do not allow control during the loading / unloading phases, where the non-compliance incidents are more likely to be happening Fig. 1.9.

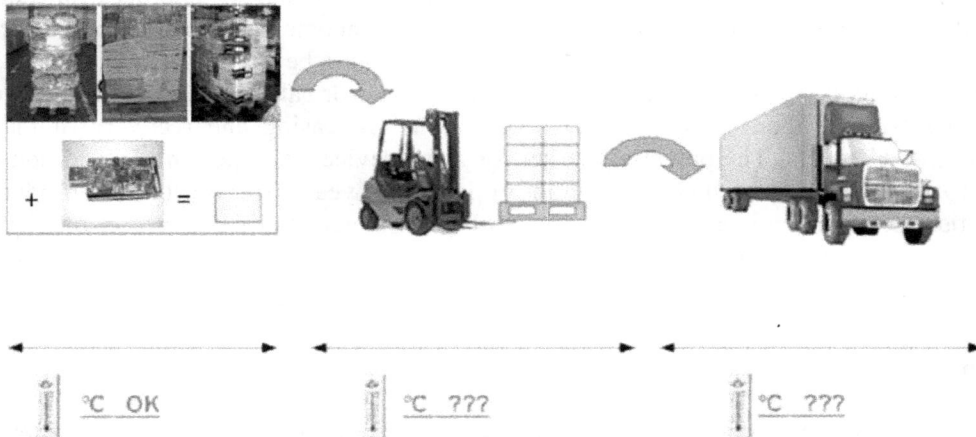

Fig. 1.9. Issues and challenges of a cold chain monitoring system.

Some solutions have been proposed to measure the temperature of the product, or at least its packaging, and throughout the transfer period between the place of manufacturer and the retail outlets. For the monitoring of positive temperatures (storage of fruits and vegetables for example), chemical indicators have been proposed which change their appearance irreversibly when a threshold is exceeded. The transposition of these devices to negative temperatures comes up against many difficulties. The only product of this type that we know consists of labels that become invisible when the cold chain has been broken for a predetermined time. In addition to a delicate implementation, since the labels once manufactured must always be kept below the threshold temperature, this system does not in practice allow intermediate control of the products during their routing. Indeed, no actor in the cold chain can afford to visualize each label when he receives a product and before sending it to the next actor. Thus, only the end user has the knowledge of whether or not the cold chain is respected, which raises the problem of the precise designation of the company responsible for the rupture.

Consequently, the actors in the chain have currently measurements in different regions of a warehouse or in a truck. The consequence of the detection of an incident in a part of the warehouse or in the truck is the destruction of products located in that region or in that truck. For one-off measurement campaigns to qualify a chain, there are portable recorders placed in or near packaging, but these are relatively bulky and costly and are not designed for systematic and extensive use ladder. Indeed, the systematic exploitation of portable temperature recorders encounters a number of problems and constraints. Such solutions are not suitable for intermediate checks because they have to be connected to a portable reader by a wired link.

This means that these recorders must be located. The cold chain monitoring system implemented by wireless sensor networks aims to solve these different problems. Since recording sensors can communicate wirelessly via a network of sensors, it becomes possible to interrogate a large number of them in a warehouse from a single reading station without having to locate them. Since multiple reading points having Internet connection could be positioned at different storage locations, it becomes possible from a central station to follow the entire logistic chain of the goods and to be warned of any problem in real time. Finally, integrating sensors with transport bins or pallets solves the problem of logistics because there is already a whole activity of leasing and recovery of these packaging elements. The potential market for these devices is important. In addition to the frozen food market, which has millions of pallets per year in the world, fresh products and fisheries, applications in other areas can be considered. For the transport of fragile products, they may be equipped with shock or moisture sensors instead of temperature transducers.

1.5.6.2. Needs and Operating Principles

As part of a cold chain monitoring application, each sensor is associated with a pallet: it records the temperatures as well as it transmits information. The sensors communicate with each other, and then with the base station which sends or receives data from the central server. In addition, an operator can communicate directly with a specific sensor, using a portable terminal, in order to initialize it, identifying, and retrieving, recording or modifying stored information. The data transmitted to the server are then stored in a database. All phases of supply chain is shown in Fig. 1.10.

- *Phase 1:* The operator targets the sensor attached to a pallet with a portable terminal in order to empty it of the previous information.

- *Phase 2:* The operator places the goods on the pallet. The pallet thus created is associated with an order number and / or pallet reference in the WMS (WMS for Warehouse Management Software) of the supplier. The operator again targets the sensor in order to assign this number to it, then, depending on the nature of the products, he adjusts the temperature according to predetermined approach and exceedance temperature thresholds. Depending on the wishes of the supplier, the operator can send the product data (barcode product, or pallet contain, deadline for sale, etc.) to the sensor. Finally, the operator enables the sensor to operate in networking and then start recording the temperatures.

- *Phase 3:* The pallet is placed in stock at an identified location. If during this period the temperature level drops or increases so as to exceed one of the set thresholds, then the sensor sends an alert to a base station (phase 7).

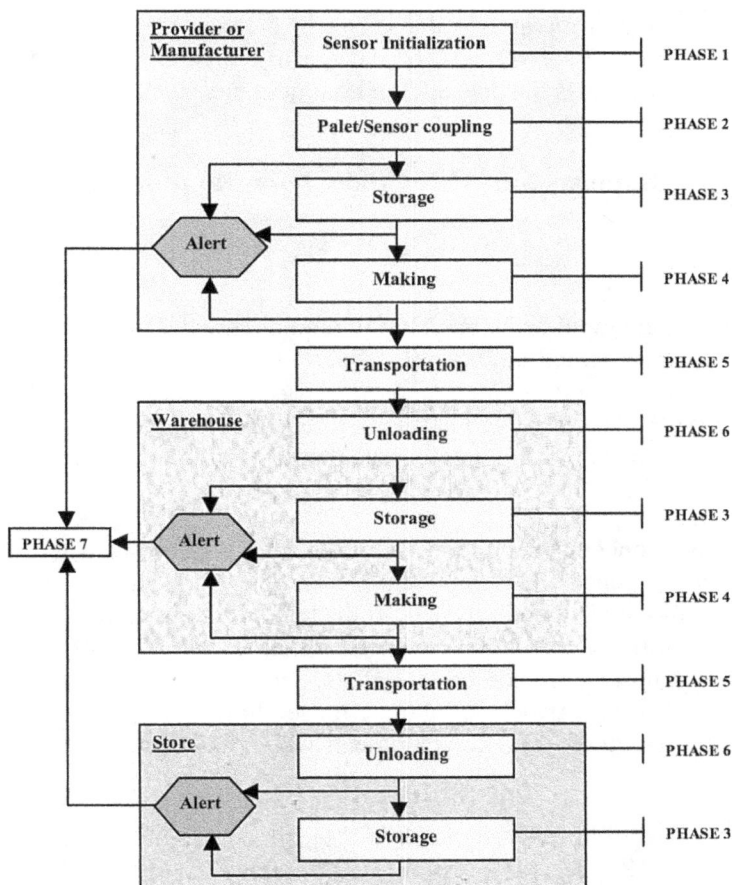

Fig. 1.10. This diagram shows operations carried out on a supply chain.

- *Phase 4:* The complete pallet is taken to the preparation area. If during this period the temperature level drops or increases so as to exceed one of the set thresholds, the sensor sends an alert to a base station (phase 7).

- *Phase 5:* The pallet is loaded into a truck and routed to the destination logistics platform. During this period, the sensor loses contact with the sender's base stations. If one of the temperature thresholds is reached, the sensor waits to reconnect to the next base station before sending an alert (except, of course, if the truck is equipped with a base station).

- *Phase 6:* The pallet is unloaded. If there has been a temperature incident during transportation, the sensor sends the alert as soon as it reconnects to a base station (phase 7). Similarly, if during the unloading phase and waiting for stocking the level temperature decreases or increases so as to exceed one of the thresholds, the sensor sends an alert to a base station (phase 7).

39

- *Phase 7:* The operator recognizes the sensor of the pallet that issued the alert thanks to its portable terminal. It verifies the data recorded by the sensor and controls the pallet. Then, only it takes the appropriate treatment measures.

1.6. Routing Mechanisms for WSN Cold Chain Monitoring System

1.6.1. Introduction

Designing a cold chain monitoring application requires special focus on at least two main phases. In [24], we presented an example of sensor network for cold chain monitoring where sensors are inside pallets. We proposed energy efficient protocols for the transport phase in which the WSN is deployed in trucks with no Base Station (BS) because it would be very expensive to install and maintain Base Stations within each truck. There are a few sensors in the truck.

The second phase concerns the product storage in a warehouse where each pallet is handling temperature sensor. This application specifically collects rare events (alarms) to ensure the proper monitoring of the system. If the temperature is over a threshold, an alarm will be generated; this "interesting event" is then sent towards the BS. Due to the size of a warehouse which hosts large number of pallets, one upon the other, the WSN can reach several hundreds of sensors which collaborate for sending data towards the BS. So, in this environment, the link quality is a key parameter which has many effects on the network performance.

In [25], we used up to 50 Moteiv Tmote Sky [7] sensors, in a small experimental platform, including a 2.4 GHz ZigBee [8, 26] wireless transceiver (chipcon's CC2420) [27]. On each packet reception, the CC2420 calculates the error rate, and produces a LQI value. To conduct experiments, we used the multiHopLQI routing algorithm [28] along with the Sensornet Protocol (SP) implementation [29]. In this algorithm, nodes sense and send "interesting events" to the BS. Based on the acknowledgement, a sensor decides to retransmit the data or not. If the acknowledgement fails, the sensor selects another node and routes data towards the BS. Under these conditions, the experimental results pointed out that the LQI based routing could have negative effects on the network performance [25].

After all, we think that the link quality might be a key parameter which some routing protocols could rely on in order to increase the network performance. Several works address WSN routing, but only few papers are related to the LQI based routing protocols. Sensors are characterized by their low energy level. Thereby load balancing traffic between different nodes, is also an essential idea to increase the lifetime of nodes and thus of the network. Our work addresses this challenge: improving the LQI based routing protocol by load balancing traffic over multiple paths.

The link quality indicator (LQI) is defined in the IEEE 802.15.4 standard [8, 26] as a measurement of the quality of packet reception between two nodes. The IEEE 802.15.4

standard does not specify the implementation of LQI, which is up to the radio manufacturer. Several works address WSN routing, but only few papers are related to LQI based routing protocols. Sensors are characterized by their low energy level. Thereby load balancing traffic between different nodes, is also an essential idea to increase the lifetime of nodes and thus the network. This work addresses two challenges: improving LQI based routing protocol by load balancing traffic over multiple paths.

When a sensor has to send data towards the Base Station, the load balancing routing consists to elect several nodes as next hop routers depending on the order of packet transmissions and the nodes previously used as the next hop routers. The idea is to involve several sensors in the routing effort to minimize the overall energy consumption and then extend the network lifetime.

The metric is a property of a route in computer networking consisting of any value used by routing algorithms to determine whether one route should perform better than another. Commonly, the route with the lowest metric is the preferred route. However, in this chapter, a metric means the local value associated with a node: for a source node, the highest value, in its neighborhood, may lead to the selection of such a node as the next hop router. For instance, the remaining energy level can be used as a metric to promote the selection of the highest powered nodes as next hop routers.

In this chapter, we propose WSN local load balancing routing mechanisms using the Wait and See (WaS) protocol [24] by comparing the following metrics: the remaining energy level, the degree of connectivity (number of neighbours), the sensor proximity with respect to the Base Station, the link quality indicator (LQI), and a hybrid metric composed of any pairs of these metrics.

The sensor networks are characterized by low energy constituting their batteries. Then energy consumption and some other performance criteria such as the Load Imbalance Factor (LIF), the average path lengths, the network lifetime and the packet loss percentage are taken into consideration to evaluate the effectiveness of routing mechanisms.

Definition of an "Achtophorous Node": we focus on homogeneous WSN where all sensors are participating together in the routing effort. Since all nodes are routers, we prefer using the term "achtophorous node" derived from Greek term *αχθοφορεω* which denotes "node handling heavy load". *For each node having data to send, its achtophorous nodes are its next hop sensors which handle the load due to the routing of its packets towards the BS. Each sensor selects among its neighbours one or more achtophorous nodes.* We will also examine the influence of increasing the number of the achtophorous nodes on the routing efficiency. The WSN deployed in a warehouse is pruned to the unreliability feature of the wireless links. Then, we present results pertaining to unreliable links impacts on the network performance.

The rest of this chapter is organized as follows. After presentation of a short background in the next part, the next one gives some topics on studied metrics. Then, we describe load balancing mechanisms and the proposed L2RP routing protocol before presenting a detailed report of L2RP performance evaluation. Finally, the last part presents the energy over-consumption induced by securing the protocol.

1.6.2. Backgrounds and Related Works

Commonly used by the TinyOS community, MultihopLQI is a routing protocol which employs the cost-based paradigm defined in [30]. Link estimation is viewed as an essential tool for the computation of reliability-oriented route selection metrics. In MultiHopLQI, the link metric is the Link Quality Indicator (LQI) which is used additively to obtain the cost of a given route. MultihopLQI avoids routing tables by only keeping state for the best parent at a given time, drastically reducing memory usage and control overhead. A new parent is adopted if it advertises a lower cost than the current parent.

Many experimental studies related to WSN, some of which are based on MultiHopLQI, [25, 31-37] have shown that high unreliability of wireless links must be explicitly taken into account when designing routing protocols. Papers [32] and [33] address load balancing embedded in reliability-oriented routing protocols and are also using MultiHopLQI.

In [38] and [39] authors address the problem of minimizing the total consumed energy to reach the destination. The performance objective of maximizing the network lifetime was considered in [40] and [41].

Several works are related to WSN and ad hoc networks load balancing routing schemes [42-47]. In [42], authors show that distributing the traffic generated by each sensor node through multiple paths instead of using a single path allows energy savings. Paper [43] defines a network optimization problem used for performing the load balancing in wireless networks with a single type of traffic. In [44], authors study wireless network routing algorithms that use only short paths, for minimizing the latency, and achieve the load balance. In [45], authors introduce collision awareness in multipath routing; while [46] propose a multipath routing protocol to address the congestion control issue in WSN. In [47], the challenge of maximizing the network lifetime by load balancing the traffic is covered. In order to balance the energy consumption among sensor nodes, they deploy multiple sinks simultaneously, which are connected through wired or wireless infrastructure. Papers [48-50] are also related to load balancing routing protocols.

The paper [51] presents a resource-aware and link quality based (RLQ) routing metric. Based on both energy efficiency and link quality statistics, the RLQ metric in [51] is intended to adapt to varying wireless channel conditions, while exploiting the heterogeneous capabilities. This protocol does not include load balancing features.

Some works are taken into account the round-robin cluster based routing [52-54], where cluster heads are selected on a round-robin fashion. In [55] authors propose a source count (packets) based weighted round-robin forwarding algorithm.

Although all these studies provide a valuable and strong contribution in WSN routing, the problems of load balancing routing mechanisms based on local metrics, with special interest on the LQI based metrics, are yet to be addressed. This is the goal of this work. To save energy, we exploit the broadcast nature of wireless links, and the fact that the weights, in our proposed L2RP protocol, are built upon the achtophorous nodes capabilities instead of the ones of the source node. This allows L2RP to avoid doing a per packet load-balancing by the source, as done in [55], where the source node sends its data without being sure that the achtophorous node is able or not to sustain the load assigned. Thus, L2RP helps in reducing packet losses. Moreover, in most of papers addressing the load balancing routing, both experimental studies and simulation models, results are validated for only small sized networks (few tens of nodes), whereas our work addresses large sized networks (several hundreds of sensors). The comparative study of different metrics in L2RP is also a contribution of this work.

1.6.3. Routes Selection Criteria

In this chapter, "metric" is used to refer to local route selection criterion. As previously defined, each time we use "Achtophorous Node" it means next hop router with respect to a specific node having data to transmit towards the BS.

1.6.3.1. Remaining Energy Level

The remaining energy of sensors could be a metric for selecting routes since a node with better battery life seems to be a better candidate for the packet routing from its neighbours. Conversely, if a sensor with low power is selected as an achtophorous node, this can lead to packet losses because it might not have enough batteries to forward packets. We consider that each node knows its energy level.

1.6.3.2. Sensor Proximity with Respect to the Base Station (Proximity-BS)

In a warehouse Fig. 1.11, depending on the nature of their respective contents (frozen foods, fresh produce, etc.), the pallets provided each with a sensor Fig. 1.12 are arranged in fixed locations Fig. 1.13 designated by the Warehouse Management Software (WMS). Thus, during the warehouse WSN initialization, sensors could be initialized with their respective positions without using the GPS technology.

So, we consider a WSN deployed with a Base Station where each node knows its exact position and that of the BS. As the main goal of the application is to send data towards the BS, it seems natural to look at the metric defined as follows:

$$ProximityBS(S_i, BS) = 1/d(S_i, BS), \qquad (1.1)$$

where $d(Si, BS)$ is the distance separating the sensor S_i from the BS. We choose inverse of the distance to promote the election of the closest sensor to the BS.

Fig. 1.11. Pallets arrangement inside a warehouse.

Fig. 1.12. Sensor plugged inside a Pallet.

Fig. 1.13. Location of a pallet: lane, location and level.

1.6.3.3. Degree of Connectivity

The degree of connectivity of a node, i.e., the number of its neighbours, is also a metric that seems interesting to study because, intuitively, the more neighbours a sensor has, the more it seems to be an appropriate candidate as an achtophorous node since a sensor with a low degree of connectivity might have little information, from its neighborhood, to forward to the BS. In the initial phase, each sensor is involved in the neighborhood information exchanges (hello protocol), which allows it to determine its degree of connectivity and the BS position.

1.6.3.4. LQI: Link Quality Indicator

In ZigBee standard [8, 26], the LQI measurement is defined as a characterization of the strength and/or quality reception of a packet. The use of the LQI result by the network or the application layers is not specified in [8, 26]. The LQI measurement is performed for each received packet Fig. 1.14, and the result is reported to the MAC sublayer as an integer ranging from 0 to 255. The minimum and maximum LQI values (0 and 255) are associated with the lowest and the highest quality IEEE 802.15.4 reception detectable by the receiver, and the LQI values in between are distributed between these two limits [8, 26].

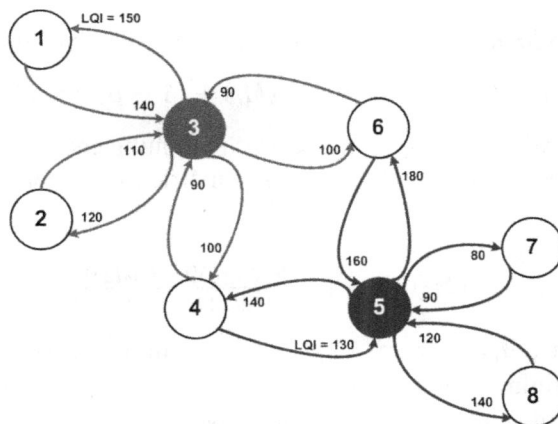

Fig. 1.14. Example of a WSN with Asymmetrical Links.

For Moteiv's Tmote Sky [7] sensors equipped with chipcon's CC2420 [27], the LQI values range from 50 to 110. Even so, we stick with the ZigBee standard [8, 26] because some manufacturers, such as SUN-SPOT [8] and WiEye [10], are still using the standard LQI values. Then, we will use the standard values (i.e., [0, 255]), instead of those of CC2420.

In this section, we define three LQI based metrics: AvgLQI, MaxLQI and MinLQI. The AvgLQI metric is the average calculated from the LQI values of all the links between the node and its neighbours. AvgLQI values give a characterization of sensors throughout their respective coverage quality. This metric might be useful in the context of the WSN

45

deployed in a warehouse which hosts a large number of pallets, one upon the other. Such an environment is proned to high unreliability of wireless links. The MaxLQI metric is the maximum LQI value which matches to the standard definition of the LQI used in the MultiHopLQI routing algorithm [25, 29]. As for the MinLQI, it is the minimum value beyond the given LQI threshold. For example Fig. 1.14, assuming that the LQI threshold for an acceptable link quality is 100, the MinLQI for node 5 is 120 (LQI of link 5-8) instead of 90 (LQI of link 5-7). Thus, Table 1.2 gives LQI metrics values for the WSN in Fig. 1.14.

Table 1.2. LQI metric values related to the WSN in Fig. 1.4.

Sensor ID	1	2	3	4	5	6	7	8
MinLQI	150	120	110	100	120	100	80	140
AvgLQI	150	120	107.5	120	125	140	80	140
MaxLQI	150	120	140	140	160	180	80	140

1.6.3.5. Composite or Hybrid Metric

In this chapter, we define the composite metric (hybrid) as follows:

$$Hybrid(LQI, M_i) = \rho * LQI + (1 - \rho) * Sc(M_i), \quad (1.2)$$

$$Hybrid(M_i, M_j) = \rho * Sc(M_i) + (1 - \rho) * Sc(M_j), \quad (1.3)$$

where $Sc(M_i)$ is the scale function, which returns remaining energy values comparable to the LQI values. This help avoiding the composite metric to be strongly influenced by the M_i component in (1.2):

$$Sc(M_i) = \alpha + \frac{\beta * \log(1 + (M_i - M_{i,min}))}{\log(1 + M_{i,max})}, \quad (1.4)$$

where M_i is the metric, $M_{i,min}$ (resp. $M_{i,max}$) is the minimum (resp. maximum) value of M_i. If M_i is the remaining energy of the node, $M_{i,min}$ represents the value under which, the sensor is considered dead (battery depletion); while $M_{i,max}$ is the initial energy value of a new battery. $\alpha = 50$, $\beta = 255$.

Like the LQI metrics definition, we can also define AvgHybrid, MaxHybrid and MinHybrid metrics depending on whether, we are respectively considering AvgLQI, MaxLQI and MinLQI as defined in Table 1.2.

1.6.4. Routing Mechanisms

1.6.4.1. Simple Routing

In the simple routing mechanism, each sensor Si selects an achtophorous node which matches the best metric in its vicinity and located between the sensor Si and the BS. For

46

each given sensor, a unique achtophorous node plays the next hop role for all its packets until the next election Fig. 1.15.

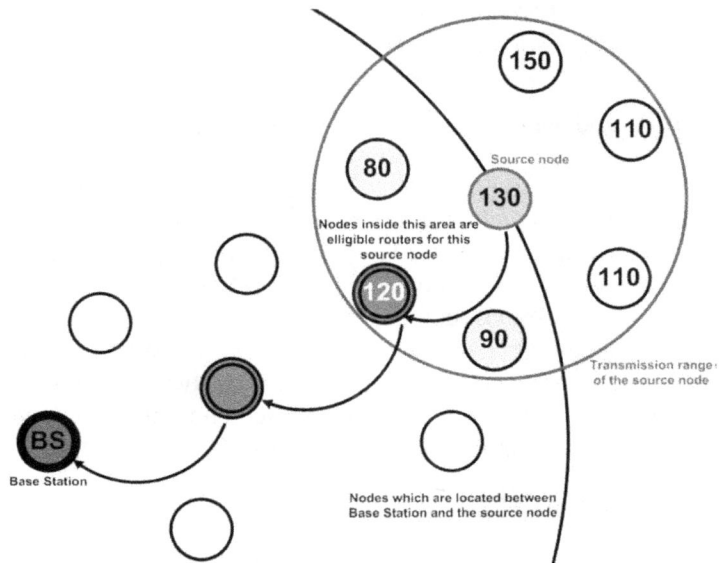

Fig. 1.15. Simple Routing from Source node 130 to the Base Station (BS).

1.6.4.2. Round Robin Routing

In the round-robin routing, each source node has to elect two or more achtophorous nodes. The source node sends data in round-robin fashion, simply taking turns which achtophorous node it routes each packet out Fig. 1.16. This routing mechanism is a per-packet load balancing routing which gives most even distribution across next achtophorous nodes.

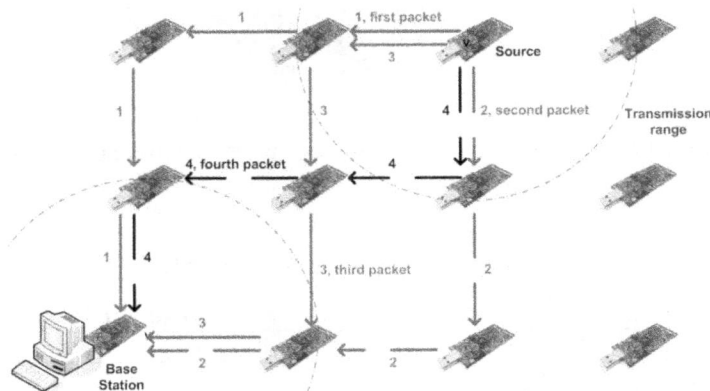

Fig. 1.16. Round-Robin Routing: Multiple routes from each source for different packets.

This per-packet load balancing method means that packets in a particular connection or flow arrive at their destination out of sequence. This does not cause a problem for most applications, but it can cause problems for the increasingly popular streaming media, both video and audio. In this chapter, only data packets are concerned within cold chain monitoring application for which the packet sequence order is not an issue.

1.6.4.3. Weighted Round-Robin Routing (W2R routing)

The weighted round-robin routing (W2R routing) is a load balancing mechanism that involves assigning a weight to each achtophorous node. Weights are proportional to metric values. In the W2R routing, each achtophorous node is assigned a value that signifies, relative to the other achtophorous nodes in the routing table, how the source node performs. The weight determines how many more (or less) packets are sent to that achtophorous node, compared to the other achtophorous nodes Fig. 1.17. The W2R routing is one way addressing some shortcomings. In particular, it provides a clean and effective way by focusing on fairly distributing the load amongst available achtophorous nodes, versus attempting to equally distribute data packets.

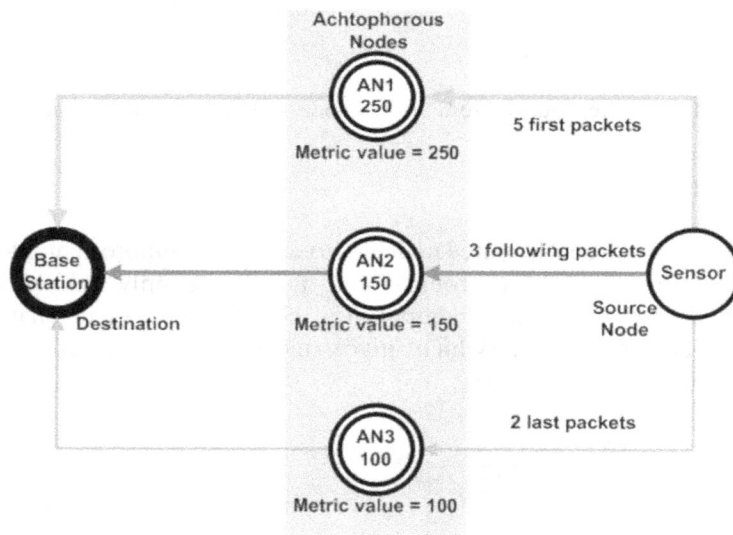

Fig. 1.17. Weighted round-robin routing (W2R routing) mechanism.

Table 1.3. Weights of the achtophorous nodes in Fig. 1.17.

Achtophorous Node	Metric	Weight	Load handled
AN1	250	0.5	50 %
AN2	150	0.3	30 %
AN3	100	0.2	20 %

For example, in Fig. 1.17, the source node routes 50 % of its packets through the achtophorous node AN1, 30 % through AN2 and 20 % through AN3. If the BS is not located within the transmission range of an achtophorous node, this one should apply the same mechanism to retransmit the packet towards the BS.

The weighted round-robin routing mechanism is computed as described in this simple Algorithm 1 which is computed each time a source node has to send a packet. The achtophorous nodes, each with its respective weight, are listed in the routing table of each source node in an ordered manner such that the first achtophorous node matches the highest weight as shown in Fig. 1.17. For each source node, the window interval is the constant length of each stream of consecutive packets to transmit. The weight of each achtophorous node is converted as an integer value based on the window interval parameter. For example, in Fig. 1.17, window = 10 consecutive packets, and weight(AN1) = 5. The use(AN) function returns the number of times the current achtophorous node AN is used during the window interval whereas packet_idx is the index of the current packet during the window interval.

Algorithm 1 : Weighted Round Robin (W2R) Routing

Require: $packet\_idx$, $window$, AN, $weight$, use

1: **if** $packet\_idx < window$ **then**
2:     **if** $use(AN) < weight(AN)$ **then**
3:         $Send\_packet\_to(AN)$
4:         $use(AN) \leftarrow use(AN) + 1$
5:         $packet\_idx \leftarrow packet\_idx + 1$
6:     **else**
7:         $use(AN) \leftarrow 0$
8:         $AN \leftarrow Next().achtophorous\_node$
9:         $\# The\ next()\ of\ the\ last\ AN\ is\ the\ first\ AN$
10:        $Send\_packet\_to(AN)$
11:        $use(AN) \leftarrow 1$
12:        $packet\_idx \leftarrow packet\_idx + 1$
13:    **end if**
14: **else**
15:     **for** *each achtophorous_node AN* **do**
16:         $use(AN) \leftarrow 0$
17:     **end for**
18:     $AN \leftarrow First().achtophorous\_node$
19:     $Send\_packet\_to(AN)$
20:     $use(AN) \leftarrow 1$
21:     $packet\_idx \leftarrow 1$
22: **end if**
23: **return** $packet\_idx$, AN, use

1.7. L2RP: The Link Reliability Based Routing Protocol

The proposed (L2RP) routing protocol Fig. 1.18 consists for a sensor having an empty routing table to elect one next hop router (case of simple routing) or more achtophorous nodes (load balancing routings) amongst its neighbours according to the following:

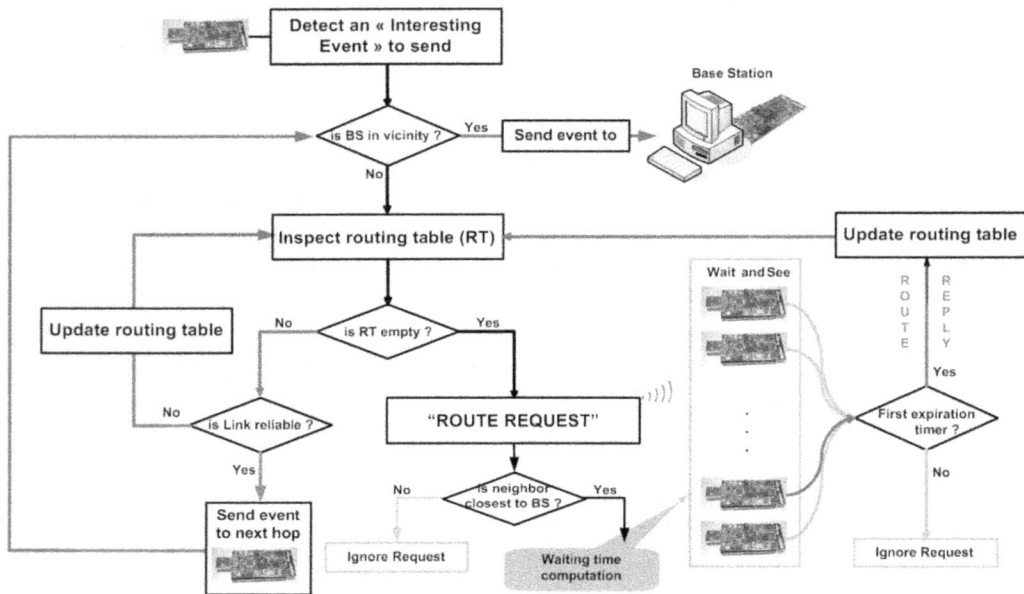

Fig. 1.18. The Link Reliability based Routing Protocol (L2RP) flowchart.

• Initial step: all sensors empty their routing tables.

• The sensors located in the vicinity (transmission range) of the BS send their data directly to it.

• A sensor, located outside of the vicinity of the BS, inspects its routing table:

  • If its routing table is not empty, it checks if the link with the next hop is reliable or not. If the link is unreliable, based on the LQI value, then:

    • Case of simple routing mechanism: it sends a "ROUTE REQUEST" to its neighbours.

    • Case of load-balancing routing: it chooses an alternate route and then checks again if the link with this next hop is reliable or not. If no link with nodes in the routing table is reliable, then it erases its routing table and it sends a "ROUTE REQUEST" to its neighbours.

- If its routing table is empty, it also sends a "ROUTE REQUEST" to its neighbours.

- Each neighbor, located between the BS and the sensor having sent the "ROUTE REQUEST", computes its own waiting time which is inversely proportional to its metric value. We use the Wait and See protocol (WaS), as in [24], where the only sensor having the best metric sends a "ROUTE REPLY" to the requester node. The other neighbours simply ignore the "ROUTE REQUEST" avoiding useless "ROUTE REPLY" packets. In the case of a load balancing routing, the number (ANs) of achtophorous nodes is a known parameter in the initialization phase of the network. This parameter is used by the WaS protocol that allows ANs sensors having highest metrics in succession to answer to the requester node, and then be elected, for this node, as achtophorous nodes.

- Upon reception of the "ROUTE REPLY" packet, the requester node updates its routing table, which remains valid until the next election. In the case of weighted round-robin routing, each "ROUTE REPLY" packet contains the metric value of the answering node, which allows the requester node to calculate weights associated with each achtophorous nodes.

- At the end of the current cycle, sensors reset their routing tables and go back to the initial step of the next cycle.

Upon receipt of a "ROUTE REQUEST" packet, a sensor S_i computes its own waiting time according to the following formula:

$$Timer(S_i) = \tau + \frac{\zeta}{1 + \log(1 + M_i + \frac{id(S_i)}{\Gamma} * M_i)}, \qquad (1.5)$$

where M_i is the metric value of the sensor S_i. τ and ζ are the nonzero positive constants. Γ is the constant, which is more large than the network size ($\Gamma = 10^6$, for example). This timer function avoids collisions between nodes having the same metric value. Since $M_i \geq 0$, if $M_i = 0$ then the sensor S_i can not be an achatophorous node.

As we can see, in this protocol the source node uses the link quality indicator (LQI) to check if the link it forms with the nominated achtophorous node is reliable or not. This helps avoiding sending the packet to an achtophorous with which it forms a link of poor quality which could lead to packet loss.

1.8. Routing Protocol Performance Criteria

1.8.1. Average Ratio of the Remaining Energy

The average ratio of the remaining energy is the ratio of the average remaining energy on the average of initial energy. Multiplied by hundred, this value represents the average battery life of sensors, in terms of percentage. The higher this value is, the more energy-efficient the routing protocol is.

1.8.2. Average Path Lengths

The average path lengths are calculated in terms of the number of hops traversed by packets before reaching the BS. A large value reflects participation of many sensors in the effort due to the routing, which may increase the overall energy consumption. A good routing protocol is recognized in this performance criterion by a relatively low value. Conversely, too small path length may lead to bad quality link.

1.8.3. LIF: Load Imbalance Factor

The load imbalance factor (LIF) is defined as the root of the squared coefficient of variation of the relative remaining energy. This shows the energy spent by communications:

$$LIF = \sqrt{\frac{Var(E_R^i)}{\bar{E}_R^2}},\qquad(1.6)$$

where E_R^i is the ratio of the remaining energy of the sensor S_i; and $\bar{E}_R$ is the average ratio of the remaining energy.

With this performance criterion, the best routing mechanism is among the ones which produce small load imbalance factor. Indeed the lowest LIF value indicates the best evenly distribution of the energy consumption between nodes.

1.8.4. Network Lifetime

In this work, we define the network lifetime as the average number of packets routed until the first time a sensor run out of battery. This could also result in network capacity. We focus on the first battery depletion, which means the instant the network stops fulfilling totally its role, because it leads to packet losses. An ideal network is a network where all packets sent by source nodes are actually transmitted to the recipient (BS). The earlier the first packet loss happened, the more ineffective the routing protocol is.

1.8.5. Average Percentage of Lost Packets

Beyond the first time battery depletion is experienced by the network, a high percentage of packet losses might reflect an unreliable network whose routing protocol is less effective.

The loss of packets is a factor aggravating the overall energy consumption because this causes poor quality of service and retransmissions to guarantee the reliability of some applications.

1.9. L2RP Routing Protocol Performance Evaluation Model

1.9.1. Energy Consumption Model

For the performance evaluation model, we will use this energy consumption model [1-3, 56, 57], Fig. 1.19.

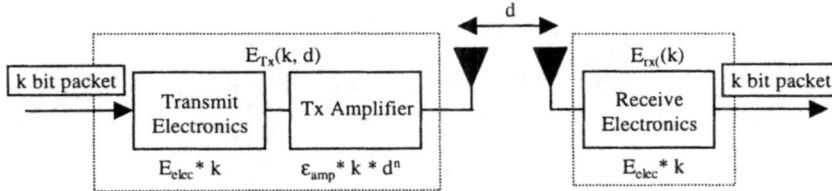

Fig. 1.19. Energy dissipation model.

Let $E_{Tx}(k, d)$ the energy [56, 57] consumed to transmit k bits message over a distance d Fig. 1.19:

$$E_{Tx}(k, d) = E_{elec} * k + \varepsilon_{amp} * k * d^2.$$ (1.7)

Let $E_{Rx}(k, d)$ the energy consumed to receive a k bits message:

$$E_{Rx}(k, d) = E_{Rx}(k) = E_{elec} * k,$$ (1.8)

E_{elec} = 50nJ/bit and ε_{amp} = 100pJ/bit/m2.

1.9.2. Network Deployment and Performance Evaluation Parameters

In the performance evaluation model N nodes are randomly (according to a uniform distribution) deployed over an area of length L=100 m, and width l=100 m (Fig. 1.20). The BS is located at the (0,0) position. Each node generates a sequence of "interesting events", which are sensed data over the temperature threshold T_{min}, following the Poisson process of parameter $\lambda = 10$. For simulation scenarios, the size of each data packet is set to k_{data} = 128 bits, and the "ROUTE REQUEST" and "ROUTE REPLY" packets of the L2RP protocol have a size of k_{rr} = 24 bits. Let us assume that each node knows its position and its remaining energy level at any time.

The initial energy amount of each node is set to a value $E_0 = (1.5 * 10^5 - \varepsilon)\mu J$, where $\varepsilon = rand(0,1) * 10^2 \mu J$. Node battery exhaustion is experienced when the remaining energy level of the node is under the given threshold $E_{min} = E_0 * 0.05$.

All nodes, including the BS, have same the transmission range R = 20 m. The main simulation parameters are listed in Table 1.4.

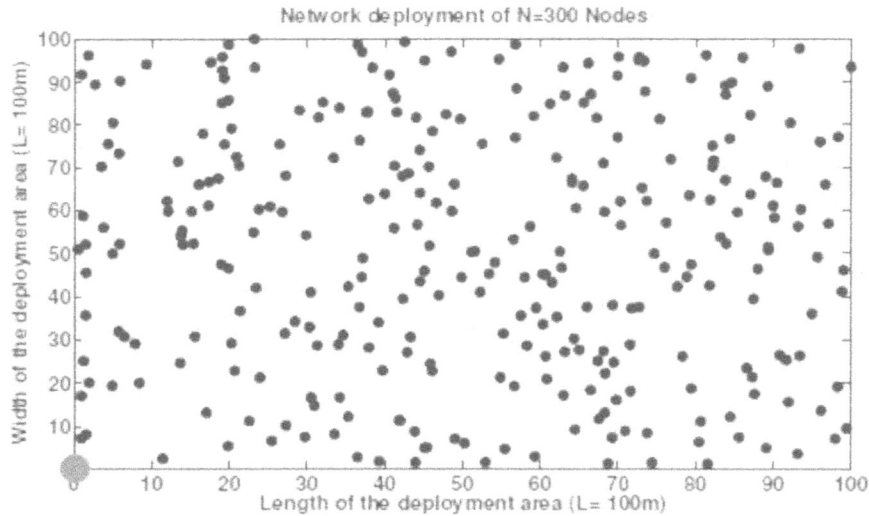

Fig. 1.20. Network deployment of N=300 sensors over an area (100 m × 100 m).

Table 1.4. Simulation parameters.

Parameter	Value
Deployment	
Area Length	$L = 100m$
Area Width	$l = 100m$
Base Station Location	$position(SB) = (0, 0)$
Radio range	$R = 20m$
Network Size (number of sensors)	$N = \{100, 200, ..., 500\}$
Poisson parameter	
Packet sent by each sensor	$\lambda = 10$
Packet sizes (bits)	
Alarms	$k_{data} = 128$
L2RP "ROUTE REQUEST"	$k_{rr} = 24$
L2RP "ROUTE REPLY"	$k_{rr} = 24$
L2RP Achtophorous Nodes	
Number of Achtophorous Nodes	$AN = 3$
Window interval for W2R	$window = 10$
LQI parameters	
Threshold for MinLQI	$LQI \geq 100$
Link Reliability (L2RP)	$LQI \geq 70$
Energy	
Initial Energy Level	$E_0 = (1.404 \times 10^5 - \varepsilon)\mu J$ $\varepsilon = rand(0,1) \times 10^2 \mu J$
Minimum Energy Level	$E_{min} = E_0 \times 0.05$

1.9.3. LQI Model for Performance Evaluation Purposes

The WSN can be modelled as a graph $G = (V, E)$, where two nodes are connected by an edge if they can communicate with each other. Let $x \in V$ be a node in the WSN. $N_1(x)$ is the neighbourhood of the node x. At each given time t, a node x forms with each $y \in N_1(x)$ a link of which the link quality indicator (LQI) value is denoted by $\ell(x, y, t) > 0$. For all other nodes: $\forall z \in V \setminus N1(x), \ell(x, z, t) = 0$. Let v be a bijective function defined in V which is a totally ordered set. The v function is defined as follows:

$$\forall x \in V, v(x) = \big(f(x), id(x)\big), \tag{1.9}$$

where $f(x)$ is the function which returns the metric value of x, and $id(x)$ returns the address of the node x. The total ordering in V is defined as follows:

$$\forall x, y \in V, v(x) > v(y) \Leftrightarrow \big(f(x) > f(y)\big) or$$
$$\big(f(x) = f(y) and id(x) > id(y)\big). \tag{1.10}$$

After the WSN deployment in the warehouse, the BS initially broadcasts a message containing its position. This information is then retransmitted to all sensors in the network. In this phase, each node knows its degree of connectivity. At each given time t, the LQI value of the link formed by any pair (x, y) of nodes is calculated by using the $\ell(x, y, t)$ function defined below:

$$\ell(x, y, t) = f(x, y, t) * g(x, y), \tag{1.11}$$

$$f(x, y, t) = 1 - \mathrm{P}_r[link(x, y, t) = Unreliable], \tag{1.12}$$

$$g(x, y) = \alpha + \frac{\beta * \log(1 + (\gamma(x,y) - \gamma_{min}(x)))}{\log(1 + \gamma_{max}(x))}, \tag{1.13}$$

$$\gamma(x, y) = \frac{1}{d(x,y)}, \tag{1.14}$$

$$\gamma_{min}(x) = \min_{y \in N_1(x)} \gamma(x, y), \tag{1.15}$$

$$\gamma_{max}(x) = \max_{y \in N_1(x)} \gamma(x, y), \tag{1.16}$$

where $\alpha = 50, \beta = 255$ and $d(x, y)$ is the distance separating y from x.

In the context of a cold chain monitoring application, the warehouse hosts hundreds of pallets, one upon the other. Each pallet is provided with a temperature sensor. This environment is subjected to the unreliably feature of the wireless links. So, in the formula (1.12), $\mathrm{P}_r[link(x, y, t) = Unreliable]$ denotes the probability that the link $\ell ink(x, y, t)$ becomes unreliable at a given time t. This probability is used in some simulation scenarios, in order to evaluate the behaviour of our L2RP protocol with respect to the unreliability aspect of the wireless links.

The choice of this model, formula (1.13) similarly to the scale function S_c defined in the composite metric, is guided by experimental results shown in [58] and [31] which stated that the LQI decreases when the distance between nodes increases in ZigBee-based WSN.

As we can see, $\ell(x, y, t) \neq \ell(y, y, t)$, because of the formulas (1.15) and (1.16). Hence, the model allows taking into account asymmetrical aspects of the wireless links.

For Moteiv's Tmote Sky [7] sensors equipped with chipcon's CC2420 [27], the LQI values range from 50 to 110. Even so, we stick with the ZigBee standard [26, 8] because some manufacturers, such as Sun-SPOT [8] and WiEye [10], are still using the standard LQI values. Then, we use the standard values (i.e. [0, 255]) increased by $\alpha = 50$, instead of those of CC2420. The use of $\alpha = 50$ allows to keep the null value, $\ell(x, y, t) = 0$, only for the two cases where the node y is not in the transmission range of the node x, or when the $\ell ink(x, y, t)$ becomes unreliable i.e. $P_r[link(x, y, t) = Unreliable] = 1$.

This LQI model is only used for simulation purposes, so sensor nodes do not compute these above formulas.

1.10. L2RP Routing Protocol Performance Evaluation

Performance evaluation, using Matlab, are run for a network size ranging from 100 to 500 nodes. The performance results presented here are obtained by averaging the results for 50 different simulations for each scenario comparing the route selection criteria. In each scenario where the three routing mechanisms are compared, 25 different simulations were run. For each simulation, a new random node layout is used.

In all simulation results presented below, $\rho = 0.5$ for the composite metric as defined in formulas (1.2) and (1.3). If it's not specified, the number ANs of Achtophorous Nodes is set to ANs = 3 for each load balancing mechanism.

In all performance evaluation scenarios, except those in Section 1.10.8, links are considered reliable, i.e.:

$$\forall t, \forall x \in V, \forall y \in N_1(x), P_r[link(x, y, t) = Unreliable] = 0. \qquad (1.17)$$

For some results, the related confidence intervals for a confidence coefficient of 95 % are computed as detailed in the Section 1.3.3 of [59].

1.10.1. Average Path Length

The Fig. 1.21 shows the average path length for the simple routing, and the Fig. 1.22 its related confidence interval.

The results of Fig. 1.21 show, in the case of simple routing mechanism, the average path lengths in terms of the average number of hops obtained with the different studied metrics when the node density is increasing in the deployment area. This result shows that routes are longer for the MaxLQI and degree of connectivity metrics. The remaining energy, AvgLQI, MinLQI and "Proximity with respect to the BS" metrics have better average path lengths. The gap is more important for the MaxLQI metric with respect to the other metrics. Moreover, for MaxLQI, the average number of hops is a monotonically increasing function of the network density. This reflects the fact that the routing according to the metric MaxLQI consists of choosing as an achtophorous node the node having the best link quality with the source node. In the absence of obstacles and other phenomena like interferences, the best link quality is determined by the shortest distance separating a node from the source node. So, routing according to the MaxLQI metric is equivalent to a multihop "step by step" routing which is characterized by a great number of hops due to small distances separating each source node and its achtophorous node.

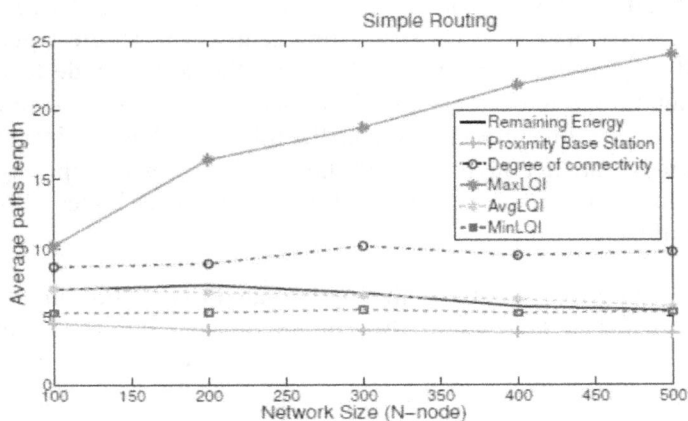

Fig. 1.21. Average path length: Comparison of metrics in simple routing mechanism.

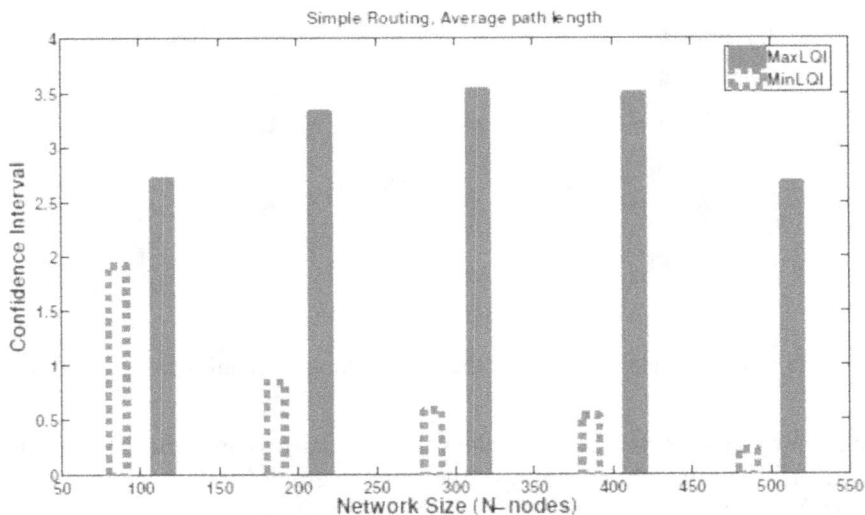

Fig. 1.22. Confidence interval, for the result of Fig. 1.21, for a confidence coefficient of 95 %.

When the network density is increasing, the distances separating sensors decrease. Thus, the distances separating each source node and its achtophorous node also decrease, as well. So, from any source node towards the base station, the number of hops of each sent

57

packet becomes increasingly high when the MaxLQI is used. By multiplying the number of hops, in this manner, the sensor network could not claim to have a good performance. This result explains the low performance of the MultiHopLQI routing algorithm which is used today in many TinyOS based empirical WSN analysis. Indeed, MultiHopLQI uses the LQI metric as defined in the ZigBee standard [26, 8], that is to say the MaxLQI metric.

Conversely, the Proximity-BS and MinLQI metric have the lowest average path lengths Fig. 1.21. For any given source node, the selected Proximity-BS based achtophorous node matches the farthest neighboring node towards the Base Station. Therefore, the routing according to the Proximity-BS metric is equivalent to the shortest geographical path routing. Accordingly, packets are transmitted from the source node to the base station requiring the minimum number of hops. This result Fig. 1.21 is also interesting for the MinLQI metric. Indeed, this metric promotes the use of the links of intermediate quality. Links of good quality are synonymous with the nearest nodes multiplying the number of hops, whereas the links of poor quality stand for lot of packet losses. This explains why MinLQI is a good metric.

The Fig. 1.23 compares the average path length related to the "Proximity with respect to the BS" metric when it is used in the simple routing and load balancing routing mechanisms.

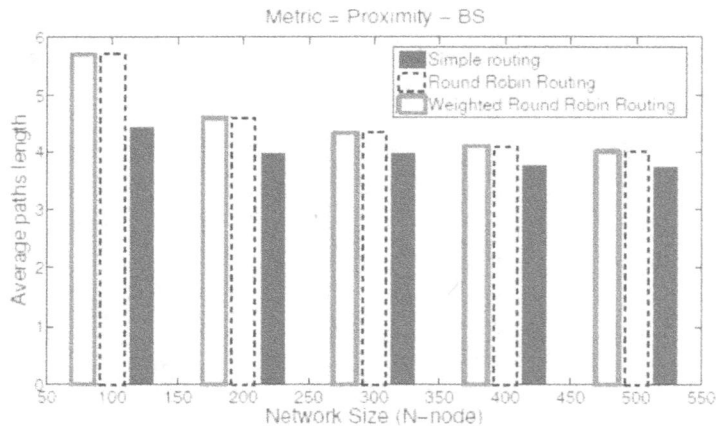

Fig. 1.23. The average path length in the three routing mechanisms (Proximity-BS).

For the Proximity-BS metric, the load balancing mechanisms have the effect of increasing the average path lengths which is almost the same average for the weighted round robin routing and the round robin one Fig. 1.23. In the case of load balancing mechanisms, each sensor has in its routing table several achtophorous nodes of which only one exactly corresponds to the achtophorous node used by the simple routing. The other achtophorous nodes are necessarily more distant from the base station. So, the average path lengths slightly increase for load balancing mechanisms with respect to the simple routing using the Proximity-BS metric Fig. 1.23. For any given source node, the selected achtophorous nodes are identical for both load balancing mechanisms, their use only differs by the weight introduced in the weighted round robin routing. This leads to an average number of hops almost identical Fig. 1.23.

Unlike the Proximity-BS and MinLQI metrics, the MaxLQI one has an average path length which is reduced by the load balancing mechanisms Fig. 1.24. In this case, the weighted round robin routing mechanism has an average number of hops closer to the one of the simple routing mechanism than the round-robin one. Indeed in the case of W2R, the achtophorous node which forms the better link quality (MaxLQI) is also the one which has the highest weight. Thus, depending on the weight value, the sensors choose to send their packets more frequently to that achtophorous node. Therefore, W2R leads to an average number of hops closer to the one of the simple routing mechanism Fig. 1.24.

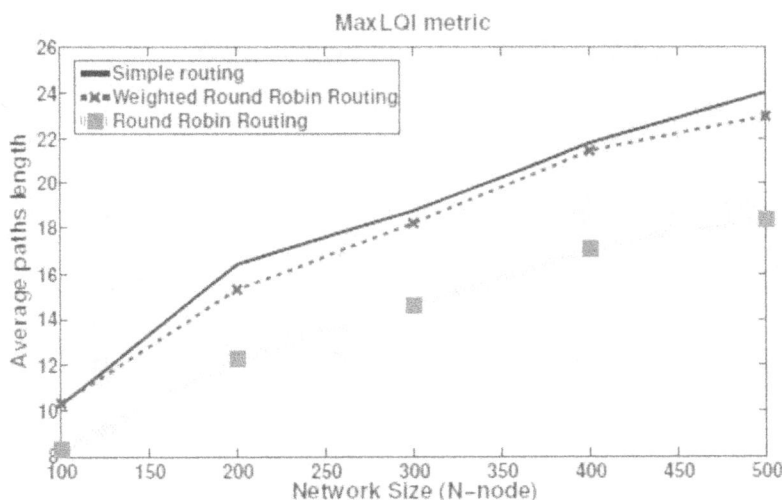

Fig. 1.24. The average path length in the three routing mechanisms (MaxLQI).

1.10.2. LIF: Load Imbalance Factor

The Fig. 1.25 shows the LIF when the "Proximity with respect to the BS" is used as metric. It displays results for the simple routing and load balancing mechanisms. The Fig. 1.26 for MaxLQI and the Fig. 1.27 for MinLQI also display the LIF for the three routing mechanisms.

The lowest LIF value indicates the best evenly distribution of the energy consumption between nodes. It would be redundant to say that the load balancing mechanisms (round robin and W2R) help evenly balancing the load. That is to say that the average LIF values are lower for load balancing mechanisms compared to the simple routing, whatever the chosen metric Fig. 1.25, Fig. 1.26 and Fig. 1.27. But the gap is more important for MaxLQI than other metrics.

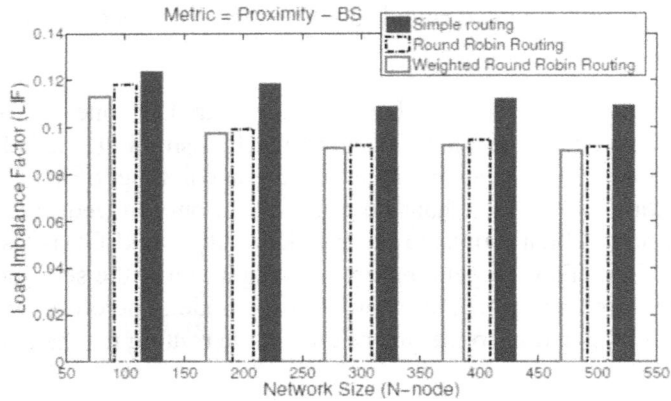

Fig. 1.25. LIF in the three routing mechanisms (Proximity-BS).

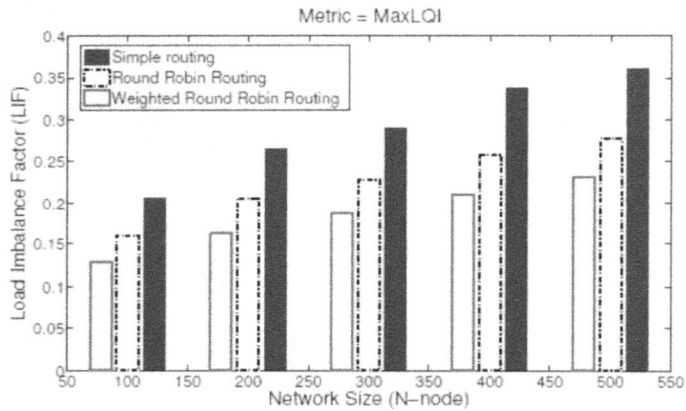

Fig. 1.26. LIF in the three routing mechanisms (MaxLQI).

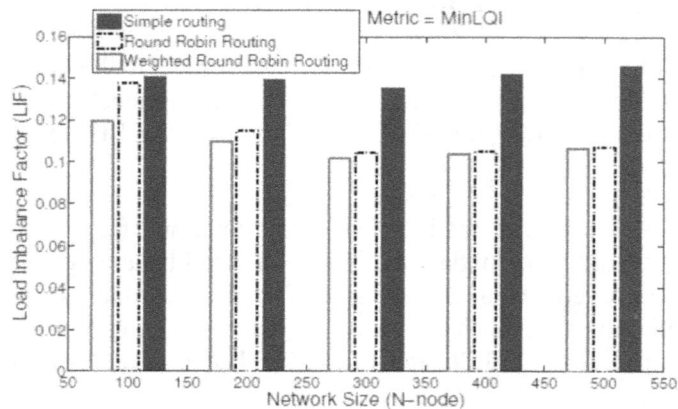

Fig. 1.27. LIF in the three routing mechanisms (MinLQI).

Moreover, when the network density is increasing, the difference between round robin and W2R tend to vanish for the Proximity-BS metric and for the MinLQI metric Fig. 1.25 and Fig. 1.27. For these two metrics, it would be more suitable in dense wireless sensor networks to use the round robin mechanism than the W2R one. Therefore, in doing so, one saves the power required, mainly by the processor, for the achtophorous weight computations Fig. 1.17 and Table 1.3.

These results confirm that load balancing mechanisms help in the distribution of the load across the nodes, because whatever the metric used: the W2R routing produces lower LIF than the round-robin routing which is followed by the simple routing Fig. 1.25, Fig. 1.26 and Fig. 1.27.

As for the result of Fig. 1.28, it compares the average LIF of different metrics in the W2R routing. The "Proximity with respect to the BS" and MinLQI metrics produce lower LIF values Fig. 1.28. The remaining energy metric has an intermediate LIF, while the degree of connectivity and MaxLQI metrics tend to imbalance the energy consumption on the network: some sensors exhaust their batteries while others have a little participation in packet routings towards the BS. This negative phenomenon is much more important for the MaxLQI metric when the network size is increasing Fig. 1.26 and Fig. 1.28.

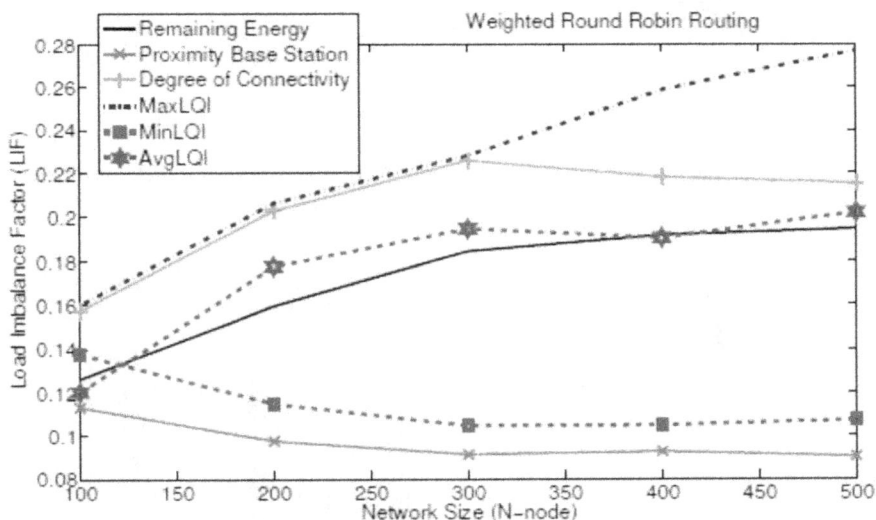

Fig. 1.28. LIF: Comparison of different metrics in the Weighted Round Robin Routing (W2R).

This reflects the fact that the degree of connectivity and the MaxLQI metrics are the ones for which packets arrive at the Base Station by routes using the largest number of hops as shown in Fig. 1.20 and explained in the last section. Thus, along each route, the WSN experiences more retransmissions and then more energy wastage due to the effects of overhead, latency and overhearing phenomena which are more important when the average number of hops is increasing.

1.10.3. Average Percentage of Packet Losses

The result of Fig. 1.29 displays, for each metric, the average percentage of packet losses experienced by the network when the simple routing is run. The three routing mechanisms are compared Fig. 1.30 using the degree of connectivity metric.

Generally, the loss percentage is quite low. This reflects the fact that, in L2RP, losses are mainly due to the node battery exhaustion. The first result Fig. 1.29 compares the different criteria in the mechanism of simple routing.

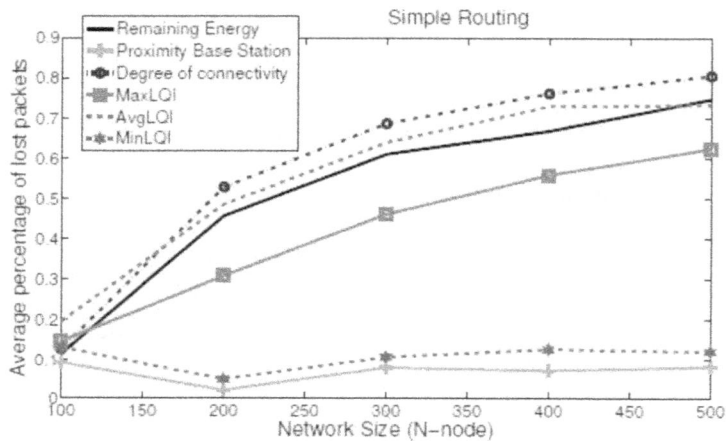

Fig. 1.29. Average percentage of lost packets: Comparison of the different metrics in the Simple Routing mechanism.

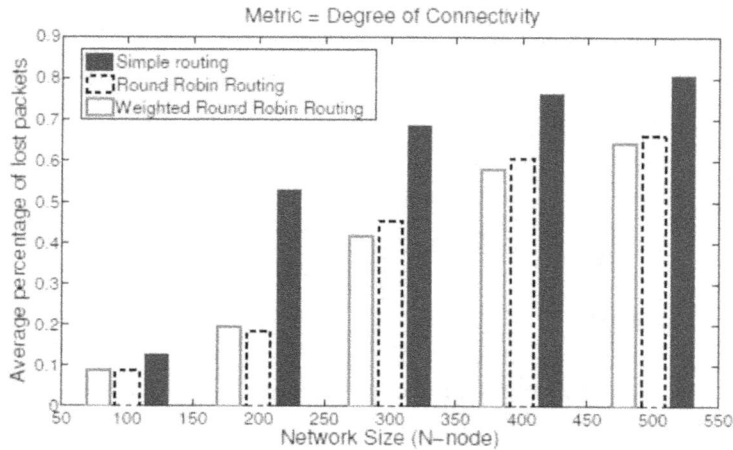

Fig. 1.30. Average percentage of lost packets: Comparison of the three routing mechanisms with the Degree of connectivity metric.

Here again, best results are produced by MinLQI and "Proximity with respect to the BS" metrics. MaxLQI has an intermediate average percentage of packet losses, while the remaining energy and degree of connectivity metrics have higher percentages. For the Proximity-BS metric this result is easy to understand, because according to the previous results Fig. 1.20, Proximity-BS is the metric which produces the shortest path lengths. Accordingly, as the overhearing phenomenon and the overhead induced by routing are reduced when the number of hops is minimal, then the node battery exhaustion occurs later (in time) leading to a low loss percentage for the metric Proximity-BS.

Conversely, the degree of connectivity metric has the highest percentage of packet losses Fig. 1.29. By choosing to route packets according to this metric, any given sensor which has data to transmit chooses its achtophorous node, in simple routing, as its neighbor which has the highest number of neighbours. Therefore whenever an achtophorous node is requested to route a packet, the overhearing phenomenon causes more energy consumption which leads to a greater packet loss percentage.

For all metrics, load balancing significantly reduces the average percentage of packet losses Fig. 1.30. Load balancing mechanisms produce lower packet losses than the simple routing; differences are more important when load balancing is run with the degree of connectivity metric, the remaining energy metric or the MaxLQI metric.

Indeed, for the degree of connectivity metric of which the overhearing phenomenon is the most important, Load balancing requires the selection of different achtophorous nodes for each source node. So, in the routing table there is exactly one node which has the highest number of neighbours: this the one used by the simple routing mechanism. Then, in load balancing when the other achtophorous nodes with fewer neighbours are used, this helps reducing the overhearing phenomenon. This justifies why load balancing reduces the percentage of packet losses compared to simple routing which always requires the highest degree of connectivity as achtophorous node Fig. 1.30.

1.10.4. Composite or Hybrid Metric

The Fig. 1.31 (simple routing) and the Fig. 1.32 (W2R routing) display the average percentage of packet losses including the hybrid metric which is a combination of the remaining energy metric and the "Proximity with respect to the BS" metric.

These results show that the hybrid metric composed of 50 % of the remaining energy and 50 % of "Proximity with respect to the BS" (i.e. $\rho = 0.5$ in Formula (1.3)) is a very good metric. It has a percentage of packet losses which is relatively low, especially when it is used with load balancing mechanisms. As we can see, there are fewer lost packets when the simple routing is run with the MinLQI metric than the W2R routing run with the remaining energy metric, MaxLQI or the degree of connectivity metric Fig. 1.31 and Fig. 1.32.

These results show that compared to the "remaining energy level" metric, the hybrid metric helps mitigating losses particularly in load balancing (W2R Routing) where the average packet loss percentage is less than 0.1 % for this metric. This kind of metric is very interesting to consider because depending on the specific WSN application purposes, it may be useful to consider several criteria for selecting routes by computing a single hybrid metric.

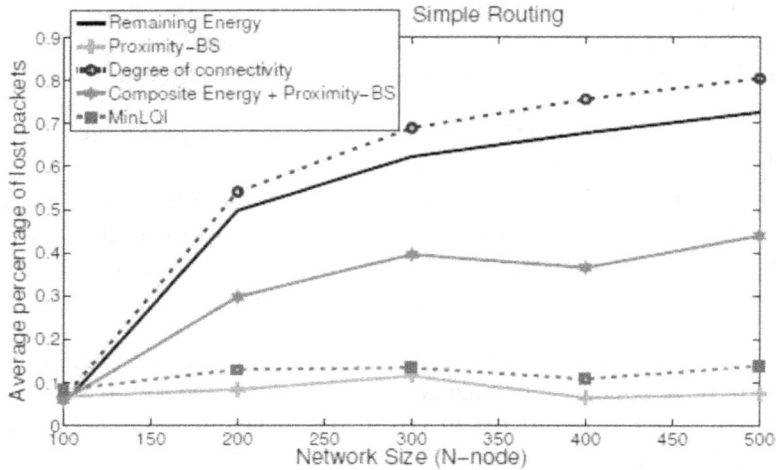

Fig. 1.31. Average percentage of lost packets: Comparison of different metrics including the hybrid metric (remaining energy level + Proximity-BS) in W2R Routing.

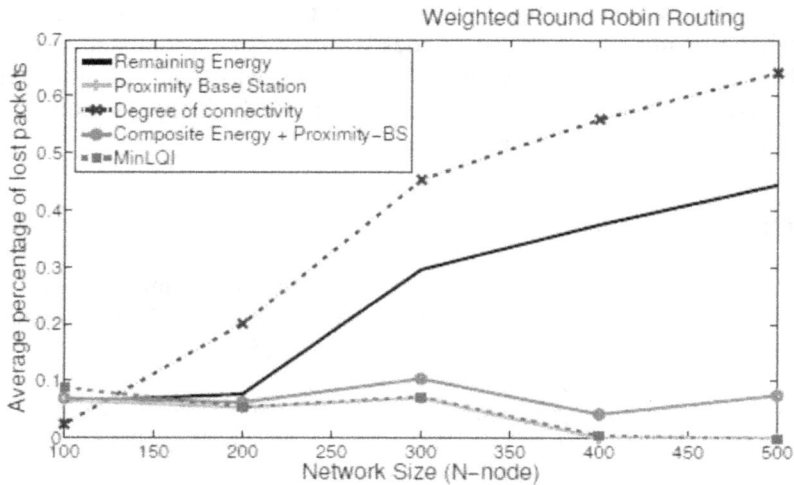

Fig. 1.32. Average percentage of lost packets: Comparison of different metrics including the hybrid metric (remaining energy level + Proximity-BS) in W2R Routing.

In this result, it is more beneficial to route jointly depending on the distance and the remaining energy than to route only along with the remaining energy criterion. This reflects the fact that the "remaining energy level" criterion is not a good metric for route selection. Because in our simulation scenario Table 1.4 each node is deployed with an initial energy level $E0$ which is randomly and slightly lower than a reference value $E_0 = (1.5 \times 10^5 - \varepsilon)\mu J$, where $\varepsilon = rand(0,1) \times 10^2 \mu J$. This scenario is very realistic, because even if the AA batteries powering the sensors are new, they also have slightly different energy levels in real world scenario.

Although the average percentage of packet losses is generally too low, the load balancing helps reducing the packet loss percentage for the hybrid metric similarly to all other studied metrics.

1.10.5. Average Network Lifetime

The Fig. 1.33 displays the average network lifetime for the simple routing, and the Fig. 1.34 shows its related confidence interval for a confidence coefficient of 95 %. The Fig. 1.35 shows the average network lifetime when MinLQI is used in each routing mechanism.

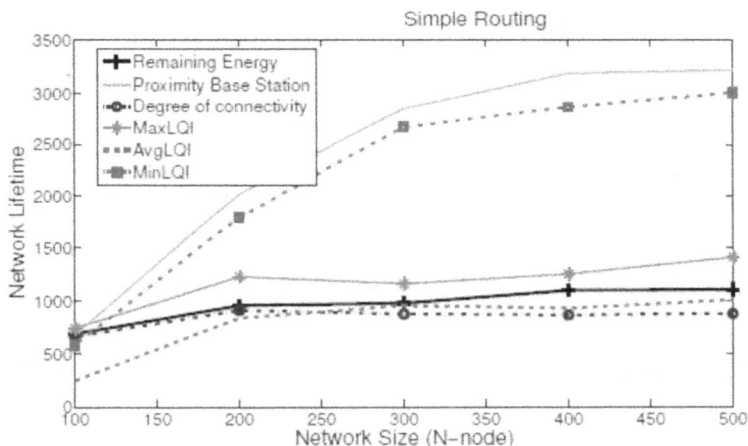

Fig. 1.33. Average network lifetime: Comparison of different metrics in the simple routing.

Firstly, these results show that more dense networks have better lifetime. The MinLQI and "Proximity with respect to the BS" metrics produce better network lifetime. MaxLQI is better than the remaining energy metric which is followed by the degree of connectivity metric Fig. 1.33. Load balancing mechanisms significantly increase the average network lifetime which is larger than the one of the simple routing with more differences for MinLQI (Figs. 1.35, 1.36) and "Proximity with respect to the BS" metrics.

The time of first packet loss occurs earlier for the degree of connectivity metric. As we explained in the previous sections, this result is also caused by the overhearing phenomenon of which effects are more important for the degree of connectivity metric with respect to other metrics. The Proximity-BS metric improves the network lifetime by minimizing the number of hops Fig. 1.33.

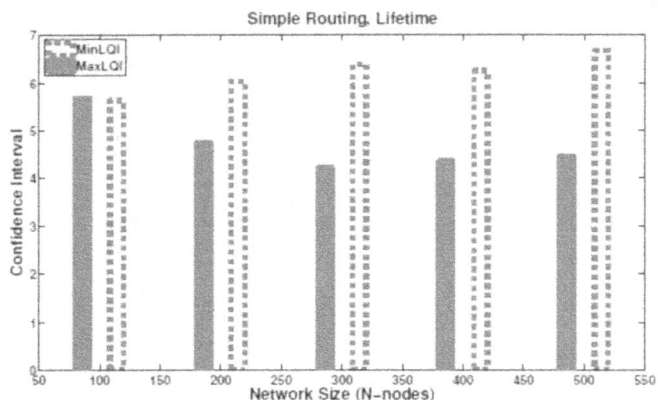

Fig. 1.34. Average network lifetime: confidence interval related to Fig. 1.33 for a confidence coefficient of 95 %.

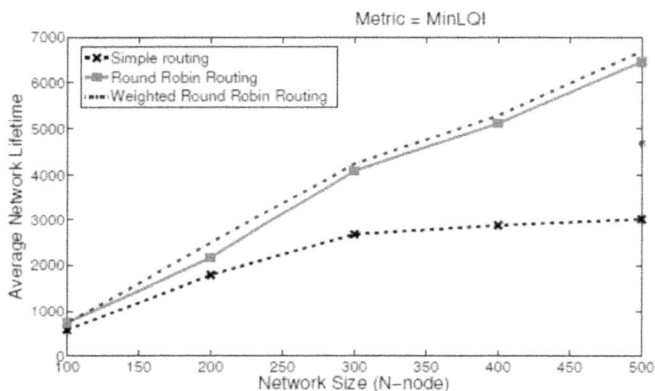

Fig. 1.35. Average network lifetime: Comparison of the three routing mechanisms using the MinLQI metric.

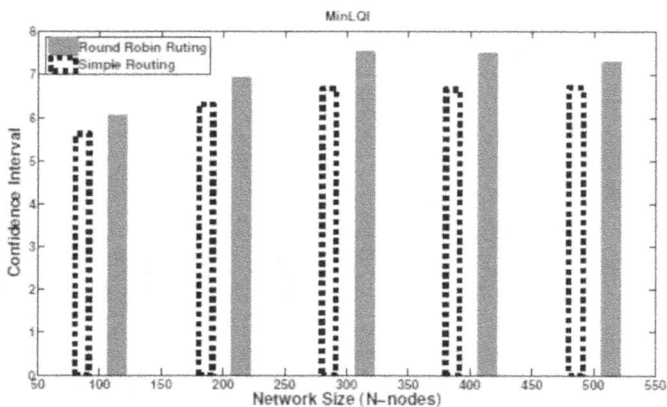

Fig. 1.36. Average network lifetime: confidence interval related to Fig. 1.35 for a confidence coefficient of 95 %.

Compared to the simple routing, the load balancing mechanisms Fig. 1.35 significantly increase the WSN lifetime. However, even if the weighted round robin routing leads to a better WSN lifetime than the round robin routing, the gap between the two load balancing mechanisms is not significant for the MinLQI metric Fig. 1.35.

By rotating the achtophorous node this helps splitting the load among different sensors. So, load balancing helps delaying the moment of the first node battery depletion of and therefore extending the lifetime of the network: Load balancing adds lifetime benefits to the WSN.

1.10.6. Average Ratio of the Remaining Energy

The Fig. 1.37 and Fig. 1.38 display the evolution average remaining energy after a complete cycle, when the network density is increasing. The cycle is constituted by: the network deployment, the detection of alarms and the data routing towards the base station where each source node uses L2RP to build its routing table. The cycle ends when all nodes have sent their alarms.

The degree of connectivity and MaxLQI metrics are the least energy efficient metrics Fig. 1.37. In contrast, Proximity-BS and MinLQI are the metrics that ensure better energy efficiency.

The weighted round-robin routing (W2R) leads to less energy consumption than the round-robin routing which is better than the simple routing whatever the metrics used. The Fig. 1.38 shows the result for the MaxLQI metric.

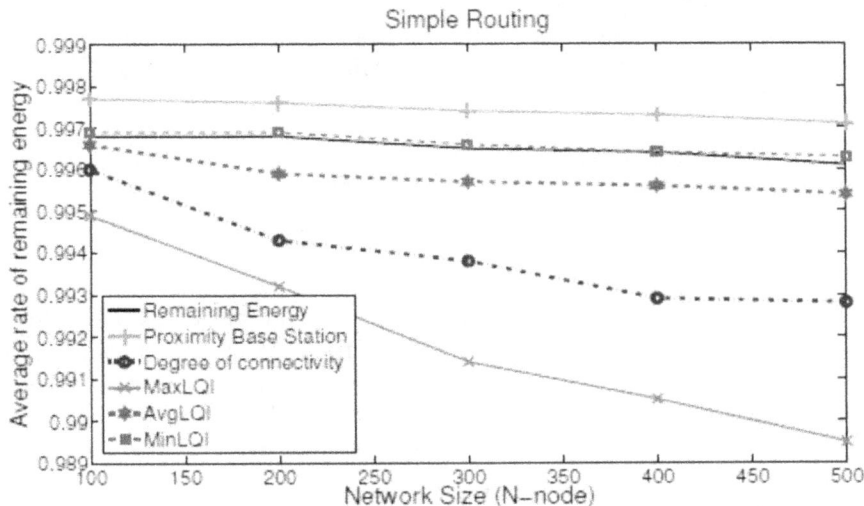

Fig. 1.37. The average ratio of the remaining energy: Comparison of the different metrics in the Simple Routing mechanism, after one cycle of which all sensors had sent their alarms towards the Base Station.

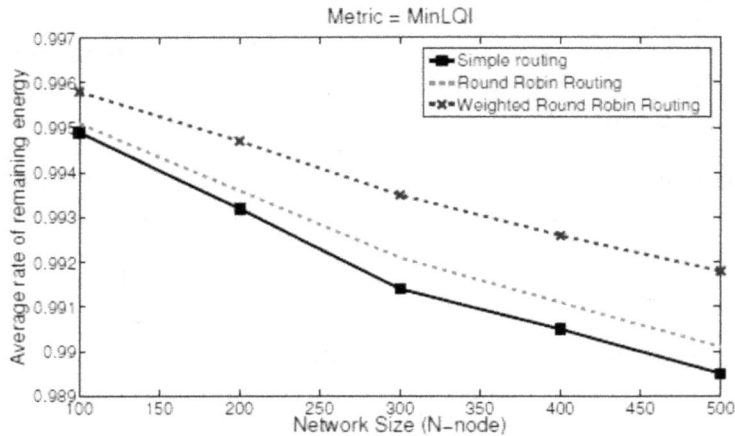

Fig. 1.38. The average ratio of the remaining energy: Comparison of the three routing mechanism with the MaxLQI metric, after one cycle of which all sensors had sent their alarms towards the Base Station.

In summary, these results are natural consequences of the previous ones. Indeed, for the MaxLQI metric of which the average number of hops (average path length) is high, the energy consumption is also large because of the increasingly overhearing, latency, and overhead phenomena.

1.10.7. Impacts of Increasing the Number of Achtophorous Nodes

The Fig. 1.39 shows the influence of the number (ANs) of the achtophorous nodes on the network lifetime performance criterion by comparing the results for *ANs = 3* and *ANs = 5*, when the remaining energy and MinLQI metrics are combined with the round-robin routing.

The Fig. 1.40 shows the influence of increasing the number (*ANs*) of the achtophorous nodes on the average percentage of lost packets by comparing results for *ANs = 3* and *ANs = 5*, when the W2R routing is run with the remaining energy and MinLQI metrics.

These two results Fig. 1.39 and Fig. 1.40 show that the average percentage of lost packets decreases for the MinLQI metric. The network lifetime increases for both metrics when the number of achtophorous nodes varies from 3 to 5. This is not obvious to predict, because increasing the number of achtophorous nodes might increase the risk of using low-energy sensors in routing process, which could cause more packet losses.

From a given number of achtophorous nodes, the result should be reversed. Nevertheless, until the value *AN = 5*, it remains within reasonable limits for a cold chain monitoring application.

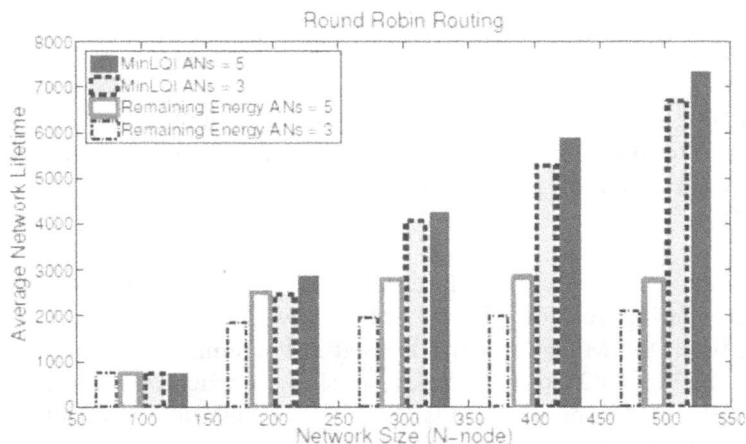

Fig. 1.39. Impacts of increasing the number of achtophorous nodes on the average network lifetime for the round-robin routing mechanism.

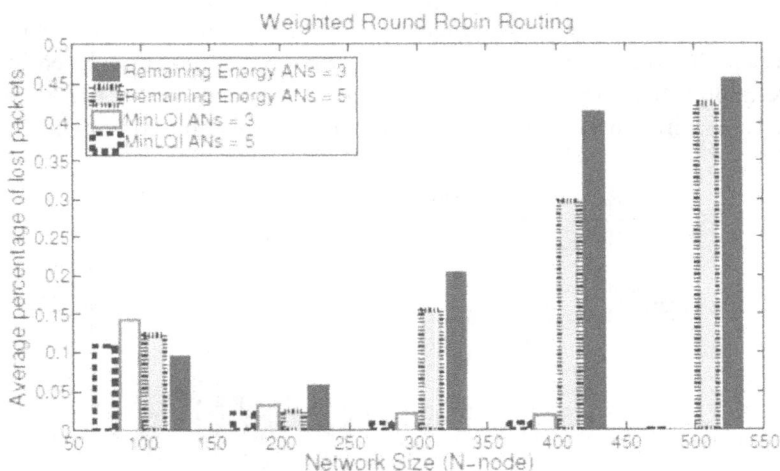

Fig. 1.40. Impacts of increasing the number of achtophorous nodes on the average percentage of packet losses for the W2R Routing mechanism.

1.10.8. Impacts of the Unreliability of Wireless Links

In the context of our application, the warehouse hosts hundreds of pallets, one upon the other. Each pallet is provided with a temperature sensor. This environment is subjected to the unreliably feature of the wireless links. In this section we take into account such a phenomenon. At any given time t, for a sensor S_i, its unreliable links $(P_r[\ell(i,j,t) = Unreliable] = 1$ in Formula (1.12)) with some neighbours are modelled by the Poisson process of parameter $\gamma(S_i, t)$ calculated as follows:

$$\gamma(S_i, t) = \frac{\mu}{\delta(S_i)},$$

(1.17)

69

where $\delta(S_i)$ is the number of nodes located between the node S_i and the BS. If $\delta(S_i) = 0$, then the node S_i has no eligible achtophorous node.

At any given time t, for each sensor S_i, $\gamma(S_i, t)$ is too small, then the Poisson process returns a series T_i of integers T_i, in which nonzero values $T_i[j]$ denote the unreliable links formed by S_i with some of its neighbours S_j i.e. ($P_r[\ell(i, j, t) = Unreliable] = 1$ in Formula (1.12).

The Fig. 1.41 shows the effect of the unreliability of the wireless links on the WSN lifetime by comparing results for μ = 0.01 (low unreliability) and μ = 0.1 (high unreliability), when the MinLQI metric is used in the simple routing and in the round-robin routing. The Fig. 1.42 (resp. the Fig. 1.43) shows impacts on the average path length (resp. on the LIF) by comparing results for μ = 0.1 (high unreliability), when MinLQI metric is used in the three routing mechanisms.

The first result in Fig. 1.41, shows that the network lifetime is smaller in high unreliable WSN (μ = 0.1). In this case, the load balancing also increases the network lifetime. Indeed, the round-robin routing in high unreliable WSN (μ = 0.1) is much better than the simple routing in low unreliable links environment (μ = 0.01), even if the simple routing produces lower average path length Fig. 1.42 than load balancing mechanisms. Even in the context of high unreliable links, the load balancing routing produces better LIF than the simple routing Fig. 1.43, which means that the load is more evenly shared between nodes.

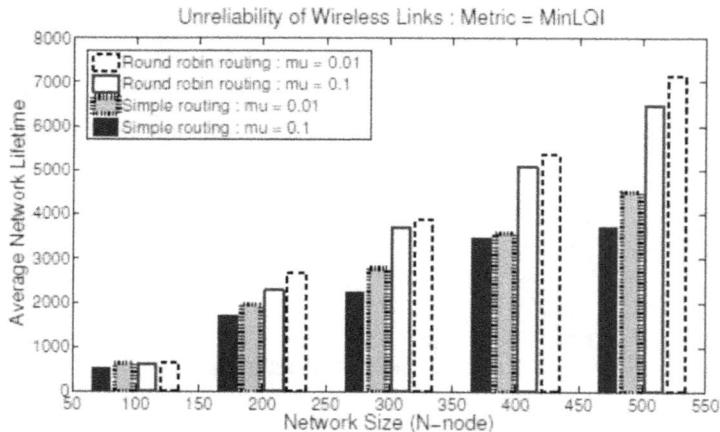

Fig. 1.41. Impacts of the unreliability of the wireless links on the average network lifetime (MinLQI, μ = 0.01 and μ = 0.1) for both simple and round robin routing mechanisms.

This result pertained to the MinLQI metric, clearly shows that the unreliability of the wireless links phenomena reduce the WSN lifetime because of more retransmissions needed in such an environment. But the key point of this result relies on the fact that load balancing mechanisms also add lifetime benefits in high unreliable environment. Indeed, in the case of simple routing, a weak link between a sensor and its achtophorous node involves the sending of a new "ROUTE REQUEST" message. In contrast, for load balancing mechanisms, each sensor has several achtophorous nodes in its routing table. If a link between a sensor and its achtophorous node were to be unreliable, the source node first examines the quality of the link it forms with the next achtophorous node listed in its routing table. So, if this link is reliable, it simply sends the packet without having to request a new route.

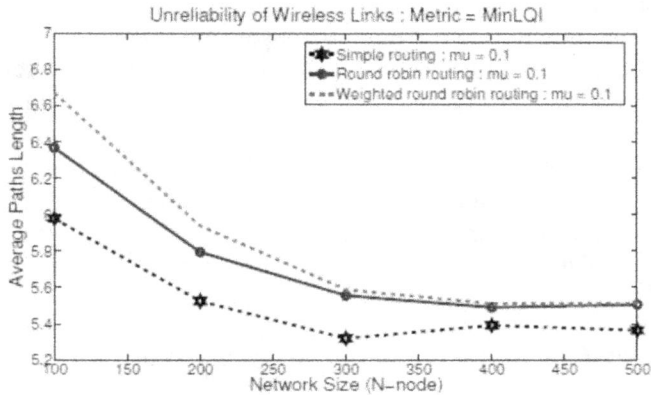

Fig. 1.42. Impacts of the unreliability of the wireless links on the average path length (MinLQI, $\mu = 0.1$) for the three routing mechanisms.

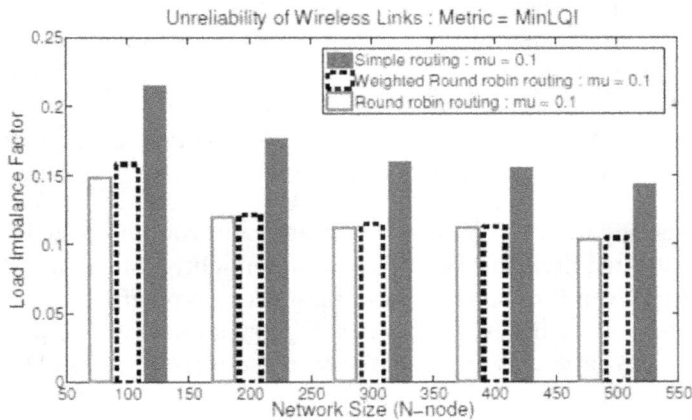

Fig. 1.43. Impacts of the unreliability of the wireless links on the Load Imbalance Factor (MinLQI, $\mu = 0.1$) for both simple and load balancing routings.

Therefore, in load balancing mechanisms, a source node has to send a new "ROUTE REQUEST" message, if and only if all the links it forms with all the achtophorous nodes listed in its routing table were to become unreliable at the same time.

In the scenarios which do not take into account the unreliability of the wireless links, the L2RP protocol leads to identical routes for the two load balancing mechanisms Fig. 1.23. The Fig. 1.42 shows the impacts of the unreliability of the wireless links on the average

71

path lengths (number of hops) in an environment subjected to high unreliable links ($\mu = 0.1$). In this result, we observe that the routes obtained with the round robin mechanism are now different from those obtained with the weighted round robin routing (W2R) Fig. 1.42. This is due by the fact that link quality parameters are very fickle and time variant and are greatly dependent on the poison parameter $\gamma(S_i, t)$ in formula (1.17).

The Fig. 1.43 plots, for $\mu = 0.1$, the impacts of high unreliability of the wireless links on the LIF criterion performance. This still confirms the effectiveness of the load balancing mechanisms in unreliable environments. Indeed, the load imbalance factor (LIF) is lower for the round robin and W2R routing compared to the simple routing. Contrary to the previous result Fig. 1.27, this one Fig. 1.43 shows that the gap between the average LIF of the simple routing mechanism and those obtained via the load balancing routings decreases as the network density is increasing. Indeed, the unreliability of the wireless links become more and more important when the density of the WSN is increasing [3]. Consequently, load balancing mechanisms gradually begin to lose some of their interest.

1.11. Energy Over-Consumption Induced by Securing Routing Operations

1.11.1. Introduction and Backgrounds

Considering the fact that sensors are often dispersed in unprotected environments, WSNs are vulnerable to several types of intrusions and attacks. Thus, the proper functioning of the network might be compromised. In such hostile environments, the information exchanged between two communicating parties can include highly sensitive data that must be secured when sent through intermediate nodes. In the state of the art, there are currently several security mechanisms to ensure a good level of data protection as it flows from one source toward one (or more) destination(s) in the network.

Security solutions are usually based on cryptographic routines which help to ensure traditional security requirements such as confidentiality, integrity, availability and authentication [61]. Even for wireless sensor networks, cryptographic techniques [62-64] are used for protection purposes. But calculating crypto keys could push down WSNs performance. Security protocols designed for wired networks (such as DES, 3DES, AES, RSA, TLS, MD5, SHA-1, etc.) are generally not suitable for WSNs, because in such wireless networks, nodes have restricted capacity calculation (CPU), scanty capacity of memorization (RAM) and small energy supply. Thereby, security solutions must be custom sized and tightly adapted for wireless environments and especially for WSNs.

In this way, related works propose several security mechanisms for WSNs without caring about energy overconsumption induced by adding cryptography keys. We agree that such security solutions could naturally decrease network performance, but the real question is to understand in what proportion adding cryptography keys could affect network lifetime. Accordingly, in this chapter, we will evaluate the energy consumption induced by increasing the size of the packet when establishing cryptographic keys for wireless sensor networks, and then estimate the decrease rate of network performance.

We call unnecessary over-consumption of energy the consumption that can be avoided to maintain the energy of a sensor node. Energy efficiency is one of the most important issues in wireless sensor networks. Whatever the TCP/IP network layer (network access, internet, transport or application) considered, we must always take into account factors of energy consumption in order to develop an energy efficient protocol for WSNs. As a matter of fact, the major causes of energy losses are [3]: Collisions, Idle listening, Overhearing, Overmitting, Control packets overhead, Network protocols, Attacks and security issues, Cryptography and security solutions, etc.

Many attacks [65] are carried out against the sensor networks in order to exhaust the batteries of the nodes and then to reduce their lifetimes, and consequently that of the network. For instance, with a very high power transmitter node, an attacker can perform Hello Flood attack by broadcasting Hello messages on a large range to certain sensors of the network. Then sensors will try to respond to the attacker node which is actually out of range. In such a case attack, an attacker uses this mechanism to consume energy.

Security mechanisms, including cryptography [66-69], are aggravating factors of energy consumption in sensor networks. Many works are concerned with the establishment of cryptographic keys without worrying about key lengths. This is the strong point of cryptographic algorithms for wired networks. However, this becomes a major disadvantage in wireless sensor networks where the safeguarding of energy is a sine qua non condition for longer node and network lifetime.

Moreover, adding integrity key to secure routing protocol exchanges should increase packet size and therefore energy expenditure. In the next section, we will try to estimate how, that's to say in what proportion, adding an integrity key will cause energy exhaustion.

1.11.2. Energy Over-Consumption Induced by Adding an Integrity Key to L2RP

For routing protocols [70], nodes compute their routing table by exchanging messages in order to compare selection criteria also called metric [1-4, 71]. Then the node which has the lowest metric is chosen as next hop router. The network convergence is the fact that all nodes have successfully completed computing their routing table. Generally, the convergence occurs when all the routers in the routing domain agree on the routes that are available. Convergence time is the time that it takes for every router's routing table to synchronize after there has been a change in the network topology. It is important to ensure that the time taken is as short as possible because while the routers disagree on the available network, they cannot route data correctly or efficiently.

So, a commonly used WSN attack is to introduce malicious nodes in the network which will play the role of next hop routers by sending lowest criterion values (after a small period of sniffing network communications). In this way, traffic could be hijack toward the attacker platform. To avoid this type of security issue, one mechanism is to add integrity keys in order to secure communications between nodes. Integrity keys help ensuring authentication and message integrity in such a way that a malicious node will not be able to perform this kind of attack. Most of algorithms providing integrity keys are designed for wired network and are not suitable for wireless sensor network operations because provided integrity keys are often too long. In this section, we aimed to show that integrity keys have non negligible energy cost and we'll estimate the rate of energy depletion induced by adding an integrity key in order to secure routing protocol processes. Firstly, we will present in the next section some types of common deployment WSN topologies we are using to run simulations.

The size of messages exchanged in the network has an effect on power consumption of transmitter and receiver nodes. Thus packet size should be neither too high nor too low. While it is small, the number of control packets (acknowledgment) increases. Otherwise, more transmission power is necessary for larger packet size.

In this work, we consider two routing different scenarios in which each sensor need to send a fixed number of packets before reaching network convergence:

- *Scenario 1:* each sensor node has to send 3 packets before network convergence. This is common to the most of the many WSN routing protocols, such as L2RP (with hello packet), in which sensors often exchange 1, 2 or 3 messages before computing its routing tables [2, 71].

- *Scenario 2:* each sensor node has to send 10 packets before network convergence. This scenario is not common to WSN routing protocols, but it's not an exaggeration if we consider all control packets needed for network auto-organization such as: initialization and authentication phase [72, 73], routing process, clustering formation [74, 75], data aggregation computation [76, 77], other security mechanisms [78-80], etc.

In each scenario, each packet has a fixed size length of 128 bits. Adding an integrity key whose length is equal to 5 % of the initial packet size, we obtain the results of the additional energy expenditure for the first scenario Fig. 1.46 to Fig. 1.48 following different deployment strategies and those of the second scenario in Fig. 1.49 to Fig. 1.51.

For all these results displayed in Fig. 1.46 to Fig. 1.51, the height of each bar highlights the energy over-consumption (in Joule) of a sensor node induced by adding an integrity key which size (7 bits) is equal to around 5 % of the initial package size (128 bits). The abscissa represents the number-id of each sensor. The energy consumption model used for simulations is the same than the Section 1.9.1 of this chapter, and can also be found in [1-3, 56, 57, 60, 81].

First of all, for the random deployment topology, it should be noted that the bars have different heights as shown in the Fig. 1.46 and Fig. 1.49. This results from the fact that each node has a number of neighbours that could be different from that of the other nodes. Because the deployment is random, some areas could be much more densely populated (so more neighbours) than others.

On the other hand, in the two scenarios, for the grid topology, the bars are grouped by several sets of bars having the same height Fig.1.47-1.51. This means that for each group, the concerned nodes have the same average power consumption. By zooming on each of the results concerning the type of grid topology, we obtain, for example, the Fig. 1.44. This will help us better understand these results.

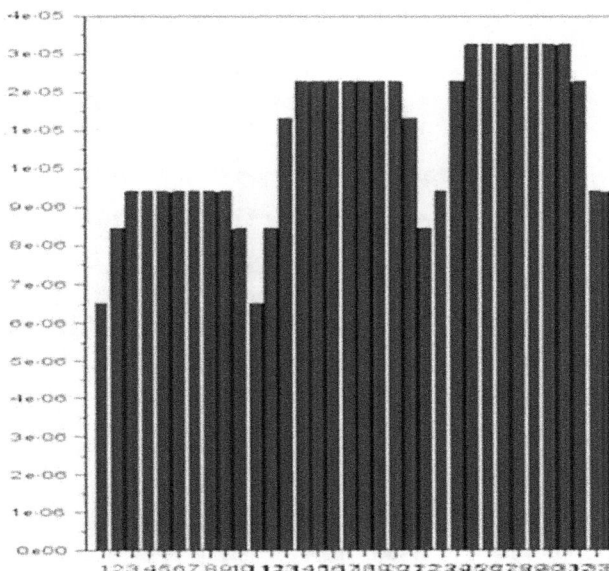

Fig. 1.44. A part of the zoom of the Fig. 1.48.

This results from the inherent properties of the uniform grid topology, where the nodes are numbered from the first row (which is defined by: $\forall x \in [0,100], y = 0$) to the last one ($\forall x \in [0,100], y = 100$), and for each row form the left ($x = 0$) to the right ($x = 100$). In fact, by observing the example illustrated in the Fig. 1.45.

- The first node and the last node of each row have the same number of neighbours. This explains why the node number 1 and the node number 11 have the same energy expenditure in the Fig. 1.44, Fig. 1.48 and Fig. 1.51.

- The second node and the penultimate node of each row have the same number of neighbours. This explains why the node number 2 and the node number 10 have the same energy expenditure in the Fig. 1.44, Fig. 1.48 and Fig. 1.51.

- For the first line, the nodes within the green rectangle (dashed line) Fig. 1.45 have the same number of neighbours. The same applies to the corresponding nodes for each of the other lines of the grid. This explains why all the nodes from the node number 3 to the node number 9 have the same energy expenditure in the Fig. 1.44, Fig. 1.48 and Fig. 1.51.

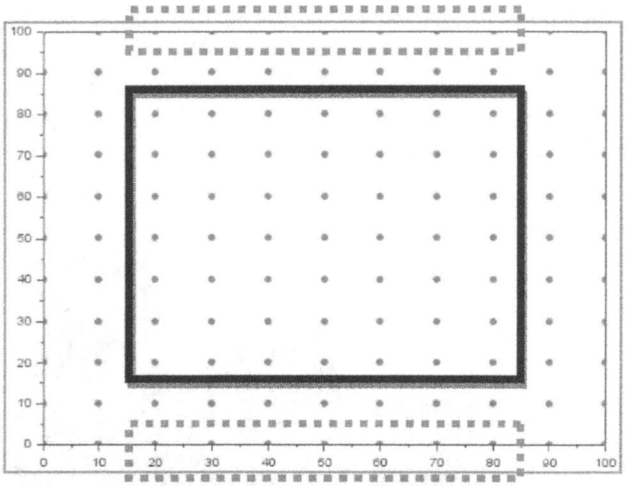

Fig. 1.45. Case of a uniform grid topology ($X_{step} = Y_{step} = 10$ m).

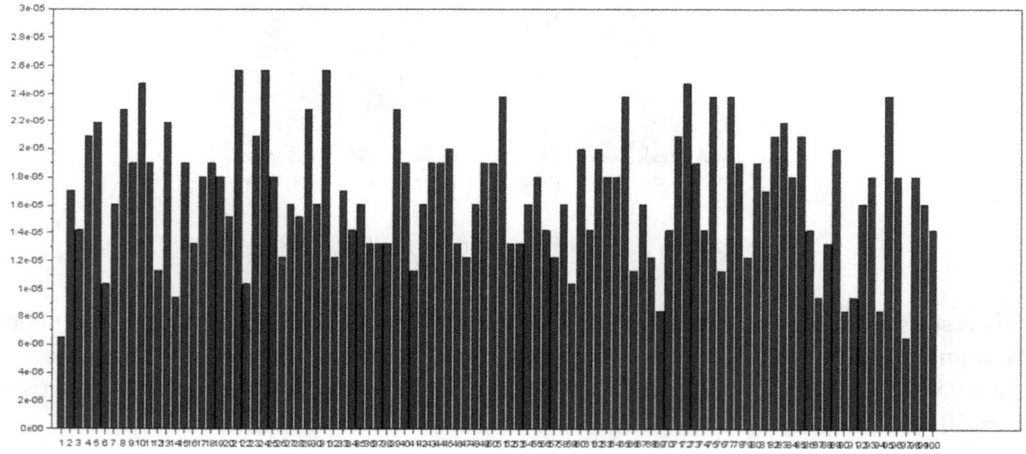

Fig. 1.46. Scenario 1: The energy over-consumption of each sensor node induced by adding an integrity key which size is equal to 5 % of the initial package size, in the case of a random deployment topology as described in Section 1.9.2. The simulation parameters are summarized in Table 1.5.

- All the nodes inside the blue rectangle (solid line) Fig. 1.45 have the same number of neighbours. Moreover, it is the nodes with the greatest number of neighbours in the grid. As a result, all lines affected by the blue rectangle (i.e. from the 3rd row to the 9th row) have the same average energy expenditure. This corresponds to the seven sets of bars having the highest height in the Fig. 1.48 and Fig. 1.51.

- The first set (and then the 11th set) of bars corresponds to the nodes of the first line (and then the 11th line) of the grid. Similarly, the 2nd set (and then the 10th set) of bars corresponds to the nodes of the 2nd line (and then the 10th line) of the grid Fig. 1.48 and Fig. 1.51.

Table 1.5. Simulation parameters for random deployment topology.

Parameter	Value
Area Length	L = 100 m
Area Width	l = 60 m
Base Station Location	position(SB) = (0, 0)
Sensor transmission radio range	R = 20 m
Initial packet size	128 bits
Length of the added key integrity	7 bits

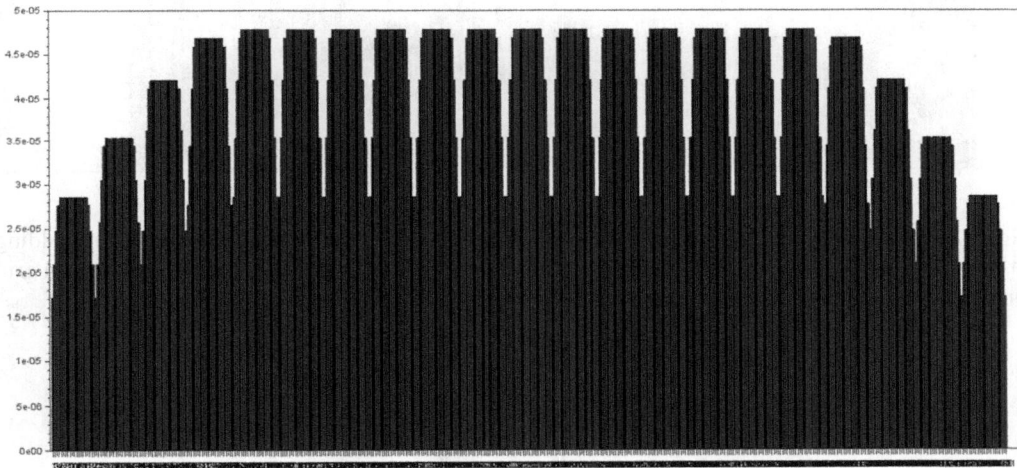

Fig. 1.47. Scenario 1: Energy over-consumption of each sensor node induced by adding an integrity key which size is equal to 5 % of the initial package size in the case of an uniform grid topology (X_{step} = Y_{step} = 5 m). The simulation parameters are summarized in Table 1.6.

Table 1.6. Simulation parameters for a uniform grid topology.

Parameter	Value
Area Length	L = 100 m
Area Width	l = 100 m
Base Station Location	position(SB) = (0, 0)
Sensor transmission radio range	R = 20 m
Initial packet size	128 bits
Length of the added key integrity	7 bits
Uniform grid topology	X_{step} = Y_{step} = 5 m

In each case of deployment topology specified in Fig. 1.46 to Fig. 1.51, we calculated the energy over-consumption of each sensor node induced by adding an integrity key which size is equal to 5 % of the initial package size. The integrity key is added to each of the 3 or 10 packets exchanged during routing process before reaching the network convergence. Results show that the average rate of energy over-consumption per sensor is high despite the smallness of the integrity key.

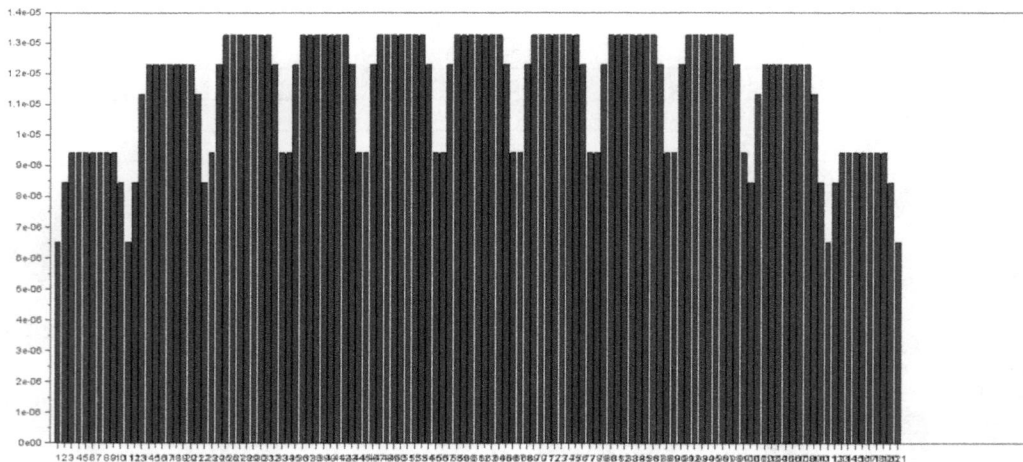

Fig. 1.48. Scenario 1: Energy over-consumption of each sensor node induced by adding an integrity key which size is equal to 5 % of the initial package size in the case of an uniform grid topology ($X_{step} = Y_{step} = 10$ m). The simulation parameters are summarized in Table 1.7.

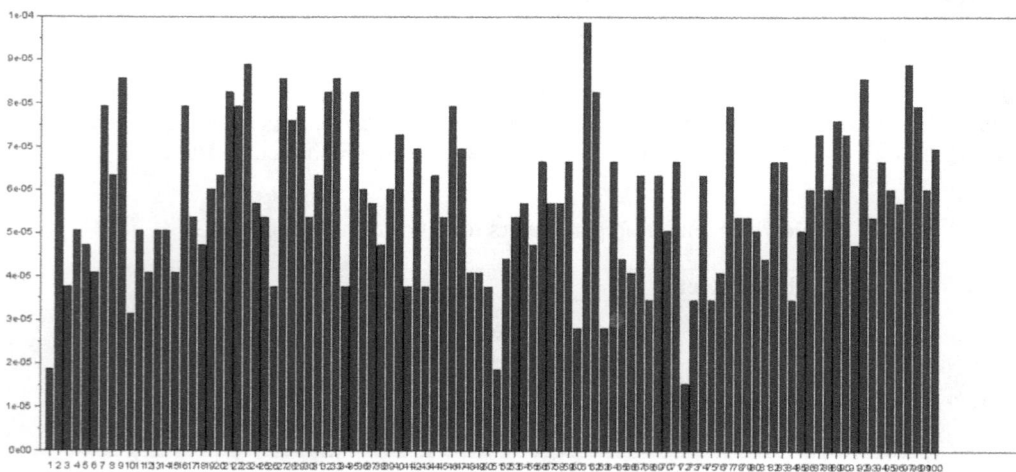

Fig. 1.49. Scenario 2: The energy over-consumption of each sensor node induced by adding an integrity key which size is equal to 5 % of the initial package size in the case of the random deployment topology. The simulation parameters are summarized in the Table 1.5.

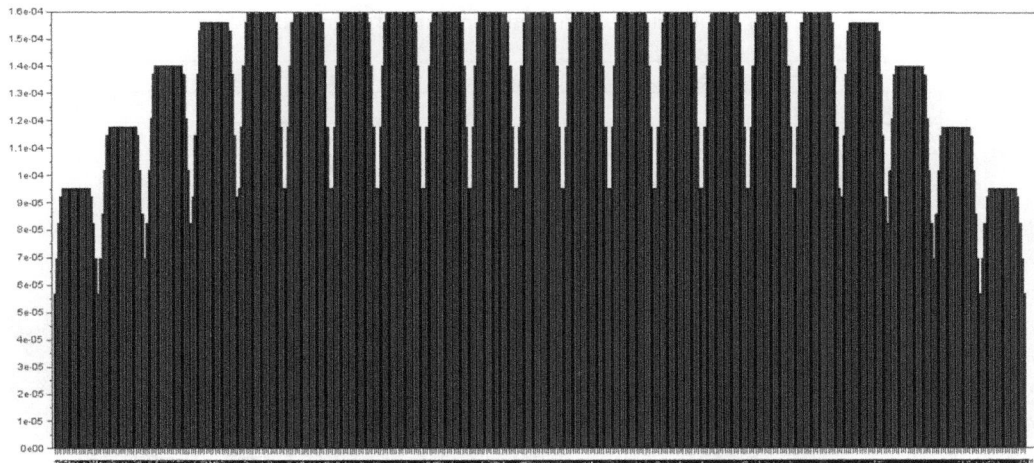

Fig. 1.50. Scenario 2: Energy over-consumption of each sensor node induced by adding an integrity key which size is equal to 5 % of the initial package size in the case of an uniform grid topology ($X_{step} = Y_{step} = 5$ m). The simulation parameters are summarizedin the Table 1.6.

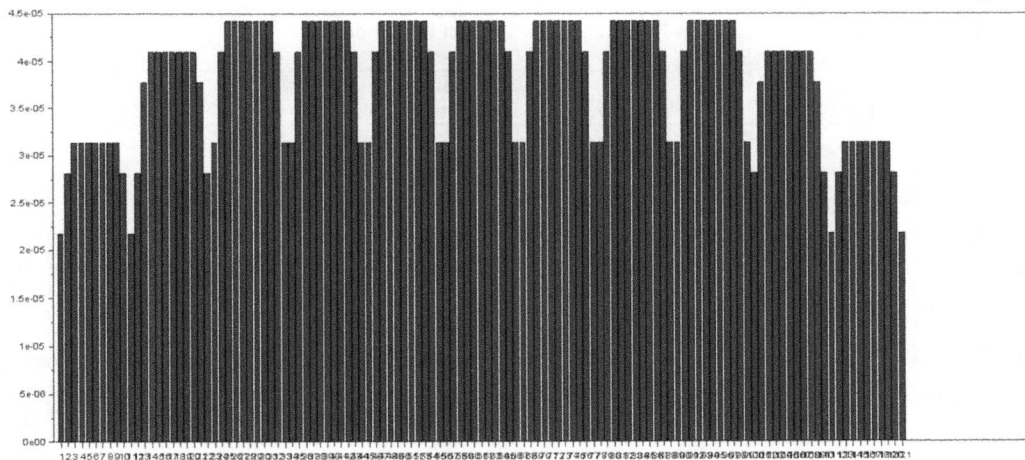

Fig. 1.51. Scenario 2: Energy over-consumption of each sensor node induced by adding an integrity key which size is equal to 5 % of the initial package size in the case of an uniform grid topology ($X_{step} = Y_{step} = 10$ m). The simulation parameters are summarizedin the Table 1.7.

Consequently, we calculated Table 1.8 the average number of additional packets each sensor would be able to send with this amount of energy over-consumption in the absence of the integrity key.

Table 1.7. Simulation parameters for a uniform grid topology.

Parameter	Value
Area Length	L = 100 m
Area Width	l = 100 m
Base Station Location	position(SB) = (0, 0)
Sensor transmission radio range	R = 20 m
Initial packet size	128 bits
Length of the added key integrity	7 bits
Uniform grid topology	Xstep = Ystep = 10 m

Table 1.8. The average number of additional packets that each sensor would be able to send if the routing protocol were not being secure with the integrity key.

Deployment topology			Scenario 1	Scenario 2
Random deployment topology		Average number of additional packets	4	16
		computed from the results displayed in the figures	Fig. 1.46	Fig. 1.49
Uniform grid topology	Xstep = Ystep = 5 m	Average number of additional packets	8	25
		computed from the results displayed in the figures	Fig. 1.47	Fig. 1.50
	Xstep = Ystep = 10 m	Average number of additional packets	4	7
		computed from the results displayed in the figures	Fig. 1.48	Fig. 1.51

These results point out that in both scenario and for the two deployment topologies, the energy required for the added integrity key is sufficient to run the routing protocol again. That's to say that it is more energy-expensive to secure the routing protocol than running it again.

The energy over-consumption of each sensor node is more important as the network density becomes larger (Fig. 1.47 and Fig. 1.48 in comparison with Fig. 1.50 and Fig. 1.51). Moreover, when the average number of neighbours is increasing in the network, as a consequence of the increase of the network density, the eavesdropping and the overhearing phenomena could then constitute aggravating factors in energy consumption. For uniform grid deployment topologies, moving from $\{X_{step} = Y_{step} = 10$ m$\}$ to $\{X_{step} = Y_{step} = 5$ m$\}$ has the effect to increase the average number of neighbours in the network. This explains the fact that the energy expenditure is almost multiplied by a factor of 2 in the results Fig. 1.47 with respect to Fig. 1.48 and Fig. 1.50 with respect to Fig. 1.51.

Thus, as we can see, providing a small integrity key, whatever the type of the deployment topology considered, greatly increases energy consumption. On the other hand, even if decreasing the node deployment density results in a decrease in average energy consumption due to Collisions, Idle listening, Overhearing, Overmitting, etc., one must still avoid in exaggeratedly swerving the positions of the nodes on the deployment area. As this may result in increased loss of connectivity due to the unreliability feature of the wireless links, which could increase the phenomenon of packets losses and retransmissions, and then annihilates quality of service while increasing energy exhaustion. So one should be careful with this, and operate to moderately swerve the positions of the sensors in a real environment.

1.12. Conclusions

In this work, we have proposed the L2RP routing protocol (Link Reliability based Routing Protocol) which takes into account the quality of the links formed by any source node with the achtophorous nodes listed in its routing table. This avoids sending data over a link disrupted, unreliable or unstable.

The L2RP protocol also includes load balancing mechanisms where the source node, based on "ROUTE REPLY" packets, is able to estimate the load sustainable by each of its achtophorous node. This property allows L2RP to avoid doing a per packet load-balancing by the source, as done in [55], where the source node sends its data without being sure of the capacity of the achtophorous node to sustain the load assigned. Thus, by doing so, L2RP helps to reduce packet losses.

Applications often have their specific objectives and constraints, so it is essential to have the choice between several possible settings when deploying a wireless sensor networks. Thus, in its design, the L2RP protocol can use any chosen metric. This allows L2RP to be able to support different applications by offering the choice of the metric which ensures the best performance in the specific context of each application.

We therefore evaluated the L2RP performance based on routing mechanisms (simple or load balancing) and then presented a comparative study of the different metrics in each routing mechanisms. This work has shown that:

The degree of connectivity metric is the metric that leads to the highest percentage of packet losses. This metric also has the lowest network lifetime. Indeed, it is the metric which is the most sensitive to the overhearing phenomenon.

The Proximity-BS metric provides better energy efficiency. With this metric, the alarms sent by any sensor reach the Base Station in fewer hops. By minimizing the number of hops, it helps in reducing energy wastefulness due to overhearing, overhead and latency.

The LQI used as a metric by considering the best link quality (the MaxLQI metric) leads to an inefficient routing regardless of the performance criterion considered. This confirms our previous experimental results obtained in [25]. The MaxLQI metric matches the standard definition of the LQI used in the MultiHopLQI routing algorithm [29]. Indeed, this metric is characterized by a relatively high average number of hops. In the absence of obstacles and any interference, the best link quality is often observed for the nodes which are located relatively close to each other. By multiplying the number of hops, the MaxLQI metric has the effect to increase energy wastefulness due to overhearing, overhead and latency.

Accordingly, despite its popularity in WSN empirical analysis based on TinyOS platforms, the MultiHopLQI routing algorithm is not suitable for WSN applications, because it uses the MaxLQI metric for route selection.

By setting a given LQI threshold, i.e. a value of acceptable LQI, and considering the lowest LQI value beyond this threshold (the MinLQI metric), we obtain an optimal LQI based metric which highly enhances the energy efficiency. As the LQI decreases when the distance between the nodes increases, the average path length is larger for MaxLQI than for MinLQI: this explains why MinLQI is more energy-efficient than MaxLQI. Then, the average percentage of packet losses is larger for MaxLQI. There is a trade-off between routes consisting of good links quality and small average path length (i.e. without too many retransmissions).

This interesting result shows that it is better for LQI based routing algorithm to promote links of intermediate quality (such as MinLQI metric) to avoid:

The better links which are synonymous of nodes located relatively close to each other and also synonymous of higher number of routing hops which are responsible for excessive energy consumption;

The bad links (low quality) which are synonymous of higher percentage of packet losses.

The load balancing mechanisms significantly improve the routing efficiency by extending the network lifetime, while minimizing the average percentage of packet losses. The load balancing also helps evenly splitting the load on all nodes in the WSN.

Increasing the number of achtophorous nodes improves the network performance: a low average of packet losses and a longer network lifetime.

The composite metric, resulting of the remaining energy metric combined with the Proximity-BS metric, offers good routing performance. This metric is interesting, as each node ignores the settings of its neighbours (such as the remaining energy, the position) when selecting its achtophorous nodes.

Since it is LQI based routing algorithm, the question that naturally arises is how L2RP behave in an environment subjected to high unreliability of the wireless links. Simulation results have shown that, such an environment slightly impacts the L2RP efficiency. Generally, packet loss percentage is relatively low because in L2RP a source node avoids sending data to an achtophorous node with which it forms an unreliable link at the moment it has data to transmit.

Embedded with load balancing mechanisms, L2RP adds lifetime benefits to the wireless sensor network. Nevertheless, it would be more profitable to combine L2RP with aggregation techniques like cluster formation and data aggregation in order to gain more scalability and lifetime. So, in [60] we have used L2RP in a cold chain monitoring application where regular sensors send alarms to their respective clusterheads, which aggregate received alarms and then forward the aggregated data towards the BS using the L2RP routing protocol. In this application L2RP is run with the weighted round robin load balancing mechanism using the "MinLQI" metric.

On the other hand, in this report, we show that how the energy over-consumption induced by securing network operations is high. We added a small integrity key (7 bits) which length is equal to 5 % of the initial packet size (128 bits) in order to secure the routing tables' computation process. Simulation results for two deployment strategies allow us to conclude that adding integrity key, for network operations security, significantly increases the energy consumption of sensors and therefore reduces network performance. Even if there is lot of works on lightweight cryptography [61] and [82], which concerns the science of small cryptography keys, it is still found that many works continue offering relatively long keys with respect to the size of the protected data without caring about energy over-consumption.

Acknowledgements

This research was supported by:

- SAMOVAR CNRS Research Lab – UMR 5157; Dept. Réseaux et Services de Télécommunications (RST); Télécom & Management SudParis, Evry, France. I would like to thank Prof. Monique Becker and Prof. Michel Marot of Télécom & Management SudParis. I am more than grateful to them for their help. (http://www.telecom-sudparis.eu/);

- Centre d'Excellence en Mathématiques, Informatique et TIC (CEA-MITIC); UFR Sciences appliquées et de Technologies (UFR SAT), Université Gaston Berger, Saint-Louis, Sénégal. (http://www.ceamitic.sn/).

References

[1]. C. Diallo, M. Marot, M. Becker, Efficiency benefits through load-balancing with link reliability based routing in WSNs, *Journal on Advances in Networks and Services*, Vol. 3, No. 3&4, 2010, pp. 430-446.

[2]. C. Diallo, M. Marot, M. Becker, Link quality and local load balancing routing mechanisms in wireless sensor networks, in *Proceedings of the 6th Advanced International Conference on Telecommunications (AICT'10)*, Barcelona, Spain, May 2010, pp. 306-315.

[3]. C. Diallo, Techniques d'amélioration du routage et de la formation des clusters multi-*sauts* dans les réseaux de capteurs sans fil., PhD Thesis, *Télécom & Management SudParis*, Evry, France 2010.

[4]. M. T. Sow, C. Diallo, Energy over-consumption Induced by securing network operations, in *Proceedings of the 2ⁿᵈ International Conference on Frontiers of Sensors Technologies (ICFST'17)*, Shenzhen, China, April 2017.

[5]. ZigBee Specificationv1.0, *ZigBee Alliance*, USA, June 2005.

[6]. TinyOs Community Forum, http://www.tinyos.net/

[7]. Tmote Sky datasheet, http://www.moteiv.com/products/docs/tmote-skydatasheet.pdf

[8]. Sun SPOT World, http://www.sunspotworld.com

[9]. Sun SPOT, smartsantander, http://www.smartsantander.eu/wiki/index.php/Main/Sunspot

[10]. EasySen WiEye Sensor Board, http://www.easysen.com/wieye.htm

[11]. Crossbow sensor platform, http://www.xbow.com/)

[12]. Micaz datasheet, http://www.openautomation.net/uploadsproductos/micaz_datasheet.pdf

[13]. Jennic Wireless Microcontrollers, http://www.jennic.com/index.php

[14]. Jennic Wireless Microcontrollers, http://www.glyn.com.au/HF_Opto_Jennic.htm

[15]. Jennic Wireless Microcontrollers, http://www.glyn.com.au/PR01062007_Jennic.htm

[16]. TinyNode Sensors, http://www.tinynode.com/

[17]. T-Node de SOWNet, http://www.sownet.nl/

[18]. TinyNode 584, http://www.tinynode.com/?q=product/tinynode584/tn-dev584

[19]. BTnode rev 3, http://www.btnode.ethz.ch/Documentation/BTnodeRev3SensorGuide

[20]. Silicon Labs, http://www.silabs.com/Pages/default.aspx

[21]. Microchip Wireless Devices, http://www.microchip.com/

[22]. WSN architecture illustration, http://vlssit.iitkgp.ernet.in/ant/ant/8/theory/

[23]. I. F. Akyildiz, S. Weilian, Y. Sankarasubramaniam, E. Cayirci, A survey on sensor networks, *IEEE Communications Magazine*, Vol. 40, Issue 8, 2002, pp. 102-114.

[24]. C. Diallo, A. Gupta, M. Becker, M. Marot, Energy aware database updating protocols for autoconfigurable sensor networks, in *Proceedings of the 8ᵗʰ International Conference on Networks GlobeNet 2009(ICN'09)*, Cancun, Mexico, Mar. 2009, pp. 138-143.

[25]. A. Gupta, C. Diallo, M. Marot, M. Becker, Understanding topology challenges in the implementation of wireless sensor network for cold chain, in *Proceedings of the IEEE Radio and Wireless Symposium(RWS'10)*, New Orleans, LA, USA, 2010, pp. 376-379.

[26]. Wireless medium access control (mac) and physical layer (phy) specifications for low-rate wireless personal area networks (wpans), IEEE Std 802.15.4-2006, *IEEE Computer Society*, 2006.

[27]. CC2420 Radio, http://www.chipcon.com redirect to http://www.ti.com

[28]. MultiHopLQI, http://www.tinyos.net/tinyos-1.x/tos/lib/MultiHopLQI

[29]. J. Polastre, J. Hui, J. Z. P. Levis, D. Culler, S. Shenker, I. Stoica, A unifying link abstraction for wireless sensor networks, in *Proceedings of the 3ʳᵈ International Conference on Embedded Networked Sensor Systems (SenSys'05)*, 2005, pp. 76-89.

[30]. A. Woo, T. Tong, D. Culler, Taming the underlying challenges of reliable multihop routing in sensor networks, in *Proceedings of the 1ˢᵗ International Conference on Embedded Networked Sensor Systems(SenSys'03)*, Los Angeles, CA, USA, 2003, pp. 14-27.

[31]. M. Becker, A.-L. Beylot, R. Dhaou, A. Gupta, R. Kacimi, M. Marot. Experimental study: Link quality and deployment issues in wireless sensor networks, in *Proceedings of the 8ᵗʰ International IFIP-TC 6 Networking Conference (NETWORKING'09)*, LNCS 5550, Aachen, Germany, 2009, pp. 140-25.

[32]. D. Puccinelli, M. Haenggi, Lifetime benefits through load balancing in homogeneous sensor networks, in *Proceedings of the IEEE Wireless Communications and Networking Conference(WCNC'09)*, Budapest, Hungary, April 2009.

[33]. D. Puccinelli, M. Haenggi, Arbutus: Network-layer load balancing for wireless sensor networks, in *Proceedings of the IEEE Wireless Communications and Networking Conference(WCNC'08)*, Las Vegas, NV, USA, March 2008, pp. 2063-2068.

[34]. D. Lal, A. Manjeshwar, F. Herrmann, E. Uysal-Biyikoglu, A. Keshavarzian, Measurement and characterization of link quality metrics in energy constrained wireless sensor networks, in *Proceedings of the IEEE Global Communications Conference(Globecom'03)*, San Francisco, USA, 2003, pp. 446-452.

[35]. J. Zhao, R. Govindan, Understanding packet delivery performance in dense wireless sensor networks, in *Proceedings of the 1st International Conference on Embedded Networked Sensor Systems (SenSys'03)*, CA, USA, 2003, pp. 1-13.

[36]. D. Son, B. Krishnamachari, J. Heidemann, Experimental analysis of concurrent packet transmissions in low-power wireless networks, in *Proceedings of the 1st International Conference on Embedded Networked Sensor Systems (SenSys'06)*, Colorado, USA, 2006, pp. 237-250.

[37]. G. Zhou, T. He, J. Stankovic, T. Abdelzaher, Rid: Radio interference detection in wireless sensor networks, in *Proceedings of the IEEE24th Annual Joint Conference of the IEEE Computer and Communications Societies (Infocom'05)*, Miami, USA, 2005, pp. 891-901.

[38]. S. Singh, M. Woo, C. Raghavendra, Power-aware routing in mobile ad hoc networks, in *Proceedings of the 4thAnnual ACM/IEEE International Conference on Mobile Computing and Networking (Mobicom'98)*, Dallas, Texas, USA, 1998, pp. 181-190.

[39]. K. Scott, N. Bamboos, Routing and channel assignment for low power transmission in pcs, in *Proceedings of the 5th IEEE International Conference on Universal Personal Communications (ICUPC'96)*, Cambridge, USA, 1996, pp. 498-502.

[40]. R. Shah, J. Rabaey, Energy aware routing for low energy ad hoc sensor networks, In *Proceedings of the IEEE Wireless Communications and Networking Conference (WCNC'02)*, Orlando, Florida, USA, March 2002, pp. 350-355.

[41]. J. Chang, L. Tassiulas, Maximum lifetime routing in wireless sensor networks, *IEEE/ACM Transactions on Networking*, Vol. 12, 609619.

[42]. F. Othman, N. Bouabdallah, R. Boutaba, Load-balanced routing scheme for energy -efficient wireless sensor networks, in *Proceedings of the IEEE Global Telecommunications Conference (Globecom'08)*, New Orleans, LA USA, 2008, pp. 1-6.

[43]. S. Toumpis, S. Gitzenis, Load balancing in wireless sensor networks using Kirchhoff's voltage law, in *Proceedings of the IEEE INFOCOM'2009*, Rio de Janeiro, Brazil, 2009, pp. 1656-1664.

[44]. J. Gao, L. Zhang, Load balanced short path routing in wireless networks, in *Proceedings of the IEEE INFOCOM 2004*, Hong Kong, China, 2004.

[45]. Z. Wang, E. Bulut, B.K. Szymanski, Energy efficient collision aware multipath routing for wireless sensor networks, in *Proceedings of the IEEE International Conference on Communications (ICC'09)*, Dresden, Germany, 2009.

[46]. L. Popa, C. Raiciu, I. Stoica, D.S. Rosenblum, Reducing congestion effects by multipath routing in wireless networks, in *Proceedings of the 14th IEEE International Conference on Network Protocols(ICNP'06)*, Santa Barbara, USA, 2006, pp. 96-105.

[47]. C. Wu, R. Yuan, H. Zhou, A novel load balanced and lifetime maximization routing protocol in wireless sensor networks, in *Proceedings of the IEEE Vehicular Technology Conference (VTC'08)*, Singapore, 2008, pp. 113-117.

[48]. K. Sha, J. Du, W. Shi, Wear: A balanced, fault-tolerant, energy aware routing protocol for wireless sensor networks, in*ternational Journal of Sensor Networks*, Vol. 1, Issue 3-4, 2006, pp. 156-168.

[49]. I. Raicu, L. Schwiebert, S. Fowler, S.K.S. Gupta, Local load balancing for globally efficient routing in wireless sensor networks, in*ternational Journal of Distributed Sensor Networks*, Vol. 1, 2005, pp. 163-185.

[50]. R. Vidhyapriya, P.T. Vanathi, Energy efficient adaptive multipath routing for wireless sensor networks, *IAENGInternational Journal of Computer Science*, Vol. 34, Issue 1, 2007.

[51]. V.C. Gungor, C. Sastry, Z. Song, R. Integlia, Resource-aware and link quality based routing metric for wireless sensor and actor networks, in *Proceedings of the IEEE International Conference on Communications(ICC'07)*, Glasgow, Scotland, 2007, pp. 3364-3369.

[52]. S. Hussain, A. W. Matin, Hierarchical cluster-based routing in wireless sensor networks, in *Proceedings of the IEEE/ACM International Conference on Information Processing in Sensor Networks (IPSN'06)*, Nashville, TN, USA, 2006, pp. 335-344.

[53]. D. Nam, H. Min, An energy -efficient clustering using a roundrobin method in a wireless sensor network , in *Proceedings of the 5thACIS International Conference on Software Engineering Research, Management & Applications(SERA'07)*, 2007, pp. 54-60.

[54]. D. Choi, J. Shen, S. Moh, I. Chung, Virtual cluster routing protocol for wireless sensor networks, in *Proceedings of the International Conference on Parallel and Distributed Computing and Networks (PDCN'09)*, Innsbruck, Austria, 2009.

[55]. M.O. Rashid, M.M. Alam, A. Razzaque, C.S. Hong, Reliable event detection and congestion avoidance in wireless sensor networks, in *Proceedings of the High Performance Computing Conference(HPCC'07)*, Houston, Texas, USA, 2007, pp. 521-532.

[56]. W.B. Heinzelman, A. Chandrakasan, H. Balakrishnan, An application-specific protocol architecture for wireless microsensor networks, *IEEE Transactions on Wireless Communications*, Vol. 1, Issue 4, October 2002, pp. 660-670.

[57]. C. Diallo, A. Gupta, M. Marot, M. Becker, Virtual base station election for wireless sensor networks, in *Proceedings of the 8thInternational Conference on New Technologies in Distributed Systems (NOTERE'08)*, Vol. 2, Lyon, France, Jun. 2008.

[58]. J. Blumenthal, R. Grossmann, F. Golatowski, D. Timmermann, Weighted centroid localization in ZigBee-based sensor networks, in *Proceedings of the IEEE International Symposium on Intelligent Signal Processing(WISP'07)*, 2007.

[59]. M. Becker, A.L. Beylot, Simulation des Réseaux (Traité IC2, Série Réseaux et Télécoms), *Hermes*, 2006.

[60]. C. Diallo, M. Marot, M. Becker, Single-node cluster reduction in WSN and energy -efficiency during cluster formation, in *Proceedings of the 9th IFIP Annual Mediterranean Ad Hoc Networking Workshop(Med-Hoc-Net'10)*, Juan-Les-Pins, France, June. 2010.

[61]. C. Manifavas, G. Hatzivasilis, K. Fysarakis, K. Ranto, Lightweight cryptography for embedded systems - a comparative analysis, in *Proceedings of the 8th International Workshop on Data Privacy Management and Autonomous Spontaneous Security* , Vol. 8247, 2014, pp. 333-349.

[62]. Y. Lu, L. Li, H. Peng, Y. Yang, An energy efficient mutual authentication and key agreement scheme preserving anonymity for wireless sensor networks, *Sensors*, Vol. 16, Issue 6, 2016, E837.

[63]. M. Turkanovi´c, B. Brumen, M. Hölbl, A novel user authentication and key agreement scheme for heterogeneous ad hoc wireless sensor networks, based on the Internet of Things notion, *Ad Hoc Networks*, Vol. 20, 2014, pp. 96-112.

[64]. R. Amin, G. P. Biswas, A secure light weight scheme for user authentication and key agreement in multi-gateway based wireless sensor networks, *Ad Hoc Netw.*, Vol. 36, Issue P1, January 2016, pp. 58–80.

[65]. J. Vijay, Daniel K.Venkatraman, G. Murugaboopathi, Various attacks in wireless sensor network: survey, international *Journal of Soft Computing and Engineering*, Vol. 3, 2013.

[66]. H. K. Patil, S. A. Szygenda, Security for Wireless Sensor Networks using Identity-Based Cryptography, *CRC Press*, 2012.

[67]. C. Alcaraz, R. Roman, J. Lopez, A survey of cryptographic primitives and implementations for hardware-constrained sensor network nodes, *Mobile Networks and Applications*, Vol. 12, Issue 4, 2007, pp. 231-244.

[68]. C. Castelluccia, A. C.-F. Chan, E. Mykletun, G. Tsudik, Efficient and provably secure aggregation of encrypted data in wireless sensor networks, *ACM Trans. Sen. Netw.*, Vol. 5, Issue 3, June 2009, pp. 1-20.

[69]. H. A. Maw, H. Xiao, B. Christianson, J. A. Malcolm, A survey of access control models in wireless sensor networks, *Journal of Sensor and Actuator Networks*, Vol. 3, Issue 2, 2014, pp. 150-180.

[70]. M. P. Singh, S. K. Singh, D. K. Singh, Routing protocols in wireless sensor network - a survey, in*ternational Journal of Computer Sciences and Engineering*, Vol. 1, Issue 2, Nov. 2010.

[71]. S. K. Singh, M. P. Singh, D. K. Singh, Routing protocols in wireless sensor networks - a survey, in*ternational Journal of Computer Science and Engineering Survey*, Vol.1, Issue 2, Nov. 2010.

[72]. Arpit A. Kumar, Optimized elliptic curve cryptography as fine balance for wireless sensor network , in*ternational Journal of Modeling and Optimization*, Vol. 1, Issue 4, 2011, pp. 348-353.

[73]. M. Khalid, X. H. Le, R. Shankar, An efficient mutual authentication and access control scheme for wireless sensor networks in healthcare, *Journal of Networks*, Vol. 6, 2011, pp. 355-364.

[74]. P. Vyas, M. Chouhan, Survey on clustering techniques in wireless sensor networks, in*ternational Journal of Computer Science and Information Technologies*, Vol. 4, Issue 1, 2014, pp. 28-32.

[75]. P. Ji, Y. Zhang, B. Zheng, J. Co, A key management method based on dynamic clustering for sensor networks, in*ternational Journal of Distributed Sensor Networks - Special issue on Algorithm and Theory for Robust*, 2015.

[76]. S. S. Maksud, A. D. Patel, Secure data aggregation using homomorpic encryption in wireless sensor networks: a survey, *Advances in Computer Science and Information Technology*, Vol. 2, 2015, pp. 13-20.

[77]. H. Alzaid, E. Foo, J. G. Nieto, Secure data aggregation in wireless sensor network : a survey, in *Proceedings of the 6th Australasian Conference on Information Security (AISC'08)*, Vol. 81, 2008, pp. 93-105.

[78]. B. Vaidya, M. Chen, J. J. P. C. Rodrigues, Improved robust user authentication scheme for wireless sensor networks, in *Proceedings of the 5th International Conference on Wireless Communication and Sensor Networks (WCSN'09)*, 2009, pp. 1-6.

[79]. K. Han, T. Shon, Sensor Authentication in Dynamic Wireless Sensor Network Environments, *International Journal of RFID Security and Cryptography (IJRFIDSC)*, Vol. 1, Issues 1/2, 2012, pp. 36-44.

[80]. S. Ruj, A. Nayak, I. Stojmenovic, Distributed finegrained access control in wireless sensor networks, in *Proceedings of the 25th IEEE International Symposium on Parallel and Distributed Processing(IPDPS'11)*, Anchorage, Alaska, USA, 16-20 May, 2011, pp. 352–362.

[81]. C. Diallo, M. Marot, M. Becker, A distributed link quality based d-clustering protocol for dense ZigBee sensor networks, in *Proceedings of the 3rd IFIP/IEEE Wireless Days International Conference (WD'10)*, Venice, Italy, October 2010.

[82]. T. Eisenbarth, S. Kumar, C. Paar, A. Poschmann, A survey of lightweight cryptography implementations, *Journal IEEE Design & Test Archive*, Vol. 24, Issue 6, November 2007, pp. 522-533.

Chapter 2
Connectivity Recovery and Augmentation in Wireless Ad Hoc Networks

Ulisses Rodrigues Afonseca, Camila Feitosa Rêgo and Jacir Luiz Bordim

2.1. Introduction

Wireless ad-hoc networks allow communication between devices without the need of fixed network infrastructure to operate. In such networks, each node is equipped with wireless capabilities and act as a mobile router. These characteristics make ad hoc networks suitable to support communications in urgent and temporary tasks, such as disaster-and-relief, search-and-rescue, law enforcement, and other special applications. In an ad hoc network, nodes typically communicate over a common channel, use omnidirectional antennas and operate on battery power. In many cases, such as wireless sensor networks, which is a special kind of ad hoc network, these devices run on non-replaceable battery [1, 2]. Hence, battery depletion is a major concern in wireless ad hoc networks as their replacement may not be feasible during operation. Also, unbalanced power consumption, which may be caused by distinct traffic load, can exacerbate battery drainage. Traffic load, in turn, may be influenced by node location, number of nodes in the vicinity and their activities, the choice of the routing algorithm, and so on [2]. Such events can lead to node failure as well as it can lead to network disconnection.

The presence of cut-vertices and cut-edges (a.k.a. articulations and bridges) are considered potential disruption points [3]. Informally, an articulation is as a vertex whose removal disconnects the network. Similarly, a bridge is a link which removal disconnect the network. Articulations may arise during deployment or during operation. Density and maximum transmission range are directly related to the connectivity and the presence of articulations. During operation, articulations usually arise due to changes in the topology such as mobility or by the unavailability of nodes, in general, by battery depletion. Articulations are required to route packets in the path from one network component to

Ulisses Rodrigues Afonseca
Instituto Federal de Goias - Campus Luziânia, Luziânia-GO, Brazil

another. Hence, such nodes may experience high load and battery consumption when contending for network resources. This, in turn, may lead to unbalanced power consumption and network segmentation [4]. Finding critical elements in the network, such as articulations and bridges, is a step in a solution to increase or recover connectivity in a wireless ad hoc network.

Given the importance of critical elements, several papers discuss how to locate articulation points and bridges on wireless ad hoc networks. Topological information can be obtained via iteration among nodes and, once such information is available, mechanisms to locate critical elements can be applied [1, 5]. The main problem of this approach is the number of messages required to gather such information. Jorgic et al. [6] proposed the use of k-hop information with the aim of reducing the input size of the algorithm with the drawback of producing false positives. Chaudhuri [7, 8], Turau [4] and Afonseca et al. [9] presented a modified version of the distributed depth first search that can locate critical elements without collecting global topology information using a reduced number of messages.

After locating critical elements, their effect on the network can be minimized or neutralized. Khelifa et al. [1] proposed a solution that can reinforce network connectivity by activating inert nodes near the critical elements or moving a mobile node to its proximity. Goyal and Caffery [5] presented a similar approach that uses coordinated mobility to bring nodes near to the critical elements. Afonseca et al. [10] proposed the use of data aggregation on nodes around articulations to reduce its energy consumption.

Cooperative Communication (CC) emerged as an alternative to increase or recover connectivity. This technique allows single antenna devices to explore spatial diversity and combine signals to increase transmission range. Zhu et al. [11] and You et al. [12] proposed the use of CC to augment connectivity by increasing the transmission range. Latter, Neves et al. [13] and Rêgo et al. [14] considered this approach in the context of wireless sensor network to improve connectivity to a sink node. Afonseca et al. [15] applied this technique to create backup-links near the articulation nodes in such a way that network connectivity can be restored in case of critical element failure.

This chapter reviews the concepts of articulations and bridges and presents theoretical models on Cooperative Communication. More precisely, this chapter details how the information on critical elements can be used to reinforce network connectivity. A review on mechanisms to augment or restore the network connectivity via cooperative communication is also presented. The remaining of this chapter is organized as follows. Section 2.2 presents the main definitions of articulation and bridges and the algorithms to identify and locate them in a wireless ad hoc network. Section 2.3 describes two cooperative communication models. Section 2.4 presents CARCC, an algorithm that uses CC and location of articulation and bridges to recovery connectivity and Section 2.5 presents the Sink-PathCC algorithm which uses CC to provide connectivity augmentation. Finally, Section 2.6 concludes this chapter.

2.2. Articulation Nodes and Bridges

Articulations and bridges are concepts widely discussed in graph theory. Informally, an articulation is a vertex whose removal (along with their adjacent edges) turns the graph disconnected. Analogously, a bridge is an edge whose removal turns the graph disconnected. Due to the importance of network connectivity, efficiency and lifetime, articulations and bridges are also known, respectively, as critical nodes and critical links [2, 3]. These elements are also related to the concept of biconnectivity in communication networks, whose importance stands out when providing resilience and/or load balancing [16]. The main feature of articulations and bridges is that they form the only path among network components (isolated cluster of nodes) and, due to that, may experience higher load and power consumption [2]. As a result, the network is prone to become disconnected prematurely. Owing to their relevance, several papers present solutions to locate these critical elements and mitigate their effects to prevent network segmentation [1-3, 5-8].

The subsequent sections review the definition of articulation and bridges as well as the algorithms to locate these critical elements in a wireless ad hoc network setting.

2.2.1. Definitions and Terms

A graph $G(V,E)$ is connected if there is a path between each pair of vertices $v_i, v_j \in V$, $v_i \neq v_j$. A graph $G'(V',E')$ is a subgraph of $G(V,E)$, if $V' \subset V$ and $E' \subset E$. Connected components are disjoint sets of elements, or subgraphs, with no connectivity to each other. In other words, a connected component $G_a(V_a,E_a)$, of an undirected graph $G(V,E)$ is a subgraph such that any vertex $v_i \in V_a$ is connected to all other vertices $v_j \in V_a$ by a path and, given two different components $G_a(V_a,E_a)$ and $G_b(V_b,E_b)$, they do not have vertices in common, that is, $V_a \cap V_b = \varnothing$. The following definition summarizes the above discussion:

Definition 2.1 *(Connected component [17]) An unconnected graph consists of a set of connected components that are maximum connected subgraphs (with as many elements as possible).*

Sedgewick et al. [17] and Cormen et al. [18] define articulation (a.k.a. cut-vertex) as a vertex $v \in V$ of a connected graph $G(V,E)$ whose removal (and, consequently, of its edges) makes the graph G disconnected. Analogously, a bridge (a.k.a. cut-edge) is an edge $(v_i v_j) \in E$ such that its exclusion turns the graph disconnected [18]. Note that when $|V| \geq 3$, the existence of a bridge implies the existence of more than one articulation in the graph. The definitions below follow from [19]. The concepts of articulation and bridge are formally presented in Definitions 2.2 and 2.3.

Definition 2.2 *(Articulation) Given a connected graph $G(V,E)$, node v_a is an articulation if $G'(V',E')$ is a disconnected graph when $V' = V \backslash \{v_a\}$ and $E' = E \backslash \{(v_a, v_i)\}$ for each $v_i \in V$.*

Definition 2.3 *(Bridge) For a connected graph $G(V,E)$, the edge $v_a v_b$ is a bridge if $G'(V,E')$ is a disconnected graph for $E' = E \backslash \{(v_a v_b)\}$.*

A graph having articulations is called a separable graph, whereas a graph without articulations is said to be non-separable or a biconnected graph [16]. In [16] and [20], several articulation properties are shown, in particular, it is demonstrated that a graph without articulations is biconnected. A biconnected component, or a "block", of G is a maximum set of edges such that any pair of edges in the set are in the same cycle [18]. Also, a biconnected component can be defined using the concept of articulation, according to Definition 2.4.

Definition 2.4 *(Biconnected component) A component $G_a(V_a, E_a)$ is said a biconnected component, or a block, when it is the biggest set (or a maximal subgraph) without articulations.*

A connected graph, or subgraph, that has only two nodes is, by definition, a biconnected graph. In [19], it is denoted that a graph or component is 2-connected if there is a cycle between any pair of nodes. A connected graph, or component, where $|V| = 2$ is a biconnected component, but it is not 2-connected. The Definition 2.5 presents formally the concept of the 2-connected component.

Definition 2.5 *(2-Connected component) A component $G_a(V_a, E_a)$ of $G(V, E)$, such that $V_a \subset V$ and $E_a \subset E$, is 2-connected if, and only if, for any pair of vertex $v_i, v_j \in V_a$, there is a cycle in E_a containing v_i and v_j.*

2.2.2. Location of Articulation and Bridges

A naive algorithm to locate articulations in a connected graph $G(V, E)$ consists of, for each node $v_i \in V$, remove it and verify if the resulting graph is connected. Removing each node of the graph has a cost of $O(|V|)$ and verifying the connectivity of a graph can be performed in $O(|V|+|E|)$ [17]. Thus, the naive algorithm is not efficient, since its complexity is $O(|V|*(|V|+|E|))$.

Tarjan [20] and Hopcroft and Tarjan [21] developed algorithms with linear complexity that apply spanning trees to detect articulations with time $O(|V|+|E|)$. In [21], a tree is produced from the original graph, using the Depth First Search (DFS) algorithm. Articulations are nodes that do not have, for each child, an edge that connects a descendant (in that branch) to an ancestor. As a specific case, if the root has more than one child, it is also an articulation. This mechanism is based on a set of lemmas whose proofs are presented in [8] and in the original work [21]. Fig. 2.1 exemplifies the location of articulations using DFS and the premises presented by Hopcroft and Tarjan [21]. Fig. 2.1 (a) presents the initial graph $G(V, E)$ with $|V| = 6$ and $|E| = 8$. The Fig. 2.1 (b) shows the tree $T(V, E')$ created from the initial graph of the Fig. 2.1 (a). Note that $E' = E \backslash \{v_1v_5, v_2v_5, v_2v_4\}$. The removed edges, or nontree edges, are represented by dashed lines. The arrows, on the edges of the tree, represent the hierarchy and its direction determines the parent node. The highlighted node v_2 is an articulation since the child node v_4 and its descendants do not have an edge for an ancestor of v_2.

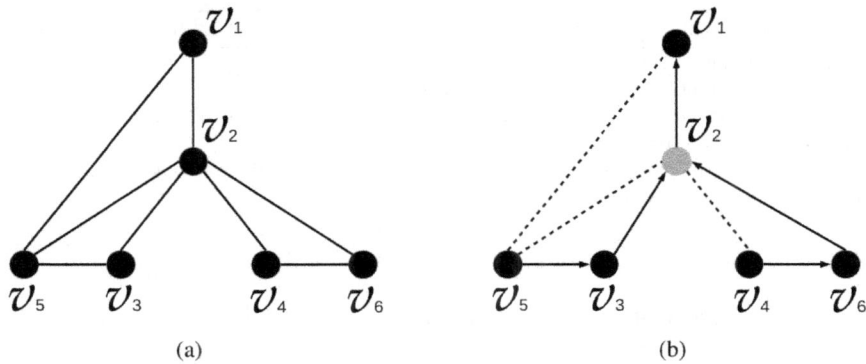

Fig. 2.1. Example of locating articulation nodes using DFS [21]: (*a*) original graph; e (*b*) tree created using DFS and identification of articulation v_2.

Tarjan [20] proposed a technique similar to [21]. It is based on the Depth First Number (DFN), the search order of the DFS. When performing the depth-first search, each node v_i is marked with a number $DFN(v_i)$ corresponding to the order in which it was added to the tree. Then, each node is assigned a value $L(v_i)$ indicating the lowest DFN value that can be reached by that node through a path that can include an arbitrary number of edges belonging to the tree or not. A node v_i is marked as articulation when one of its children v_j has a $L(v_j)$ value greater than or equal to $DFN(v_i)$, or if this node is the root and has more than one child.

2.2.3. Locating Articulations and Bridges in Wireless Ad Hoc Networks

The mechanisms to locate critical elements in wireless ad hoc networks can be classified as centralized, localized and distributed. Centralized mechanisms must collect information about the entire topology to locate critical elements. Localized mechanisms utilize partial information of the topology on each node while distributed algorithms are based on the Distributed Depth First Search to locate critical elements. This section enumerates the solutions presented in the literature for each alternative.

Khelifa et al. [1] and Goyal and Caffery [5] present centralized solutions for the location of articulations. Topology information is collected and sent to a node that reconstructs the overall knowledge of the topology and is responsible for running the algorithm and locating the articulations. In [5], the authors use a modified version of DFS to determine articulations. This method verifies, when creating the tree, whether there is a path back to the root. When no such path exists, the node is marked as an articulation. In [1], the proposal uses a centralized algorithm that performs a DFS search on the graph such that nontree edges are used similarly to the algorithm of Hopcroft and Tarjan [21] to check if the node is an articulation. The location of the articulations has complexity $O(|V|+|E|)$, whereas the collection of topology information requires a quadratic time as a function of the number of nodes [6]. The Algorithm 1 enumerates the steps of the centralized solution, similar to the one used in [1] and [5].

Algorithm 1 Centralized location of articulations in wireless ad hoc networks
---
1: Collect information about all vertex and edges of the network;
2: Construct a graph $G(V,E)$ based on the collected information;
3: Compute, using DFS, a tree $T(V,E')$ from the graph $G(V,E)$, such that $E' \subset E$, using an arbitrary node v_r as root;
4: Node $v_i \in V$ is an articulation if, for any child node v_j of v_i, there is not and edged from a descendant of v_j to and ancestor of v_i;
5: Node v_r is an articulation if it has more than one child.

In Wireless Sensor Networks (WSNs), where nodes have limited capacity, centralized methods become impractical since they require a large number of messages to collect information about the network topology. Jorgic et al. [6] shows that centralized solutions require a quadratic number of messages (in terms of number of nodes) to update the information of links when the topology changes. An alternative is the use of localized algorithms that execute at each node and use information about the topology limited to a certain number of hops in relation to its neighbours. These alternatives try to reach the global objective using local behavior [3, 6].

Jorgic et al. [6] proposed a solution that assume that nodes collect information at k-hops using a special message called "*HELLO*". The topology information at k-hops is propagated to the neighbours at $(k-1)$-hops allowing nodes to be aware of part of the network topology. According to the authors, the topological information at k-hops corresponds to the subgraph of k-neighbours, including all existing links among k-hop and $(k-1)$-hop neighbours. This set of information does not strictly determines whether the neighbours at k-hops are connected. In the method proposed by the authors, a node is k-critical (or articulation using information at k-hops) if the subgraph of neighbours at k-hops is disconnected when considering the removal of such node. Nodes must individually collect the topological information and run the algorithm to determine, through partial representation, whether they are articulations. This solution can use positional information, where the topology of the network is determined by the position of the nodes and their transmission radius. The authors also propose verifying critical edges at k-hops based on articulation information.

Fig. 2.2 presents an example of the use of localized knowledge, on nodes v_4 and v_7. In Fig. 2.2(b), nodes v_4 and v_7 determine that they are articulations by applying information corresponding to $k = 1$. The demarcated areas represent the knowledge obtained by each node to perform the calculations. Note that all nodes are articulations for 1-hop knowledge. In Fig. 2.2(c), when using topological information at 2-hops, node v_4 determines that it is an articulation, whereas v_7 determines that it is an ordinary node. The highlighted areas represent the localized knowledge of nodes v_4 and v_7. For v_4, it is not possible to find a cycle (multiple paths) through localized knowledge and, therefore, the node determines that it is an articulation. In this case, note that v_4 produces a false positive since, in fact, it is not an articulation.

Localized algorithms allow the decentralization of the processing and partial representation of the graph. Although this methodology does not reduce the complexity of the algorithm, the requirements for processing and storage are reduced along with the input size (number of vertex and edges of the graph). In contrast, it can detect false positives. An alternative are the distributed algorithms that use, as knowledge, only the adjacent edges and the directly connected nodes. In general, these algorithms locate articulations using a distributed version of the DFS algorithm that represents partially, in the memory of the nodes, the tree produced by the algorithm and the information about the cycles in the graph.

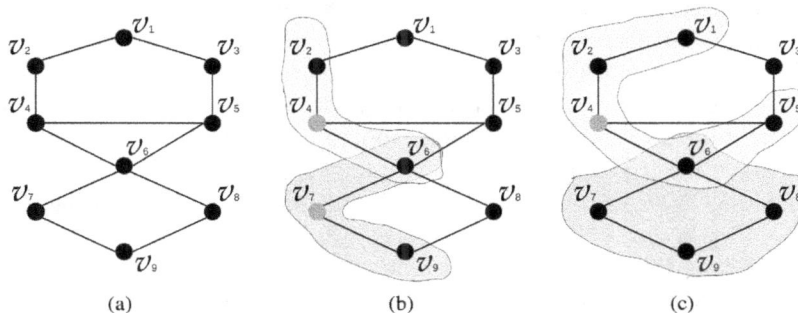

Fig. 2.2. Localized algorithm to detect articulation nodes: (*a*) graph representing the network $G(V,E)$, (*b*) running the algorithm in node v_4 and v_7 with $k = 1$; and (*c*) running the algorithm in v_4 and v_7 with $k = 2$.

Chaudhuri [7] presented a distributed solution to determine the bridges and biconnected components of a graph. A similar solution to determine the articulations of the graph, optimal in time and number of messages was presented in [8]. Nodes compute if they are articulations verifying the same assumptions of the algorithm proposed by Hopcroft and Tarjan [21]. The algorithm requires that nodes knows only the neighbours directly connected and the articulations are computed using $O(|V|)$ messages in $O(|V|)$ time. Given a connected graph and an arbitrary node as root, this mechanism calculates the articulations in two steps. In the first step, a distributed tree is built by keeping in memory of each node the information of its parent node and its children. Then, a list of ancestor nodes is propagated to each node. In the second step, the list of nontree edges (edges that do not belong to the tree) is propagated from the leaves to the root and the articulations are calculated. The tree is created by means of a special message called *SEARCH*. This message is propagated from the root to the leaves and carry a list of previously visited nodes. By message exchange, the destination is added as the child node by the source, while the destination node defines that the source is its parent node. To determine that a segment has been completely visited, a *SEARCH* message is sent back to the parent node. Initially, the return message is sent by the leaves of the tree that identify, through the list of visited nodes, that there are no more nodes to visit in that branch. After the tree is created, the root sends a special message called *TERMINATE* for each child. This message carries the list of ancestor nodes that is updated and propagated to the rest of the tree. In the second phase, a *NONTREE* message is sent from the leaves to the ancestors propagating the list of edges that are not part of the tree (nontree edges). This solution uses a maximum of $4 \times |V|$ messages.

Fig. 2.3 presents an example of the execution of the distributed solution proposed by Chaudhuri [8]. The Fig. 2.3 (a) represents the topology of the wireless network. The Fig. 2.3 (b) shows the tree construction using node v_1 as the root. Nodes are visited sequentially through the *SEARCH* message. The order in which the nodes are included in the tree is v_1, v_2, v_4, v_3, v_5, and v_6. Note that, because the process is sequential, the *SEARCH* message is sent, for example, to the node v_3 only when v_4 sends a *SEARCH* message stating that the search has already been completed in that branch. The node v_4 determines that there are no more nodes to search using the contents of the *SEARCH* message sent by node v_2. In this figure, the tree structure is shown by means of arrows in the edges whose direction represents the parent node. In Fig. 2.3 (c), the ancestors of each node are defined using *TERMINATE* messages whose propagation is performed from the root. Messages are sent in the order of the previous DFS. Each node, upon receiving this message, must store the information, add its identifier and send it to its descendants. In Fig. 2.3 (d), edges that are not part of the tree (and which constitute cycles) are propagated from the leaves to the root using *NONTREE* messages. When the nodes receive the messages, they check whether or not they are articulations. The nodes v_2 and v_3 (shown in red) are articulations. Note that v_3 receives from v_6 a message stating that there are no cycles in that branch. Therefore, node v_3 confirms that it is an articulation. Node v_2 receives from node v_3 its nontree list of edges and checks that there is no edge in this set that links to one of its ancestors. Thus, node v_2 also determines that it is an articulation.

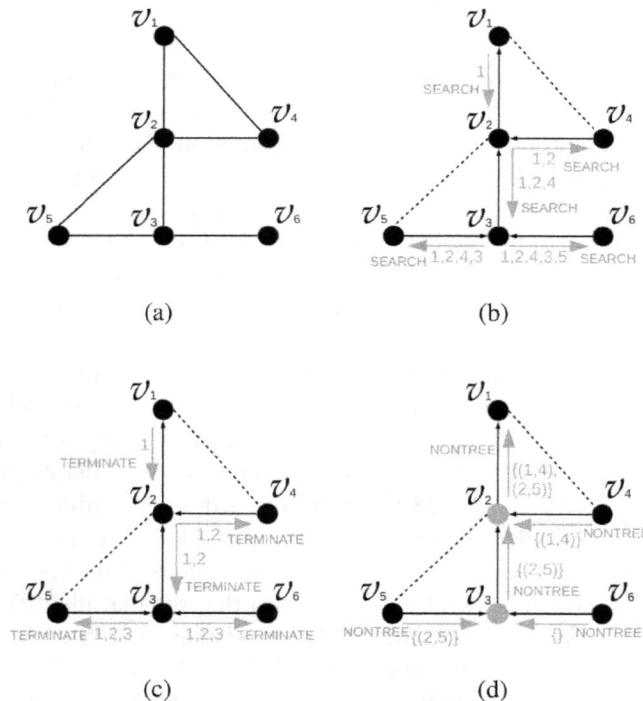

Fig. 2.3. Example of running the distributed algorithm of Chaudhuri [8]: (*a*) network topology; (*b*) building the tree; (*c*) propagation of ancestors; (*d*) propagation of cycles and identification of articulations (nodes v_2 and v_3).

Turau [4] presents another distributed solution for locating articulations. It differs from other approaches by using the methodology proposed by Tarjan [20]. This alternative allows bridges and 2-connected components to be located. Thus, as in the original algorithm, each node v_i is assigned a *DFN* value that determines the sequence of nodes visited in the DFS and a value L corresponding to the lowest *DFN* value that can be obtained by the nodes through a path to its ancestors. Articulations are nodes v_i that have children v_j such that $L(v_j) \geq DFN(v_i)$. These values are obtained by special messages in a process similar to Chaudhuri's proposal [8]. The authors show that the solution is optimal and uses $O(|V|)$ time units and messages to determine articulation, bridges, and 2-connected components. The algorithm uses at most $4 \cdot |V|$ messages.

Afonseca et al. [9] proposed improvements to the solution presented in [8]. The algorithm, called Distributed Articulation Search (or DAS), takes as input a connected graph $G(V,E)$ and a *root* parameter that indicates the node used as the root of the distributed tree. The root of the tree can be chosen arbitrarily. Each node must have a unique identifier and know its direct neighbours. The algorithm creates a distributed tree using a special *SEARCH* message that loads: (*i*) a list *a* of ancestor nodes; (*ii*) a list *v* of all nodes visited; and (*iii*) a list *nt* of nontree edges, or edges that do not belong to the tree. The message can be sent in two directions: to the children or to the parent node. The *SEARCH* message sent to the children is used to add nodes not yet visited to the tree and represent this information in memory. When it is sent to the parent node, it determines that a subtree has been fully scanned and propagates the list of edges that do not belong to the tree, found by the node itself and by its descendants. Upon receiving a *SEARCH* message from a child, nodes can determine if they are articulations. The algorithm terminates when the root receives a return message from all children. At the end of execution, each node knows, through a local variable, whether it is an articulation. Algorithm 2 presents the details of the DAS.

During node initialization of DAS, variables are created to store the identification of the parent node, set of child nodes, set of ancestors, set of edges that do not belong to the tree, nodes already visited, and those that have not yet been visited (lines 1-4). When the node is the root, a neighbor is chosen, added to the list of children and already visited nodes, and the first *SEARCH* message is sent (lines 5-11). The *SEARCH* messages received by the nodes are the trigger for executing the remainder of the algorithm. When a node receives a message, it updates the list of already visited nodes and visits it through the message content (lines 12-13). If the node is not the root of the tree and has not yet been visited (*parent$_{vj}$* = 0), then the node that sent the message is assigned as the parent node and the list of the ancestors is updated via the content of the message received (lines 15-17). If the message was sent by a child node, lines 18-31 are executed. In this case, if the node that received the message is the root of the tree, it defines itself as an articulation if it has more than one child (lines 19-23). If the node receiving the message is an ordinary node, it checks to see if it is a link through the list of nontree edges (received in the message content) and the list of ancestors (lines 24-28). Then it updates the list of edges that do not belong to the tree, applying the message content (line 29). After processing the lines 12-31, the nodes must continue building the distributed tree. If it has not been visited, one of them is removed from the list, added to the set of children, and then a *SEARCH* message is sent to the selected node (lines 32-36). If the node has no neighbor to add to the tree, it constructs its list of nontree edges and sends it to the parent node (lines 37-43).

Algorithm 2 - Afonseca et al. [22]: DAS(root)

Input: Each node v_i must know its direct neighbors $N[v_i]$.
Output: Each node knows, using its logical local variable $articulation_{v_i}$, if it is an articulation node.

\# Initialization of node v_i
1: $parent_{v_i} \leftarrow 0$;
2: $children_{v_i} \leftarrow ancestor_{v_i} \leftarrow nontree_{v_i} \leftarrow \emptyset$;
3: $visited_{v_i} \leftarrow \{v_i\}$;
4: $tovisit_{v_i} \leftarrow N[v_i]$;
5: **if** $v_i = root$ **then**
6:    $destination_{v_i} \leftarrow$ first node in $tovisit_{v_i}$;
7:    $children_{v_i} \leftarrow children_{v_i} \cup \{destination_{v_i}\}$;
8:    $tovisit_{v_i} \leftarrow tovisit_{v_i} \setminus \{destination_{v_i}\}$;
9:    $visited_{v_i} \leftarrow visited_{v_i} \cup \{destination_{v_i}\}$;
10:    send message $SEARCH(\{v_i\}, \{v_i\}, \emptyset)$ to $destination$;
11: **end if**

\# Node v_j receive a $SEARCH(a, v, nt)$ message from v_i
12: $visited_{v_j} \leftarrow visited_{v_j} \cup v$;
13: $tovisit_{v_j} \leftarrow tovisit_{v_j} - visited_{v_j}$;
14: **if** $parent_{v_j} = 0$ **and** $v_j \neq root$ **then**
15:    $parent_{v_j} \leftarrow v_i$;
16:    $ancestor_{v_j} \leftarrow ancestor_{v_j} \cup a$;
17: **end if**
18: **if** $v_i \in children_{v_j}$ **then**
19:    **if** $v_j = root$ **then**
20:      **if** $|children| > 1$ **then**
21:        $articulation_{v_j} \leftarrow TRUE$;
22:      **end if**
23:    **else**
24:      **if** $\nexists (v_x v_y) \in nt \,|\, x \in ancestor_{v_j} \textbf{ or } y \in ancestor_{v_j}$ **then**
25:        $articulation_{v_j} \leftarrow TRUE$;
26:      **else**
27:        $dualLink_{v_j} \leftarrow v_i$;
28:      **end if**
29:      $nontree_{v_j} \leftarrow nontree_{v_j} \cup nt$;
30:    **end if**
31: **end if**
32: **if** $tovisit_{v_j} \neq \emptyset$ **then**
33:    $destination \leftarrow$ first node in $tovisit_{v_j}$;
34:    $tovisit_{v_i} \leftarrow tovisit_{v_i} - destination_{v_i}$;
35:    $children_{v_j} \leftarrow children_{v_j} \cup destination_{v_j}$;
36:    send message $SEARCH(ancestor_{v_j} \cup \{j\}, visited_{v_j}, \emptyset)$ to $destination_{v_j}$;
37: **else**
38:    **if** $v_j \neq root$ **then**
39:      **for all** $v_k \in \{neighbour_{v_j} \setminus (children_{v_j} \cup \{parent_{v_j}\})\}$ **do**
40:        $nontree_{v_j} \leftarrow nontree_{v_j} \cup \{v_j v_k\}$;
41:      **end for**
42:      send $SEARCH(\emptyset, visited_{v_j}, nontree_{v_j})$ to node $parent_{v_j}$;
43:    **end if**
44: **end if**

The Fig. 2.4 illustrates the execution of the DAS algorithm in a graph with six nodes. Node v_1 is used as the root of the tree, that is, *root* = v_1. Fig. 2.4 (a) shows the exchange of the first three messages. The search is performed until a leaf is found (there are no neighbours to add to the tree). The edges v_4v_1 and v_4v_2 are marked as nontree. In the Fig 2.4 (b), return messages are sent until a node that still has neighbours not added to the tree is found. In the figure, node v_2 is not marked as an articulation, since the message origin has edges for one of its ancestors. The process of adding nodes to the tree continues in Fig. 2.4 (c) to leaf node v_5. Then, as shown in Fig. 2.4 (d), the node v_5 determines that the edge v_5v_2 is nontree and propagates this information up to v_6. Node v_2 identifies that is an articulation since it receives a message from the child node v_6 with a nontree list of edges that does not include one of its ancestors. During propagation of messages to the parent node (message 4, 5, 8, 9, 10), the list of nontree edges received is added to the local knowledge of the node and it is propagated to the ancestors. When node v_1 receives the message number 10, the search finishes since it is the root and there are no nodes to add to the tree. During the construction of the tree, there may be two edges from the articulation, which lead to a same biconnected component. This occurs when an articulation is found and some node of that component has not yet been visited. To make easier the process of defining the biconnected components, the algorithm maintains (in the articulation) a variable called *dualLink*, containing the identification of the child node that composes the edge. Line 27 of the algorithm stores this information when the node is an articulation and receives a *SEARCH* message from a child, but specifically for that subtree, the node does not constitute an articulation (lines 25-27). In other words, the articulation received the message of child who has a nontree edge for one of his ancestors.

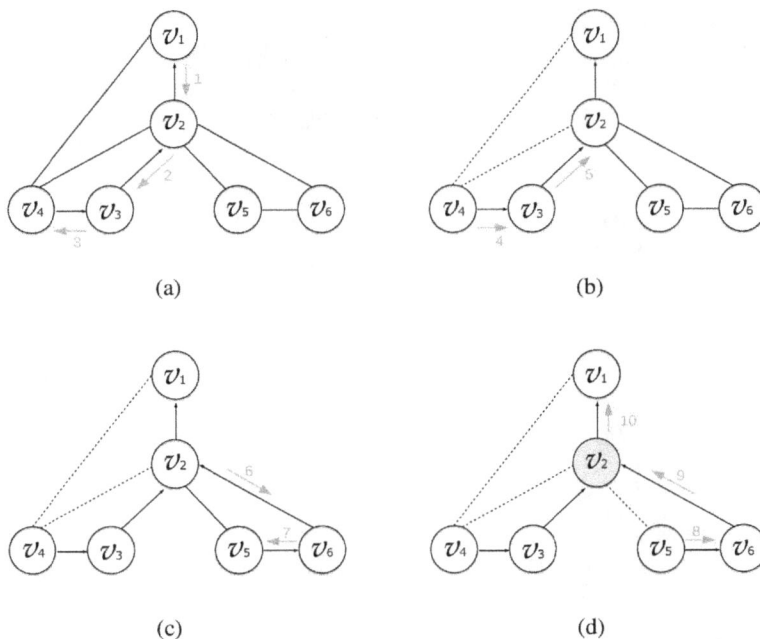

Fig. 2.4. Example for the algorithm DAS.

The complexity of the DAS algorithm can be calculated in relation to the number of messages. Given a graph $G(V,E)$, the algorithm creates a distributed tree $T(V,E')$. There are $|V|-1$ edges in T. Note that, for each edge of the resulting tree, exactly two *SEARCH* messages are transferred, one search and one return. It can be seen that only the edges that are part of the tree are used to carry the search and return message of each node. Since a node is not included twice in the tree, a total of $2\cdot|E'|$ messages are used. Regarding the number of nodes, $2\cdot(V-1)$ messages are sent. Thus, for the graph $G(V,E)$, the complexity of the DAS algorithm is $O(|V|)$.

Afonseca et al. [22] proposed two algorithms that use the DAS produced information to locate bridges and 2-connected components. The search for bridges is done by the Distributed Bridge Search (or DBS). The algorithm requires the knowledge of the tree $T(V,E')$, $E' \subset E$, represented in distributed form in the memory of the nodes, as well as the identification of the node used as root during its construction with the DAS. The general idea is to mark as bridges the links that: (*i*) join a leaf node of the tree to its ancestor, when it does not have a nontree tree edge; and (*ii*) connect an articulation to an ancestor node which is also an articulation. The algorithm runs on all nodes of the network and the trigger for execution is the definition of a node as an articulation or as a leaf. Thus, the exchange of messages is carried out only by nodes that can effectively constitute bridges. Upon starting, a special message *BRIDGE* is sent by articulations and leafs that do not have nontree edges to their parents. Upon receiving the message, the node send a *BRIDGE* response message if it verifies that the edge is a bridge. The details of this mechanism are presented in the Algorithm 3.

Algorithm 3 - Afonseca et al. [22]: DBS(root)

Input: DAS internal structures $parent_{v_i}$, $children_{v_i}$, $nontree_{v_i}$, $articulation_{v_i}$ and $ancestor_{v_i}$;

Output: Each node knows, using its logical local variable $bridge_{v_i}$, if it participates in a bridge and, the list of nodes participating in the bridges using the local array $bridgePair_{v_i}$.

```
# Initialization of node v_i
 1: bridge_vi ← FALSE;
 2: bridgePair_vi ← ∅;
 3: if children_vi = ∅ and nontree_vi = ∅ then
 4:     send message BRIDGE(vi) to node parent_vi;
 5: end if
 6: if articulation_vi = TRUE and vi ≠ root then
 7:     if ∄ vxvy ∈ nt | x ∈ ancestor_vi or y ∈ ancestor_vi then
 8:         send message BRIDGE(vi) to node parent_vi;
 9:     end if
10: end if

# Node vj receive a BRIDGE(vi) message from node vi
11: if vi = parent_vj then
12:     bridge_vj ← TRUE;
13:     bridgePair_vj ← bridgePair_vj ∪ {vi};
14: else
15:     if (vi ∈ children_vj and articulation_vj = TRUE) or vj = root then
16:         bridge_vj ← TRUE;
17:         bridgePair_vj ← bridgePair_vj ∪ {vi};
18:         send message BRIDGE(vj) to node vi;
19:     end if
20: end if
```

The location of biconnected components is performed by the algorithm Distributed Search for Biconnected Components (or DDBC), that requires the local variables produced by the algorithms DAS and DBS as the parameter and the root of the distributed tree. The details of the DDBC is shown in Algorithm 4.

Algorithm 4 - Afonseca et al. [22]: DDBC(root)

Input: DAS internal structures $parent_{v_i}$, $children_{v_i}$, $articulation_{v_i}$, e $dualLink$; DBS internal structures $bridge$ and $bridgePair_{v_i}$

Output: Each node knows, using its local array $component_{v_i}$ the list of bi-connected component it belongs.

\# Initialization of node v_i
1:   $component_{v_i} \leftarrow \emptyset$;
2:   **if** $v_i = root$ **then**
3:      **if** $articulation_{v_i} = TRUE$ **then**
4:         **for all** $v_j \in children_{v_i}$ **do**
5:            send message $COMPONENT(id(v_j))$ to v_j;
6:            $component_{v_i} \leftarrow component_{v_i} \cup \{id(v_j)\}$;
7:         **end for**
8:      **else**
9:         $component_{v_i} \leftarrow component_{v_i} \cup \{id(v_i)\}$;
10:       send message $COMPONENT(id(v_i))$ to the only element in $children_{root}$;
11:      **end if**
12: **end if**

\# Node v_j receive a $COMPONENT(c)$ message from node v_i
13: $component_{v_j} \leftarrow component_{v_j} \cup \{c\}$;
14: **if** $articulation_{v_j} = FALSE$ **then**
15:     **for all** $v_k \in children_{v_j}$ **do**
16:         send messagem $COMPONENT(c)$ to v_k;
17:     **end for**
18: **else**
19:     **for all** $v_k \in children_{v_j} \,|\, v_k \neq dualLink_{v_j}$ **do**
20:         $component_{v_j} \leftarrow component_{v_j} \cup \{id(j)\}$;
21:         send message $COMPONENT(id(v_j))$ to v_k;
22:     **end for**
23:     **if** $dualLink_{v_j} \neq 0$ **then**
24:         send message $COMPONENT(c)$ to $dualLink_{v_j}$;
25:     **end if**
26: **end if**

The root of the distributed tree, provided as a parameter, is used to start the identification of biconnected components. In addition, all nodes know if there are incident bridges and which neighbours make up these bridges (resulting from the algorithm DBS). As output, each node holds a local variable that stores an ID of the biconnected component to which it belongs. The ID of the biconnected components uses non-sequential integer values that correspond to the identifier of the first node added to the component. Nodes in disjoint biconnected components have different biconnected component identifiers. DDBC is based on a special message called *COMPONENT*. This message propagates a component identifier across all nodes of the network using the distributed tree structure. The general idea of the solution is to propagate the biconnected component identifier of the node to its children and this value should change when an articulation is found. From that point, all nodes must have different component ID. The first message is sent by the root of the tree and the following messages are sent from a parent node to the child nodes. During the initialization process, if the root is an articulation, it sends to its children its node ID, which is used as the component ID (lines 3-7). If the root is not an articulation, it sends its ID as component identifier to its only child (lines 8-11). When an ordinary node receives a message, it uses its content as component ID (line 13) and propagate this information to their child nodes (lines 14-17). If it is an articulation, the received value is added to the list of components (line 13) and a message is sent to all children with the children node ID as component ID (lines 13-26).

2.3. Cooperative Communication

Cooperative Communication (CC) is a technique in which devices with a single antenna may obtain the benefits of spatial diversity by transmitting simultaneously the same packet. In CC, nodes create a virtual antenna array without deploying multiple antennas in a single device [23, 24]. This technique, besides mitigating the signal fading, allows to increase the transmission range of nodes [23]. CC research is driven by the fact that some devices, such as wireless sensors, cannot accommodate an antenna array due to size or power constraints.

In a traditional simplified wireless communication model, the communication between nodes v_i and v_j can be simplified in terms of the transmission power, the distance between the nodes and the rate of signal fading. A node v_i can adjust its transmit power P_i with values in the range $[0, P_{MAX}]$. When $P_i = 0$, the transceiver is turned off and when $P_i = P_{MAX}$, the transceiver operates at full power. Thus, a source node v_i can communicate directly with the destination node v_j only if its transmission power complies with the following relation:

$$P_i(d_{i,j})^{-\alpha} \geq \tau \qquad (0 \leq P_i \leq P_{MAX}), \tag{2.1}$$

where: α is the exponent of signal fading, usually between 2 and 4, representing the rate of loss of the signal with increasing distance; $d_{i,j}$ is the Euclidean distance between nodes v_i and v_j; and τ is the receiver sensitivity to correctly receive a packet, i.e., the threshold of the received power so that node v_j can correctly decode the signal and obtain the original message. An edge between two neighboring nodes v_i, v_j, complying with Eq. (2.1), is termed a *direct link* and is represented as $\overline{v_i v_j}$. The direct neighboring set of node v_i is denoted as $N(v_i)$. The weight of a direct edge between nodes v_i, v_j is defined as:

$$W(\overline{v_i v_j}) = \tau(d_{i,j})^{\alpha}. \tag{2.2}$$

In the Cooperative Communication model adopted by Zhu et al. [11] and Yu et al. [12], a node v_i cooperatively transmit signals, in synchronism, with a set of nodes $v_1, v_2, ..., v_m$ to a particular destination node v_j. A cooperative transmission is successful if and only if the combined received signal at v_j is greater than or equal to τ. For latter reference, let $H_{i,j} \subseteq N(v_i)$ denote the *helper set* nodes of a source node v_i that cooperatively transmit packets to destination v_j. The received signal power P^c necessary for v_i and its helper nodes to communicate cooperatively with a destination node v_j can be computed by:

$$P^c_{i \cup H_{i,j}} = \left(\sum_{k=1}^{m} P_k \cdot d_{k,j}^{-\alpha} \right) \geq \tau \qquad (0 \leq P_k \leq P_{MAX})$$

(2.3)

A Cooperative Communication link (CC-link) between nodes v_i and v_j is denoted as $\widetilde{v_i v_j}$. As customary, the network topology is represented as graph $G(V,E)$, which denotes the union of all the direct and CC links. Let $\bar{E}$ and $\tilde{E}$, represent the set of direct and CC links, respectively. Thus, $E = \bar{E} \cup \tilde{E}$. That is, if $v_i v_j \in E$, then $v_i v_j = \overline{v_i v_j}$ if $v_i v_j$ is a direct link and $v_i v_j = \widetilde{v_i v_j}$ if $v_i v_j$ is a CC-link. Similarly, the graph of direct and CC communication is denoted by $\bar{G} = (V, \bar{E})$ and $\tilde{G} = (V, \tilde{E})$, respectively.

As in [9, 11, 12], cooperating nodes expend the same amount of power when establishing a cooperative link. Therefore, the weight of a CC-edge $\widetilde{v_i v_j}$ can be defined as:

$$W(\widetilde{v_i v_j}) = W_D(H_{i,j}) + (|H_{i,j}| + 1) W_{CC}(H_{i,j}),$$

(2.4)

where $W_D(H_{i,j})$ is the minimum power consumption of the node v_i to communicate with the farthest node in $H_{i,j}$, and $W_{CC}(H_{i,j})$ is the minimum power consumption of the node v_i to communicate directly to v_j, together with its helper nodes in $H_{i,j}$.

In the CC model considered in [11, 12], the transmission occurs in two phases. In the first phase, node v_i sends its data to its helper nodes in $H_{i,j}$. In the second phase, node v_i and its helpers must simultaneously send the same data to v_j. Thus, the weight of a CC-link comprehends the sum of communication cost of these two steps. The cost for the first stage of communication is equivalent to $W_D(H_{i,j})$, while the cost of individual nodes to transmit data using CC is $W_{CC}(H_{i,j})$.

CC was primarily used to overcome the effects of signal fading as well as to extend the transmission range of nodes in order to improve network connectivity [11, 12]. Fig. 2.5 illustrates how CC can be used by node v_i to send data to node v_j that is outside of its maximum transmission range R_{MAX}. The source node v_i selected the closest node as its helper. The nodes in black that are inside R_{MAX} could also have been selected to be included in the helper set. The dotted line represents the direct communication of the first phase, between the source and the helper. The solid arrows represent the second phase, where the source and its helper send data cooperatively to the destination.

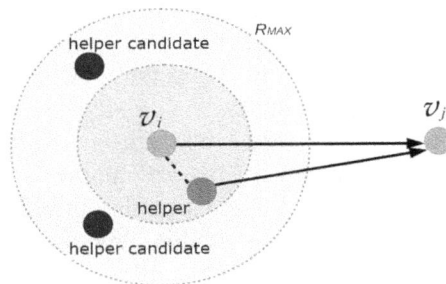

Fig. 2.5. Example of a traditional Cooperative Communication.

103

Rêgo et al. [25] proposed an alternative Cooperative Communication model aiming at reducing the power consumption of the source node in a CC-link. The first phase is similar to the previous discussed CC model, where the source node sends the data to the helper nodes using the minimum transmission power necessary to reach the farthest helper node. In the second phase, however, only the nodes in the helper set send the data to the destination node, i.e., source node does not participate in the second phase. The transmission will occur only if the sum of the transmission power of the helper nodes satisfy the following relation:

$$P^c_{H_{i,j}} = \left(\sum_{k=1}^{m} P_k \cdot d_{k,j}^{-\alpha} \right) \geq \tau \qquad (0 \leq P_k \leq P_{MAX})$$

(2.5)

Therefore, the weight of a CC-edge $\widetilde{v_i v_j}$ is modified to exclude the participation of the source node in the second moment of CC as shown below:

$$W(\widetilde{v_i v_j}) = W_D(H_{i,j}) + (|H_{i,j}|)W_{CC}(H_{i,j}),$$

(2.6)

where $W_D(H_{i,j})$ is the minimum power consumption of the node v_i to communicate with the farthest node in $H_{i,j}$, and $W_{CC}(H_{i,j})$ is the minimum power consumption of each helper node uses to communicate with v_j. The factor $W_{CC}(H_{i,j})$ is multiplied for the number of helper nodes, i.e., $|H_{i,j}|$.

2.4. Connectivity Recovery

Mitigating the effects caused by articulations becomes important to prolong the lifetime of the network and the creation of mechanisms to recover the connectivity ensure the success of the application. Section 2.2 presented the mechanisms in the literature to locate articulation nodes. This section explores existing solutions to mitigate the effect of the presence of articulations in the network using coordinated mobility, activation of inert nodes and the creation of cooperative links to reinforce the network connectivity.

Khelifa et al. [1] present a solution that, after detecting articulation nodes, reorganizes the network to improve its connectivity. The solution assumes the existence of inert nodes, previously implanted, which are called "redundant nodes". In addition, mobile nodes may exist. The proposed topology reorganization consists of: (*i*) locating the articulations; (*ii*) activate a redundant node near the articulation, if they exist; and (*iii*) if there are no redundant nodes near the articulation, the algorithm selects a redundant node from a neighbor and moves it to a coordinate close to the articulation. Fig. 2.6 provides an example of this solution. Fig. 2.6 (a) illustrates the graph of the initial state of the network. The redundant nodes are represented in white, while the articulations v_4 and v_7 are depicted in gray. In Fig. 2.6 (b), the inert node near the articulation v_7 is activated, and an inert node, close to v_6 is moved to the vicinity of the articulation v_4 and activated. Note that the nodes v_4 and v_7 are no longer articulations in this new network topology. The authors demonstrate that, with a sufficient number of inert and mobile nodes, it is possible to eliminate the presence of articulations in a wireless ad hoc network.

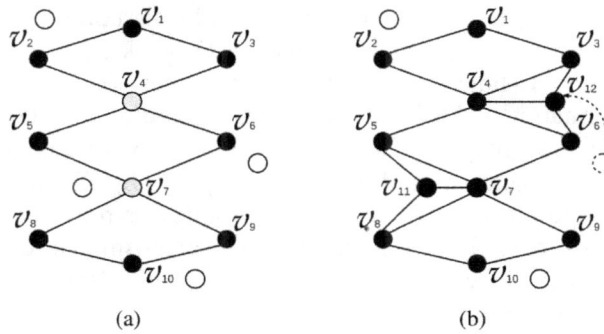

Fig. 2.6. Network reorganization using the mechanism proposed by Khelifa et al. [1]: (*a*) graph of the initial network topology; e (*b*) activating inert nodes and moving node to the articulation proximity.

Goyal and Caffery [5] proposed the postponement of network partitioning using a set of mobile nodes, called auxiliary nodes. Initially, bridges are identified through a modified version of the DFS algorithm that maps the potential quantity of network components. Then, articulations request an auxiliary node, which is moved to its proximity, enhancing the connectivity. The message used in the algorithm contains the location information where the auxiliary node should move to and the criticality of the articulation in terms of the size of the partition. The work does not define how the position of the nodes can be obtained. The message is carried out by broadcast and is limited to 2-hops to avoid over consumption of resources. Simulation data indicates that approximately 40% of the articulation lose this characteristic when using the aforementioned solution.

In [10], the authors proposed the use of data aggregation techniques to reduce energy consumption in articulations in WSNs. The solution requires that each node to check whether or not it is an articulation. Distributed Depth First Search based algorithms must be used to locate articulations and the sink node should be the root of the tree. Thus, from any node, it is possible to determine the direction to the sink. Each articulation informs its neighbours at k-hops, in the opposite direction to the sink, that they must perform aggregation. The aggregation of data is accomplished by concatenating the data of disjoint packets in such a way that the volume of the headers is reduced. To control the delay, a time limit is set that each packet can wait in buffer until another arrives. Packages are aggregated up to the maximum transfer limit. The solution was evaluated using an energy model that considers the amount of data received and sent. This mechanism of aggregation allowed the reduction in the energy consumption of the articulation and postponed the partitioning of the network. Fig. 2.7 illustrates the network topology. The red-colored node is an articulation, and it requests its neighbours at 1-hop to perform data aggregation. This way, the amount of data reaching the articulation is reduced.

Afonseca et al. [15] envisioned the use of Cooperative Communication to recover connectivity. They propose a localized mechanism to establish Cooperative Communication links to prevent network partition on articulation failure. The proposed solution, called Connectivity Auto-Recovering via Cooperative Communication

(CARCC, for short) operates by identifying possible disruption points and selecting appropriate neighboring nodes to establish backup CC-links. In case of articulation or bridge failure, backup CC-links become operative to reestablish network connectivity. Using localized information, CARCC explores the use of directional CC-links to provide ways to reestablish connectivity even when bidirectional links are not possible. Although CARCC shares some characteristics of the works proposed by Khelifa et al. [1] and Yu et al. [12], it does not require the use of additional nodes or previous knowledge of the network topology. Indeed, CARCC is a distributed mechanism that relies on a simple and yet effective heuristic to select appropriate nodes with the least amount of energy to recover network connectivity.

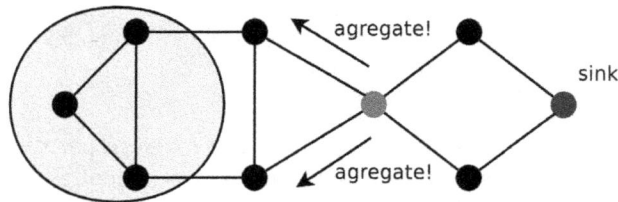

Fig. 2.7. Example of reducing the energy consumption on articulations using data aggregation.

CARCC comprises two algorithms: a helper set selection routine, called HSS, and the connectivity recovery routine, the CARCC itself. The HSS takes as input the source node v_i, the articulation node v_a, and the estimated distance $d_{a,j}$ between node v_a and v_j. HSS add nodes to the helper set $H_{i,j}$ until the destination v_j can be reached by a cooperative link. In case the nodes in $N(i)\backslash v_a$, are exhausted and a $\widehat{v_i v_j}$ link cannot be established, HSS returns an empty set. As shown in Algorithm 5, CARCC is composed of three main tasks, which are performed in sequence. The first task (lines 1-3) is to compute a suitable distance estimation to create CC links able to connect nodes in neighboring blocks (see Definition 2.4). If the articulation node fails, these blocks may still be connected via CC-links. To compute the CC-links, CARCC uses the HSS, which demands for a distance estimation so that the necessary helper nodes can be selected at each block. On the next step, the minimum distance to reach nodes in the farthest blocks is computed. The two largest values, D_1 and D_2, are informed to each node $v \in N(v_a)$. Once the distance estimation is available to the neighboring nodes of the articulations, these nodes execute routine HSS using the estimate D_1 or D_2. If $d_{a,u} = D_1$, then $u \in N(v_a)$ uses estimate D_2, otherwise u uses estimate D_1. This is necessary since u is at distance D_1 and the estimated distance to reach the second most distant node in a different block is D_2. Those nodes having sufficient helper nodes to create a CC-link for the given estimated distances report the CC-link cost back to the articulation v_a. On the final step, CARCC select the backup node having the least cost. At this stage, each network block has a single backup node whose cost is the least among the nodes in that block.

Optionally, CARCC can be succeeded by an optimization of the CC-link transmission power. More precisely, after selecting backup nodes in each block B_i, $1 \leq i \leq l$, these nodes may execute a routine to optimize the transmission power. Nodes in a neighboring block, upon receiving a message via CC-link, may inform the source of the message whether the transmission power can be reduced or not. The Algorithm 6 presents the power optimization for CARCC.

Algorithm 5 - Afonseca et al. [15]: CARCC(v_a)

Input: Articulation point v_a;
Output: $\forall H_{i,j} \neq \emptyset$, a CC-link $\widetilde{v_i v_j}$ connecting $\{v_i, v_j\} \in N(v_a)$ such that $v_i \in \mathscr{B}_k$ and $v_j \in \mathscr{B}_m$, $k \neq m$;

\# Computing the min distance to reach the farthest block
1: $\forall u \in \mathscr{B}_m$, $1 \leq m \leq l$, v_a computes $D_m = \min(d_{a,u})$;
2: v_a retrieves D_1 and D_2 satisfying $D_1 \geq D_2 \geq \ldots \geq D_m$;
3: v_a broadcasts (D_1, D_2) to all $v \in N(v_a)$;

\# Upon receiving D, check if a CC-link can be set up;
4: $\forall v \in N(v_a)$ compute $HSS(v, v_a, D_1 | D_2)$;
5: **if** $H_{v,j} \neq \emptyset$ **then**
6:      v sends $< W_{CC}(H_{i,j}), v >$ back to node v_a;
7: **end if**

\# Selecting a suitable backup node in each block $\mathscr{B}_m$;
8: Node v_a collects the messages and computes $\min_{v_i \in \mathscr{B}_m} (W_{CC}(H_{i,j}), v)$ for each block $\mathscr{B}_m$;
9: Node v_a informs to node $v_i \in \mathscr{B}_m$ that it holds the minimum cost among the nodes in that block. Node v_i becomes the backup node for block $\mathscr{B}_m$;

Algorithm 6 - Afonseca et al. [15]: CARCC_PwOpt

\# Backup nodes adjust its $W_{CC}(H_{i,j})$ based on P^c;
1: The backup node in each block B_i sends a CC message M containing $W_{CC}(H_{i,j})$, $SrcBlockID$, $SrcNode$;
2: For each received message M, node $v_z \in V$ records a tuple t with the $(P_z^c, SrcBlockID, DestBlockID, SrcNode)$;
3: Node v_z sends a message T with the collected information back to the $SrcNode$ through node v_a;
4: For each block $\mathscr{B}_i$, node v_a computes
     $P_{\mathscr{B}_i}^c \leftarrow \max(P_z^c)$, $z \in \mathscr{B}_i$;
5: Node v_a inform the $P_{\mathscr{B}_i}^c$ to $SrcNode$ in block $\mathscr{B}_i$;
6: On receiving the $P_{\mathscr{B}_i}^c$, the $SrcNode \in \mathscr{B}_i$ adjust its transmission power $W_{CC}^{opt}(H_{i,j}) \leftarrow \frac{W_{CC}(H_{i,j})}{P_{\mathscr{B}_i}^c}$.
     β, $(\beta > 0)$;

Simulation results show that CARCC obtains results matching those provided by centralized solutions. In addition, CARCC reduces the network requirements to create suitable CC-links. In fact, CARCC allows the selection of suitable CC-links in $O(\Delta(G) \times \Lambda)$ time, using $O(\Delta(G))$ messages, whereas a centralized approach takes $O(V^2 \times \Lambda)$ time, where $\Delta(G)$ is the degree of the underline network graph $G(V,E)$ and Λ is the computational cost of selecting helper nodes. In case of failure of an articulation point, network connectivity can be successfully obtained in more than 91 % in the evaluated scenarios, which is comparable to the results obtained by a centralized algorithm.

Fig. 2.8 shows a graph composed of one articulation and two network blocks $\mathcal{B}_1 = \{v_1, v_2, v_3, v_4\}$ and $\mathcal{B}_2 = \{v_4, v_5, v_6, v_7, v_8\}$. Solid and dotted arrows represent respectively, messages used to select a backup node in cluster $\mathcal{B}_1$ and cluster $\mathcal{B}_2$. In Fig. 2.8 (a), the articulation computes the distance estimation, which is sent to its direct neighbours. In Fig. 2.8 (b), each node checks if the CC-link can be established to overcome the estimated distance. The articulation point selects, for each block, the node with the least cost to be its backup node in that block. Fig. 2.8 (c) represent the message used by the articulation to define the backup node. Fig. 2.8 (d) presents the CC backup link created to overcome network disruption.

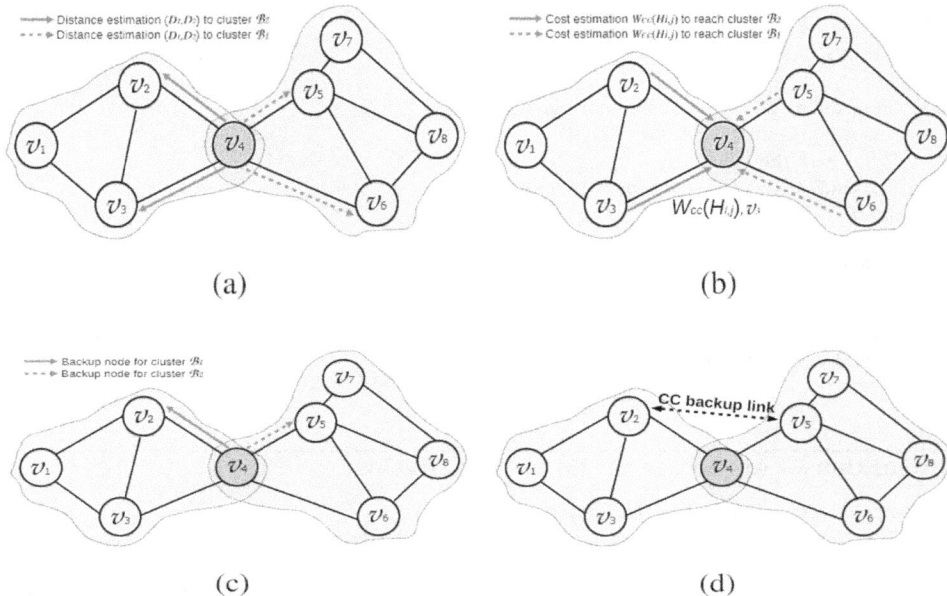

Fig. 2.8. Example of CARCC operation.

CARCC was evaluated with the same network model and parameters of [11] and [12], which are as follows: the maximum transmission power $P_{MAX} = 4900$, the receiver threshold $_\tau = 1$ and the Cooperative Communication model is defined by Eq. 2.3 resulting on a maximum transmission range of 70 *m*. Nodes were disposed randomly on an area of 300×300 *m* with a density varying from 4×10^{-4} to 12×10^{-4}. CARCC is compared to global and local strategies employing exhaustive search to compute the best "global" and "2-hop" cooperative links. Simulation results for CARCC are presented on Fig. 2.9. The "y" axis represents the transmission power and the "x" axis represents the node density. Fig. 2.9 (a) presents the individual transmission power for CARCC and CARCC with power optimization (CARCC PwOpt). The figure shows that CARCC without power optimization demands higher amount of energy from cooperating nodes as compared to global and 2-hop strategies. On the other hand, it provides several cooperative links that can be explored to find different paths with better results for total energy consumption or total delay. CARCC with power optimization can reduce energy expenditure considerably. In fact, CARCC with power optimization, in terms of energy, is quite close to the best global and 2-hop cooperative links. It should be noted, however, that these results do not consider the amount of energy expend to gather global and 2-hop information. As mentioned before, CARCC requires a reduced number messages to operate. This may not be the case for collecting and maintaining global and 2-hop topology information. Fig. 2.9 (b) presents the total energy consumption requited by source and helper nodes. That is, the figure shows the amount of power necessary to reach the nodes on the helper set and the transmission power required to cooperatively send data to the destination node. As shown in figure, the total transmission power decrease with higher node density, which comes from the reduced transmission power required to reach the helpers. The transmission power required by the optimized CARCC is $\approx 10 \%$ higher than the best global and $\approx 5 \%$ higher than the 2-hop link.

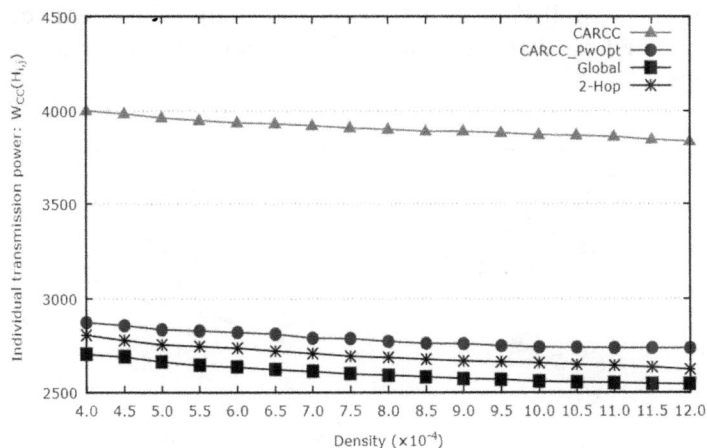

(a)

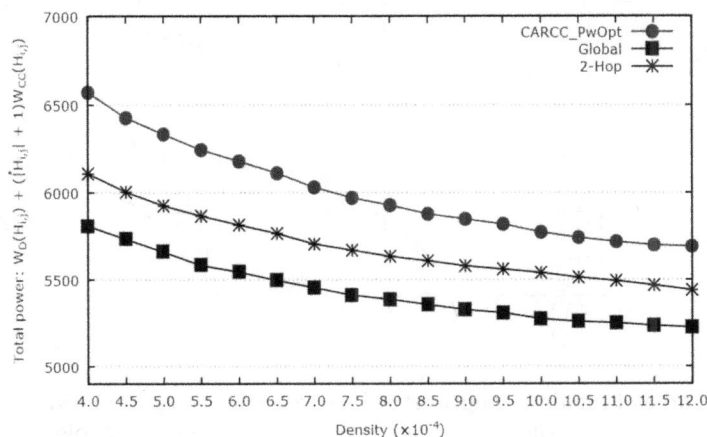

(b)

Fig. 2.9. Simulation results for CARCC.

2.5. Connectivity Augmentation

The possibility to increase network connectivity is of interest in wireless networks as it may provide alternative path to a destination while reinforcing network connectivity. The use of cooperative communication as alternative to improve network connectivity has been reported in the literature [11-14]. Yu et al. [12], proposed a topology control technique, called CoopBridges, aimed at increasing network connectivity and reducing transmission power by minimizing the number of cooperative communication links among disconnected networks. The authors also present a greedy heuristic to select helper nodes, termed Greedy Helper Set Selection (GHSS), to reduce the transmission power of the source node in the first step of CC using the Cooperative Communication model defined in Eq. 2.3. As shown in Fig. 2.10, CoopBridges initially create all possible cooperative links among each node in each network component (isolated cluster of nodes). A topology control is applied to prune costly cooperative links connecting cluster of nodes. Then, only the lowest cost containing direct and cooperative links are kept in the resulting topology.

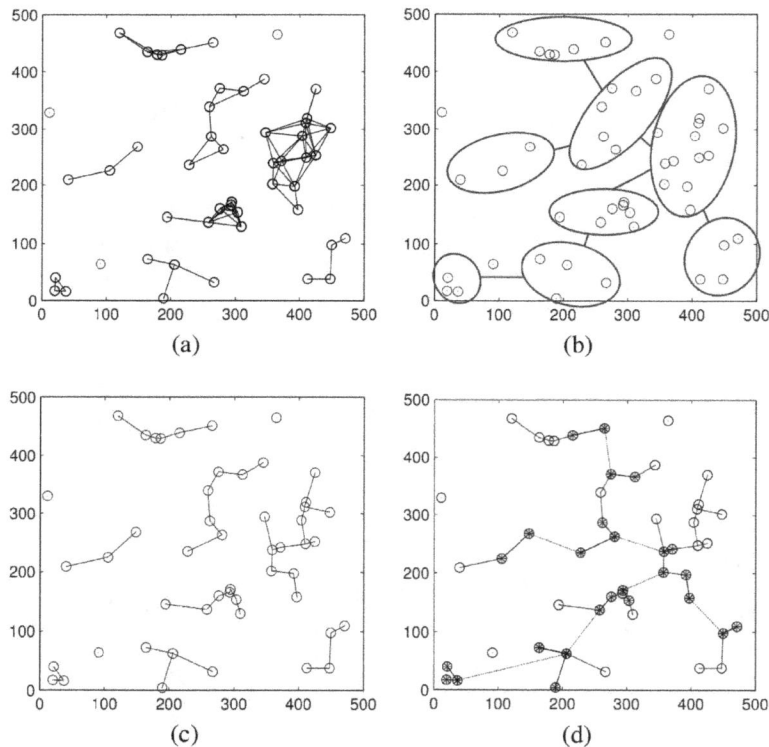

Fig. 2.10. Example of running CoopBridges [12]): (*a*) initial graph; (*b*) resulting edges after topology control to prune CC-links; (*c*) resulting edges after applying topology control to prune direct links; (*d*) resulting graph after the union of direct and cooperative links.

Neves and Bordim [13] proposed a technique in the context of WSNs, termed CoopSink, which uses the CC model defined in Eq. 2.3, to increase connectivity to the sink node and topology control to minimize energy consumption in the routes in the path. CoopSink also uses the GHSS heuristic to select the appropriate helper set. This solution runs in four steps: (*i*) generate the topology using the maximum transmission power; (*ii*) create all possible cooperative links; (*iii*) apply topology control considering the most efficient routes to the sink node; (*iv*) adjust the transmission power, keeping only the best direct and cooperative links.

CoopSink tends to select the neighboring nodes closest to the source node to be part of the helper set to decrease the burden of the source node in the first step of CC. However, this may cause an increase of the cost of the second step of CC as mentioned in Section 2.3. Although CoopSink provides more efficient routes to the sink node compared to CoopBridges, it does not consider the number of hops to reach the sink node neither the energy balance among source and helper nodes. Rêgo et al [14] proposed a technique, termed Sink-PathCC that addresses such issues. The Sink-PathCC consists of three steps, which are executed in sequence, as shown in Algorithm 7 [14]. In the first step, Sink-PathCC establish all the possible direct edges by employing the maximum available transmission power P_{MAX}. In the second step, a heuristic is proposed to select suitable helper nodes to improve network connectivity while promoting energy balance. In the third step, a shortest path algorithm is used to select the shortest path for each node v_i to the sink node. After computing the shortest path, a topology control strategy can be employed to prune unnecessary CC-links from the resulting communication graph.

Algorithm 7 - Rêgo et al. [14]: Sink-PathCC

Step#1:
Each node pair $(v_i, v_j) \in V$, where $v_i \neq v_j$, compute the direct edge $\overline{v_i v_j}$ by employing the maximum available transmission power P_{MAX};
Step#2:
Each node pair $(v_i, v_j) \in V$, where $v_i \neq v_j$ and $v_j \notin N(v_i)$, compute the CC-link $\widehat{v_i v_j}$ by employing a heuristic to select the suitable helper set to improve network connectivity and provide energy balance;
Step#3:
Compute the shortest path to the sink for each $v_i \in V$. Maintain in the graph only the links belonging to the shortest path.

Sink-PathCC was evaluated with the same network model and parameters of [11, 12, 15]. However, SinkPath-CC uses the cooperative communication model proposed in [25] in order to reduce the energy burden of the source node. SinkPathCC is also compared to CoopSink in terms of connectivity, cost of source nodes, cost of helper nodes and energy balance. As shown in Fig. 2.11 (a), both CoopSink and Sink-PathCC increase connectivity with the sink node when compared to the original topology. Since the source node does not participate in the second step of CC, the increase of the transmission range using CC is limited to number of helper nodes. Thus, the Sink-PathCC provide less connectivity gain than CoopSink. However, as the network density increases, the connectivity provided by both techniques increases as their difference diminishes.

Using SinkPath-CC, the cost of the source nodes is virtually independent to number of nodes as shown in Fig. 2.11 (b). On average, Sink-PathCC reduces the cost of the source node in 35 % compared to CoopSink. However, since the source node does not participate in the second step of CC in Sink-PathCC, the cost of the helper nodes is higher as observed in Fig. 2.11 (c). In fact, the helper nodes in Sink-PathCC uses, on average, 7 % more energy. Nevertheless, SinkPath-CC is closer to ideal energy balance than CoopSink as shown if Fig. 2.11 (d). Namely, the source node and the helper nodes spend almost the same amount of energy, which may help improve network lifetime.

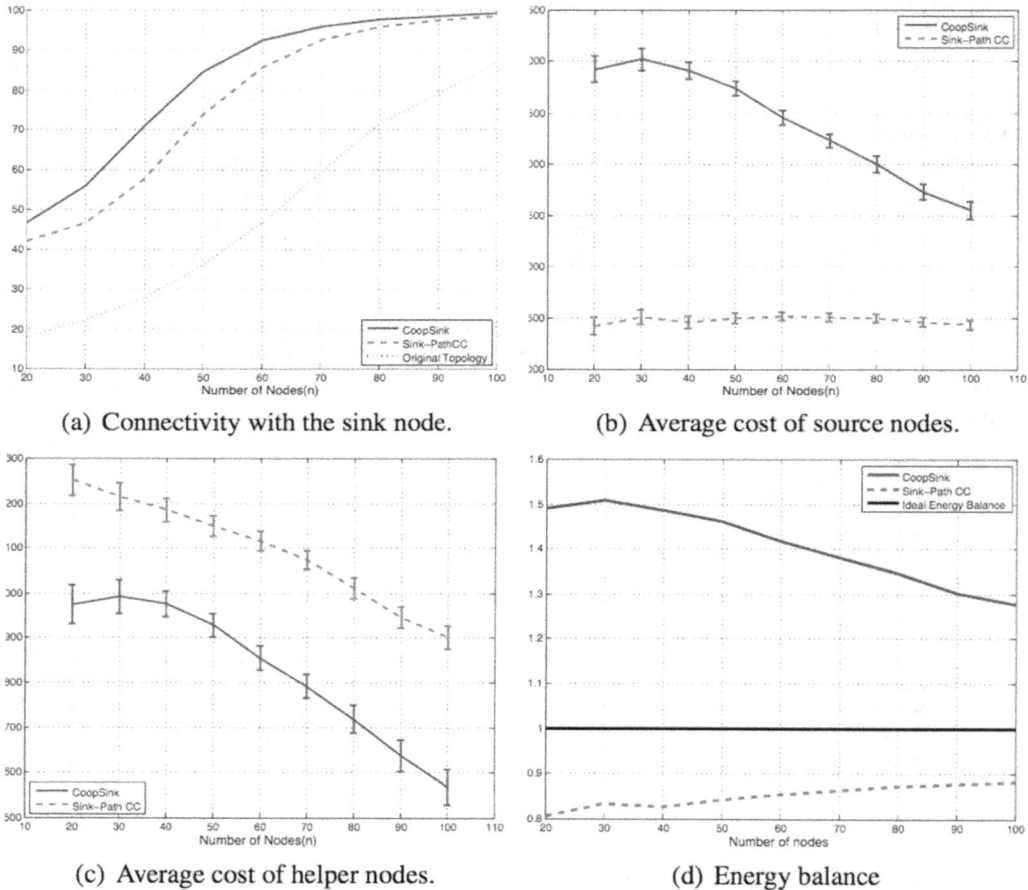

(a) Connectivity with the sink node.

(b) Average cost of source nodes.

(c) Average cost of helper nodes.

(d) Energy balance

Fig. 2.11. Simulation results for Sink-PathCC.

2.6. Conclusion

Wireless, multi-hop, ad hoc networks operate by routing data through intermediate nodes to reach a gateway node or sink. The success of applications in such scenarios, especially in critical activities such as search and rescue operations, depends on the initial connectivity as well as the capacity to maintain or recover network connectivity. Articulation nodes and bridges, a.k.a. cut-vertices and cut-edges, are considered as a potential point of network disruption. Hence, the task of locating them is of great relevance to enforce network connectivity. This chapter reviewed the techniques to locate articulations and bridges. Also, the main concepts and models of Cooperative Communication (CC), which can be used for connectivity augmentation and recovery are discussed. At first, CC was envisioned as a solution to reduce the effects of signal fading but it is also used to guarantee the initial connectivity of the network by creating cooperative links among disjoint network components (cluster of isolated nodes). However, CC can also be used to regain network connectivity when the information on potential disruption points is available.

Algorithms to locate articulation and bridges are presented in detail in this chapter. Once such nodes and links are identified, mobile nodes could be moved or activated to reinforce connectivity. Also, such information can be explored to establish backup CC-links that can be used to recover network connectivity [15]. This chapter discusses on alternatives to achieve such a goal. Furthermore, CC can be explored to provide connectivity augmentation, allowing the establishment of better routes to a gateway or sink node [14]. Although experiments in connectivity augmentation through Cooperative Communication are in the early stages, the results of proposed solutions reviewed in this chapter provide a fruitful ground for further investigation.

References

[1]. B. Khelifa, H. Haffaf, M. Madjid, D. Llewellyn-Jones, Monitoring connectivity in wireless sensor networks, in *Proceedings of the IEEE Symposium on Computers and Communications, (ISCC'09)*, 2009, pp. 507-512.

[2]. T. Yong, G. Z. Hu, The articulation nodes modeling of wireless sensor networks, in *Proceedings of the International Conference on Computer Application and System Modeling (ICCASM'10)*, Vol. 12, 2010, pp. 12-75.

[3]. I. Stojmenovic, D. Simplot-Ryl, A. Nayak, Toward scalable cut vertex and link detection with applications in wireless ad hoc networks, *Network,* Vol. 25, Issue 1, 2011, pp. 44-48.

[4]. V. Turau, Computing bridges, articulations, and 2-connected components in wireless sensor networks, in *Proceedings of the International Symposium on Algorithms and Experiments for Sensor Systems, Wireless Networks and Distributed Robotics (ALGOSENSORS'06)*, 2006, pp. 164-175.

[5]. D. Goyal, J. Caffery, Partitioning avoidance in mobile ad hoc networks using network survivability concepts, in *Proceedings of the 7th International Symposium on Computers and Communications, (ISCC'02)*, 2002, pp. 553-558.

[6]. M. Jorgic, M. Hauspie, D. Simplot-Ryl, I. Stojmenovic, Localized algorithms for detection of critical nodes and links for connectivity in ad hoc networks, in *Proceedings of the 3rd IFIP Mediterranean Ad Hoc Networking Workshop (MED-HOC-NET'04)*, Bodrum, Turkey, June 2004, pp. 360-371.

[7]. P. Chaudhuri, An optimal distributed algorithm for computing bridge-connected components, *The Computer Journal*, Vol. 40, Issue 4, 1997.

[8]. P. Chaudhuri, An optimal distributed algorithm for finding articulation points in a network. *Computer Communications*, Vol. 21, Issue 18, 199, pp. 1707-1715.

[9]. U. R. Afonseca, T. F. Neves, J. L. Bordim, Sinkpath: a cooperative communication protocol to restablish WSN connectivity, in *Proceedings of 13th International Conference on WWW/Internet*, 2014, p. 299.

[10]. U. R. Afonseca, P. H. Azevedo Filho, J. L. Bordim, P. S. Barreto, Localização e Redução do Consumo de Energia em Pontos de Articulação em Redes de Sensores Sem Fio, in *Proceedings of the II Workshop de Sistemas Distribuídos Autonômicos*, 2012, pp. 21-24.

[11]. Y. Zhu, M. Huang, S. Chen, Y. Wang, Energy-efficient topology control in cooperative ad hoc networks, *IEEE Transactions on Parallel and Distributed Systems,* Vol. 23, Issue 8, 2012, pp. 1480-1491.

[12]. J. Yu, H. Roh, W. Lee, Topology control in cooperative wireless, *IEEE Journal on Selected Areas in Communications*, Vol. 30, Issue 9, 2012, pp. 1771-1779.

[13]. T. F. Neves, J. L. Bordim, Topology control in cooperative ad hoc wireless networks, in *Proceedings of the XXXIX Latin American Computing Conference Electronic Notes in Theoretical Computer Science (CLEI'13)*, 2014, pp. 29-51.

[14]. C. F. Rêgo, R. R. Bezerra, J. L. Bordim, Improving energy balance and network connectivity in wireless sensor networks employing cooperative communication, in *Proceedings of the IEEE 12th International Conference on Wireless and Mobile Computing, Networking and Communications (WiMob'16)*, Oct 2016, pp. 1-8.

[15]. U. R. Afonseca, C. F. Rêgo, J. L. Bordim, CARCC: connectivity auto-recovering via cooperative communication, *Wireless Communications and Mobile Computing*, Vol. 14, 2017.

[16]. E. M. Reingold, J. Nievergelt, N. Deo, Combinatorial Algorithms: Theory and Practice, *Prentice Hall College Div*, 1977.

[17]. R. Sedgewick, K. Wayne, Algorithms, Fourth edition, *Pearson Education*, 2011.

[18]. T. H. Cormen, C. Stein, R. L. Rivest, C. E. Leiserson, Introduction to Algorithms, 2nd edition, *McGraw-Hill Higher Education*, 2001.

[19]. R. Merris, Graph Theory, Wiley Series in Discrete Mathematics and Optimization, *Wiley*, 2011.

[20]. R. Tarjan, Depth-first search and linear graph algorithms, *SIAM Journal on Computing*, Vol. 1, Issue 2, June 1972, pp. 146-160.

[21]. J. Hopcroft, R. Tarjan, Algorithm 447: efficient algorithms for graph manipulation, *Commun. ACM*, Vol. 16, Issue 6, June 1973, pp. 372-378.

[22]. U. R. Afonseca, T. F. Neves, J. L. Bordim, Employing cooperative communication to recover network connectivity in ad hoc networks, *International Journal of Networking and Computing*, Vol. 4, Issue 2, 2014, pp. 336-354.

[23]. C.-C. Jay Kuo, Y.-W. Peter Hong, Wan-Jen Huang, Cooperative Communications and Networking: Technologies and System Design, *Springer Science+Business Media*, 2010.

[24]. M. Uysal, Cooperative Communications for Improved Wireless Network Transmission: Framework for Virtual Antenna Array Applications, *Information Science Reference*, 2010.

[25]. C. F. Rêgo, R. R. Bezerra, U. R. Afonseca, J. L. Bordim, Cooperative communication models towards energy balance in WSN, in *Proceedings of the XLII Latin American Computing Conference (CLEI'16)*, Oct. 2016, pp. 1-9.

Chapter 3
QoS Routing in Ad Hoc Network

Tiguiane Yélémou

3.1. Introduction

In the current context of ubiquitous computing, the contribution of ad hoc network is strongly awaited. Due to erroneous nature of wireless links, interference and limited bandwidth capacities of radio channel, wireless networks offer a lower QoS compared to wired ones. Packet losses are significant in these networks. The Binary Error Rate (BER) is in the order of 10^{-3} in these wireless networks against 10^{-9} in the wired networks [1]. To address the QoS requirements of multimedia applications, several algorithms aiming at choosing better paths for data transmissions were proposed. Many take into account quality of links in route choice process. Several metrics have been proposed and incorporated into routing protocols. The provided performances have been mixed. In this chapter, we analyze the effectiveness of usual metrics and QoS methods used to improve routing in ad hoc network when considering the erroneous-nature of radio links. First, we present the context of routing in ad hoc network. Then, we carry out detailed study on usual metrics in QoS approach routing protocols. Particularly, we are interested in costs of estimating quality of the links, accuracy of measured value, relevance of the metric contribution in the choice of better paths. Finally, we analyze different QoS approaches of routing in ad hoc network.

3.2. Routing in Ad Hoc Network

An ad hoc wireless network is a dynamic collection of devices (which we call nodes), autonomous, connected without fixed infrastructure, which can be highly mobile. A node can communicate directly with other nodes within its range or serve as a relay. The relays allow two nodes out of radio range to communicate. These networks are practical and interesting communication solution, offering mobility, flexibility, robustness and low cost of deployment. The ever-increasing need to communicate anytime and anywhere and the popularization of multimedia services have contributed in recent years to a considerable

Tiguiane Yélémou
Polytechnic University of Bobo-Dioulasso, BP 1091, Bobo-Dioulasso, Burkina Faso

development of this solution. It often appears as the only solution for communication in complex environments without fixed infrastructure: rescue operations, military operations, coverage of exceptional events, vehicles in motion, etc.

These wireless networks are characterized by limited resources (bandwidth, energy autonomy) and a shared medium (radio medium). The radio wave undergoes fluctuations and attenuation in its propagation medium. Three major phenomena negatively influence the quality of the signal at the reception. These are ambient electromagnetic noise, obstacles in the propagation medium with induced multi-path effects, and node mobility, which induces a highly variable topology of the network. These situations, coupled with interference due to simultaneous transmissions, cause the characteristics of the propagation channel to change frequently. Packets losses are then important. Data packets routing of in such context is a real challenge and the ability of these networks to cope with the diverse requirements of service quality of multimedia applications is hypothecated. Several QoS techniques, applicable at various levels of Open System Information (OSI) reference layers, are available to system designers. For our study, we focused on the cross-layer approach.

In this chapter, we discuss the problem of achieving QoS at the routing level. Work presented in this section focus on the two main routing approaches: proactive approach and reactive approach. In proactive routing approach, nodes construct routing tables using periodically broadcast topology control messages. A route is thus available for each destination of the network. As an example of protocol at this level, we have Optimized Link State Routing (OLSR) protocol. In reactive routing approach, to transmit its data, the source node must first find a route through a route request process. Ad hoc On-demand Distance Vector (AODV) protocol is an example of protocol at this level. In a context of decentralized control and erroneous links, routing is a real challenge. These routing protocols used in this context of ad hoc wireless communication are, for the most part, adaptations of those already used in the wired domain.

Standard well known protocols like AODV and OLSR (their basic layouts) take into account only *number of hops* criterion in their path selection processes. Indeed, AODV chooses first-built paths that, in practice, means the paths with lowest delays and accordingly one with a low number of hops (but not necessarily the lowest). Nevertheless, some links of these paths can have a bad quality that allows small-sized messages such as route requests and replies packets to be transmitted, but not large ones. On such links, several transmissions can be mandatory to ensure data packet transmissions which induces more delay. Therefore, the chosen paths appear not as relevant as expected.

The routing computation mechanism used by the basic layouts of OLSR is also based on the minimization of the number of hops. The Multi Point Relays (MPR) computation mechanism used in this protocol is heavily based on the number of 2hop-neighbors that MPR candidates can reach [2]. This MPR mechanism can lead to a bad network capacity, since a node have a partial knowledge of the network [3]. This can result in low Packet Delivery Ratio (PDR) and important end-to-end delay due to retransmissions.

During recent years, efforts have been made to take into account link quality in the route choice process. Several methods are proposed with different QoS metrics including bandwidth, delay, packet delivery ratio. Khaled et al. [4] propose an on-demand routing protocol based on path robustness. Authors state that intermediate nodes must examine lifetime and delay of RREQ packets and compare them to specific limitations before to forward it in order to increase the robustness of each path.

Some authors use additive or multiplicative metric to enhance routing in wireless network context. For example, Kim et al. [5] have modified AODV protocol and particularly the RREQ mechanism in order to discover the best path. Authors states that all viable routes should be considered into the RREQ mechanism. To achieve this objective, authors used Expected Transmission Time (ETT) [6] which gives an indication on link quality to choose one path over another. According to Kim et al., this method allows to double the rate of received packets. However, this conclusion is based on simulations that rely on optimistic assumptions such as ideal propagation model.

Some variants of OLSR algorithm try to take into account quality of selected links in MPR selection mechanism. Munaretto et al. [7] and Ge et al. [8] propose a modification of the selection process by selecting MPR among nodes with good quality links in terms of bandwidth. In the same idea, Ingelrest et al. [9] suggested to qualify links with the probability of correct reception to reflect fluctuations due to attenuations caused by obstacles. To mathematically model this probability, the lognormal shadowing model is used. The probability of good reception is therefore considered during MPR calculation. [10-12] also present heuristics addressing the issue of QoS in wireless network routing protocols.

The hybrid Zone Routing Protocol (ZRP) is also enhanced with different metrics such as bandwidth [7, 8], delay [6], packet delivery ratio (PDR) [9]. For example, Mungara et al. [11] proposed a technique to reduce end-to-end delay and perform a better throughput. Their algorithm is based on selective bordercasting where route reconfiguration is started from the destination failure reporting node instead of beginning from the source. This allows a quick construction of a path and thereby reduces control overhead packets and end-to-end delay. As ZRP is the merge of proactive and reactive routing approaches, improvements that are proposed for AODV and OSLR may be applied to it.

3.3. Usual Metrics

To take into account the quality of the links in the choice of route, several metrics have been designed. In this section, we analyze some metrics commonly used in wireless networks. Their features and limits are detailed.

3.3.1. Hop-Count Metric

Hop-Count metric is the natural metric used in most of native multi-hop routing protocols. This metric privileges paths having the minimum number of hops. It is a very stable metric and has the isotonic characteristic [13]. Its measurement generates additional routing load. In some unstable contexts like mobility situations, it may be more efficient than metrics inducing long paths. The weakness of this metric is that it does not address interference, channel diversity, varying load on the link and capacity of the link. Thus, during the route establishment, algorithms implementing this metric do not take into account the characteristics of the network. They treat all the links identically.

These algorithms are likely to generate overloads in the center of the network because the shortest path passes through this center. Thus, in the case of multicommunication, interferences may be very frequent.

3.3.2. Delay-Based Metrics

Delay-Based metrics are also questionable. Delay at each node is composed of input queuing, processing, output queuing, transmission, propagation, and retransmission ones. Most of QoS-based delay metrics focus only on transmission delay at MAC layer [14], [15], while the other components of delay take a significant portion of the total hop-to-hop delay. Li et al. [16] consider queuing delay at network layer, but their estimation method is complex. In practice, it is not easy to obtain the number of packets waiting in network-layer buffer. Delay is closely related to packet loss rate. Packet loss that induces retransmissions grows significantly delay and also network congestion. These network performance parameters depend on the quality of used links and ambient flow. Delay and link loss ratio are often subject to high variation. End-to-end delay changes with network load as interface queue lengths vary. This can cause routes to oscillate away from a good path once the path is used. This increases the number of route changes and can lead to instability in communication and degrades performance.

3.3.3. ETX Metric

Expected Transmission Count (ETX) routing metric is one of the most popular classes of packet-loss-based metrics [17]. It is developed to improve the performance of routing in static wireless mesh networks where hop count is not suitable. The ETX of a link is calculated using the forward and reverse delivery ratios of the link. These delivery ratios are measured using probe packets. For two adjacent nodes X and Y, X measures probe delivering rate by determining the ratio between the numbers of probes received from Y and the number of expected ones. When X sends a probe, it includes the calculated ratio in the message. Y does the same. Hence, each node knows the ratio in both directions of a link (one is calculated, the other is provided by the neighbor). The metric is then obtained by:

$$ETX = \frac{1}{PDR_{X \to Y} * PDR_{Y \to X}}. \tag{3.1}$$

We note that, although ETX distinguishes two PDR values for respectively upstream and downstream direction, the obtained link metric is the same for both directions. ETX is therefore symmetric. We consider this point as a drawback of the approach. Indeed, if a link is asymmetric, we think that this link should be used but only for traffic in the reliable direction. Only Acknowledge (ACK) messages should be sent in the unreliable direction, since these messages are small and consequently are more likely to be transmitted correctly. Besides, this metric is independent on network load. Detailed analysis of OLSR [18] with the original hysteresis [19] and ETX routing metric revealed that the original hysteresis performs better than ETX-based protocols in a large dense mesh network. An analysis was then carried out on the ETX protocols. It revealed that in realistic networks, the predicted losses using the ETX algorithm are twice the actual losses that are experienced even in ideal lab conditions for 802.11 [18]. Shi et al. [20] present the design and selection of appropriate routing metrics as the principal issue to guarantee efficient routing in self-organizing networks. They attempt to analyze, compare and summarize traffic-based routing metrics in the ETX family. Several studies [21-23] have been proposed to improve the metric, but its fundamental limits remains.

3.3.4. BER-Based Metric

In this section, we, first, present a design of BER-metric, then, we present some limits of BER-based metrics. Finally, we show the relationship between BER criterion and expected number of retransmissions needed for a successful transmission.

The BER criterion characterizes a network at the lowest level (physical layer) of the transmission chain. Measuring the error rate at physical layer provides a more refined estimation of quality of radio links. It allows study of physical phenomena that influence the quality of communication. This link quality criterion has a direct impact on packet delivery rate and average communications delay. BER-based metric has been used in QoS routing as additive metric [24, 25]. Yelemou et al. in [26] formally prove that this BER criteria can be used as additive metric.

In [24], the authors used a BER-based approach to improve delay performance of on-demand routing protocol (not specified) but the used BER metric is too simplistic. BER depends on the Signal-to-Noise Ratio (SNR) at the receiving node. For simplicity, it has been assumed that the transmitted signal is affected only by free space loss. Reserve-and-go (RESGO) MAC protocol [27] was used. It is very simplistic and is based on the assumption of immediate relaying at intermediate nodes, without any anti-collision mechanism. RESGO MAC protocol is known as a low-delay MAC protocol and is relatively weak in reducing the inter-node interference. This performance of the proposed approach is limited to low interference wireless networking scenarios. The measurement of this metric has not been detailed.

Delahaye et al. [28, 29] use a ray-tracer propagation model CRT for a better estimate of the radio channel in Network Simulator (NS). The BER used in [25], [30] is the result of simulation of this realistic channel model. The use of this metric in MANET routing protocols (OLSR, AODV, ZRP) has significantly improved Packet Delivery Ratio (PDR) and delay. However, this metric has many drawbacks in actual implementation. Indeed, the BER metric is quite hard to measure in practice. A first method consists in injecting probe packets in the network. Knowing every binary elements that a packet should contain, the receiver is able to evaluate the bit error rate by counting how many bits are erroneous. Nevertheless, the packet should be large enough to allow a precise measure of BER but its size is in practice limited to the maximal transfer unit of the network. Note that control packets are too small and cannot be used to evaluate BER. So this method generates an additional load for network [31]. Another approach consists in sending impulses and measuring the impulse response associated with a transmission. The main drawback is that this method requires an adapted physical layer. An estimation of all these disadvantages is presented in [32].

Moreover, using BER as an additive metric induces long end-to-end transmission paths [30]. These long paths with an overall good BER value would potentially permit a better packet delivery ratio, but they generate long delays and induce poor throughput. Indeed, first, long paths increase intra-communication interference. Second, they also increase the vulnerability of established routes, particularly in mobility or dense networks and multi-communication contexts. For all these reasons, the BER-based metrics remain theoretical.

Against these BER metric limits, authors in [45] invested a new metric based on the number of retransmissions required to make a data transmission over a link successful. We can note that the number of packet retransmissions is highly related to the bit error rate. If we suppose a multimedia stream with constant packet size of n bits, the Packet Error Rate (PER) is $PER = 1 - (1 - BER)^n$.

Furthermore, the expected number of transmissions to get a successful packet can be computed as the mathematical expectation of the stochastic variable PER, that is $\frac{1}{(1-PER)}$.

Therefore, the expected number of transmissions is equal to:

$$nb_{transmissions} = \frac{1}{(1-PER)^n}. \tag{3.2}$$

Table 3.1 shows how the expected number of transmissions depends on the BER, for 512-byte-long packets (n = 4096).

Table 3.1. Expected number of transmissions depending on BER.

BER	Number of transmissions
10^{-5}	1.05
10^{-4}	1.51
2.10^{-4}	2.27
3.10^{-4}	3.42
4.10^{-4}	5.15
5.10^{-4}	7.76

We see (Table 3.1) that when the BER equals 4.10^{-4} or above, the expected number of transmissions is beyond the number of attempts that a default MAC layer allows to successfully deliver a packet. If possible, these links should not be used. We therefore propose a new metric based on the number of transmissions and more precisely the number of retransmissions (that appear when the first attempt is not successful). As shown in Table 3.1, this metric is highly related to BER, but, it does not require to be measured. It appears as a low-level but effective measure of the quality of links. This metric only requires that each MAC layer computes a mean value of the number of transmissions required to send packets to each neighbor, including the large ones. It is therefore not a costly measure. The next sub-section is devoted to this metric.

3.3.5. Retransmission-Based Metric

In this sub-section, first, the choice of route when intra communication interference (different transmissions for the same communication) is taken into account is discussed. In a second step, the design of number of retransmissions-based metric is presented. In a third step, this metric is compared with the ETX metric. In this retransmissions-based metric, the estimated cost of retransmission, compared to the cost of the first attempt, must be evaluated, and delay seems a convenient way to evaluate it. Let us evaluate the transmission time between a source S and its neighbor D. Let's consider a given constant time $t1$ corresponding to a successful first transmission. If transmission fails, the additional time for each retransmission is $t2$. For more details on different timing at MAC level see [33, 34]. To simplify, $t1$ is supposed to include processing time to pass from routing level to MAC level, Request To Send / Clear To sent (RTS/CTS) mechanism [35] time and propagation time, and $t2$ includes additional ACKnowledge (ACK) packet waiting timeout, RTS/CTS mechanism time and propagation time (hence $t2 > t1$).

Thereby, the delay is:

$$t = t1 + (n - 1) \times t2,$$

where n is the total number of transmissions. This equation is normalized to get the new metric (called PR for Packet Retransmission) as follows:

$$PR = \frac{t}{t1} = 1 + (n - 1) \times a,$$

with $a = \frac{t1}{t2}$.

Note that this metric appears as the number of hops penalized by a weighted number of retransmissions $a \times (n - 1)$. It equals 1 if no retransmission is needed, but it can have a greater value if retransmissions occur. This value can be seen as an equivalent (but not integer) number of intermediate hops. PR is therefore an additive metric, since equivalent number of hops can be cumulated. In a sense, it is an alternative to the simple number of hops metric: this metric is based on the number of intermediate nodes to access a recipient, but unlike the standard number of hops, it takes into account the quality of links.

To evaluate this metric, the number of packet transmissions must be determined. This information is available at the MAC level (it is a part of the communication statistics at the MAC layer) and, by a cross layer approach, is operated at routing level. There is no need to use special probes contrary to what is required in most metrics estimation processes. When the used packet size is small (such as hello packet), the number of transmissions is almost always 1 (no retransmission) when the used link exists. On the contrary, large packets allow a better estimate of the quality of a link with this metric. All kind of packets (control and data packet ones) are taken into account.

Note that *a* is a mean value that represents retransmission cost. To calibrate the value of a, statistical approach pay be used. A realistic propagation model taking into account the obstacles, with data packets in a multi-communication context, allowed to find the value *1.65* for a with *0.1* as standard deviation. In-depth study could better refine the value of *a*. This parameter may vary depending on the nature (dense or less dense) and congestion level of the studied network.

To test the effectiveness of this new metric, authors in [45] have been incorporated it in OLSR protocol as the metric used for path selection. At each node, the metric is calculated from the number of retransmissions required to make data transmissions successful over a given link. The obtained information is recorded as a new field in the record of neighbours and is disseminated through the network thanks to Topology Control (TC) messages. As it is an additive metric, a path length is computed as the sum of the metric of each of its links. The results obtained are satisfactory [45].

3.4. Taking into Account the Quality of Links in the Choice of Route

In ad hoc networks, routing is still an issue. It remains hazardous to guarantee any QoS for such networks. Route choice criterion are critical part of routing protocols. They are used to quantify paths connecting communicating nodes. Most QoS algorithms compute paths by relying on a selected metric. This allows them to compare different paths and find the best. Routing metric must take into account all specific characteristics of ad hoc wireless networks such as instability of links, interference, etc. Indeed, in this erroneous-links context, a node may need several attempts to transmit data successfully. Unfortunately, retransmissions imply additional delays, decrease throughput and increase communication overhead in the network. In critical cases, the communication fails after several attempts. The choice of the metric, the cost of its estimation and the efficiency of the algorithm allowing its exploitation are determinant for the achievement of good performance of a network.

In this section, we make a critical overview of the most commonly encountered QoS-based routing approaches. We analyze the cost (in terms of routing load and additional time) generated by link quality measurement processes and uses of these metrics in routing protocols.

Packet loss in Mobile Ad hoc NETwork (MANET) [36] is due to many factors. Among them, buffer overflow, transmission loss and link breakages are the most dominant. In addition, a received packet whose delay is over tolerable threshold is also treated as a lost packet. Loss caused by over-threshold delay can only be monitored at the receiver, requiring a feedback message be sent to the source for QoS purpose. Packet loss caused by buffer overflows and maximum retransmissions exceeding, are the only information that can be obtained from intermediate nodes. Successful design of a metric that takes into account all of these components is very delicate.

Many approaches measure the packet loss rate by injecting probing packets into a network. A large number of sample packets are required to accurately estimate a highly variable link. Shi et al. [32] evaluate the number of probing packets needed to get an accurate result as follows: $N = \frac{1-p}{m^2 * p}$ where p is the packet loss probability and m the coefficient of variation. According to this formula, we see that this active measurement scheme is not suitable for MANETs. For example, for a link with 10 % mean packet loss rate ($p = 0.1$), 900 samples must be sent on that link to get a measurement result where standard deviation is within 10 % of the loss probability (i.e., $m = 0.1$). When each node should send probe packets, these can cause a large overhead in MANETs, thus skewing the obtained results. Furthermore, it takes some time for measurements. For example, if one sample packet is sent every 1 second, 15 minutes are needed to send 900 samples. Which shows that the active measurement scheme is obviously not suitable for a wireless network particularly in mobility context. In order to overcome this dilemma (amount overhead), Link Quality Ranking (LQR) [37] uses the following trade-off: Instead of estimating a link-layer metric for each link, LQR performs a pairwise comparison of the physical-layer metrics and selects the best link. One problem faced when broadcasting probe-based estimators such as ETX, is that they decouple link estimation from data traffic: if a link goes bad and packets are lost, the link estimate will not reflect this change until the next routing beacon is dropped [38].

The average rates of link packet loss are commonly used. Link quality of a route is evaluated by summing the metric value of each link on the route [39, 17]. This way of using this metric is questionable. The average or sum of link quality measurements along one route may ignore the worst link. Indeed, if the quality of a link among one route is rather bad, the packets can not be delivered successfully although the average or sum value is rather good. Some works, such as [4], use a link metric optimal value for the path selection. This choice may not provide the best path. For example, considering number of expected retransmissions metric, a path with link metric optimal value m, is preferred to any other containing just one link whose metric value is upper than m even if its other links are better. m value of a path is assumed to be the maximum link metric value, considering all links of the path.

The Packet Delivery Ratio (PDR) metric is often used as a multiplicative metric [9]. A blind multiplication applied to this metric strongly favors long paths. In this case, inter-hop interference may be significant. Indeed, the intermediate node can not simultaneously receive a packet from a neighbor upstream and send another to a downstream neighbor. Additional delay due to intra-communication interference is often not taken into account. It has an impact on throughput and delay but not necessarily on packet delivery ratio.

Often, feasible paths under QoS metric requirement are based on a blacklisting method [42]. Link estimators consider only links with quality above a certain threshold. This minimizes the potential costs for low quality link estimation that should not be used for routing. However, a blacklisting policy could filter routing options, severely limiting the efficiency of the routing algorithm if an improper threshold is chosen [43]. Some work like [44, 16, 15] use a composite metric. It is shown that taking into account several metrics simultaneously is NP-complete [41].

In addition to the sensitivity of the link quality criterion measurement, many authors have questioned the use of these QoS values. In [32, 34, 40, 41, 4], the authors highlight the complexity and exorbitant cost (overhead and computing time) of route-discovery approach with admission control processes. Technique used in [41] to take into account intermediate node lifetime for robust network, performed five examinations and the size of the packet is raised at each intermediate node. Thus, this algorithm induces more weakness in the routing protocol than advantage because it increases control messages size and delays.

In order to guarantee a certain level of QoS, routing protocols should be smart enough to pick a stable and good quality communication route in order to avoid retransmission and packets loss. To be efficient, QoS metric-based algorithms must use a light and immediate way to evaluate link quality. It should not induce significant additional load.

3.5. Conclusion

Because of the ease of deployment and low deployment cost, ad hoc wireless networks are attracting renewed interest in a socio-economic context where it is necessary to communicate at any time and anywhere. However, the radio channel used for communication in these networks is not a safe medium. It undergoes the effect of ambient electromagnetic noise, obstacles in the propagation medium, interference. Packet loss rates are high. These networks struggle to satisfy the QoS constraints imposed by multimedia applications.

In this context of error-prone medium, it is very important that routing algorithms are smart enough to contain the routing load and establish reliable routes. In this chapter, we discuss the problem of QoS in routing by the cross-layer approach. Several studies have focused on this theme. Complexity of estimating the quality of the links and complexity of the implemented solutions mean that the expected results are mixed under realistic conditions.

We retain that, to be efficient, metric used to quantify link must not induce more instability in the routes used for data transmission. In the context of mobile ad hoc networks, it must not lead to too long paths. The number of hops must be judiciously exploited.

When node speed is very high (too dynamic network topology), QoS results are not satisfactory. These allow us to say that the main issue in high mobility context is not to take into account link quality, but rather control of neighborhood information. A node should more frequently inventory its links and routes.

Better use of the cognitive radio approach in ad hoc network would make possible to optimize the exploitation of the bandwidth offered by the radio channel.

Acknowledgements

The author would like to thank Prof. Philippe Meseure for his considerable contribution to the ideas and design of this chapter.

References

[1]. A. S. Tannenbaum, Computer Networks, 4th edition, *Prentice Hall*, 2002.

[2]. T. Clausen, P. Jacquet, C. Adjih, A. Laouiti, P. Minet, P. Muhlethaler, A. Qayyum, L. Viennot, Optimized Link State Routing Protocol - OLSR, *Tech. Rep. RFC3626*, 2002.

[3]. T. Yélémou, P. Meseure, A.-M. Poussard, A new BER-based approach to improve OLSR protocol, in *Proceedings of the 8th IEEE and IFIP International Conference on Wireless and Optical Communications Networks (WOCN'11)*, Paris, France, May, 2011.

[4]. K. Al-Soufy, A. Abbas, A path robustness based quality of service routing for mobile ad hoc networks, in *Proceedings of the International Conference on Internet Multimedia Services Architecture and Application (IMSAA'10)*, December 2010, pp. 1-6.

[5]. C. Sunghyun, K. Seongkwan, L. Okhwan, L. Sung-Ju, Comparative analysis of link quality metrics and routing protocols for optimal route construction in wireless mesh networks, *Ad Hoc Networks*, Vol. 9, September 2011, pp. 1343-1358.

[6]. P. Esposito, M. Campista, I. Moraes, L. Costa, O. Duarte, M. Rubinstein, Implementing the expected transmission time metric for OLSR wireless mesh networks, in *Proceedings of the Wireless Days 2008 (WD'08)*, 2008, pp. 1-5.

[7]. K. Al Agha, A. Munaretto, H. Badis, G. Pujolle, Qolsr: Routage avec qos dans olsr, in *Proceedings of the 5ème Rencontres Francophones sur les Aspects Algorithmiques des Télécommunications (AlgoTel'03)*, May 2003.

[8]. T. Kunz, Y. Ge, L. Lamont, Quality of service in ad-hoc networks using olsr, in *Proceedings of the 36th Hawaii International Conference on System Sciences (HICSS'03)*, 2003.

[9]. F. Ingelrest, D. Simplot-Ryl, Maximizing the delivery of MPR broadcasting under realistic physical layer assumptions, *Journal of Computer Science and Technology*, May 2008, Vol. 23, Issue 3, pp. 451-460.

[10]. K. Oudidi, N. Enneya, M. Elkoutbi, Enhancing delay in MANET using OLSR protocol, *International Journal of Communications, Network and System*, Vol. 5, 2009, pp. 392-399.

[11]. M. SreeRangaRaju, J. Mungara, Performance evaluation of ZRP in ad-hoc mobile wireless network using QualNet simulator, in *Proceedings of the Signal Processing and Information Technology (ISSPIT'10)*, December 2010, pp. 457-466.

[12]. I. A. Saroit, I. M. A. Fahmy, L. Nassef, S. Ahmed, QoS parameters improvement for the hybrid zone-based routing protocol in MANET, in *Proceedings of the International Conference on Informatics and Systems (INFOS'10)*, March 2010, pp. 1-9.

[13]. M. Siraj, A Survey on Routing Algorithms and Routing Metrics for Wireless Mesh Networks, *World Applied Sciences Journal*, 2014, pp. 870-886.

[14]. R. Draves, J. Padhye, B. Zill, Routing in multi-radio, multi-hop wireless mesh networks, in *Proceedings of the 10th Annual International Conference on Mobile Computing and Networking (MobiCom'04)*, New York, NY, USA, 2004, pp. 114-128.

[15]. Y. Yang, Designing routing metrics for mesh networks, in *Proceedings of the IEEE WiMesh*, 2005.

[16]. H. Li, Y. Cheng, C. Zhou, W. Zhuang, Minimizing end-to-end delay: a novel routing metric for multi-radio wireless mesh networks, in *Proceedings of the IEEE Conference on Computer Communications (INFOCOM'09)*, Apr. 2009, pp. 46-54.

[17]. D. S. J. de Couto, D. Aguayo, J. Bicket, R. Morris, A high-throughput path metric for multi-hop wireless routing, *Wireless Networks*, Vol. 11, July 2005, pp. 419-434.

[18]. D. Johnson, G. Hancke, Comparison of two routing metrics in olsr on a grid based mesh network, *Ad Hoc Netw.*, Vol. 7, Mar. 2009, pp. 374-387.

[19]. T. Clausen, P. Jacquet, Optimized Link State Routing Protocol (OLSR), *IETF RFC 3626*, October 2003.

[20]. F. Shi, D. Jin, J. Song, A survey of traffic-based routing metrics in family of expected transmission count for self-organizing networks, *Computers and Electrical Engineering*, Vol. 40, No. 6, 2014, pp. 1801-1812.

[21]. P. Esposito, M. Campista, I. Moraes, L. Costa, O. Duarte, M. Rubinstein, Implementing the expected transmission time metric for OLSR wireless mesh networks, in *Proceedings of the Wireless Days (WD'08)*, November 2008, pp. 1-5.

[22]. R. Draves, J. Padhye, B. Zill, Routing in multi-radio, multi-hop wireless mesh networks, in *Proceedings of the 10th Annual International Conference on Mobile Computing and Networking (MOBICOM'08)*, 2004, pp. 114-128.

[23]. A. Cerpa, J. L. Wong, M. Potkonjak, D. Estrin, Temporal properties of low power wireless links: Modeling and implications on multi-hop routing, in *Proceedings of the International Symposium on Mobile Ad Hoc Networking and Computing (ACM MobiHoc'05)*, 2005, pp. 414-425.

[24]. N. Wisitpongphan, G. Ferrari, S. Panichpapiboon, J. Parikh, O. Tonguz, QoS provisioning using BER-based routing in ad hoc wireless networks, in *Proceedings of the Vehicular Technology Conference (VTC'05)*, Vol. 4, May 2005, pp. 2483-2487.

[25]. T. Yélémou, P. Meseure, A. M. Poussard, A new ber-based approach to improve olsr protocol, in *Proceedings of the 8th International Conference on Wireless and Optical Communications Networks (WOCN'11)*, May 2011, pp. 1-5.

[26]. T. Yélémou, P. Meseure, A.-M. Poussard, Taking binary error rate into account to improve OLSR protocol, *International Journal of Computer Networks and Communications Security*, 2016, pp. 276-285.

[27]. G. Ferrari, O. Tonguz, MAC protocols and transport capacity in ad hoc wireless networks: Aloha versus PR-CSMA, in *Proceedings of the Military Communications Conference (MILCOM'03)*, Vol. 2, October 2003, pp. 1311-1318.

[28]. R. Delahaye, Y. Pousset, A.-M. Poussard, C. Chatellier, R. Vauzelle, A realistic physic layer modeling of 802.11g ad hoc networks in outdoor environments with a computation time optimization, in *Proceedings of the 11th World Multi-Conference on Systemics, Cybernetics and Informatics (WMSCI'07)*, Orlando, Florida, USA, 2007.

[29]. R. Delahaye, A.-M. Poussard, Y. Pousset, R. Vauzelle, Propagation models and physical layer quality criteria influence on ad hoc networks routing, in *Proceedings of the 7th International Conference on ITS Telecommunications (ITST'07)*, June 2007, pp. 1-5.

[30]. T. Yélémou, P. Meseure, A. M. Poussard, Improving ZRP performance by taking into account quality of links, in *Proceedings of the IEEE Wireless Communications and Networking Conference (WCNC'12)*, April 2012, pp. 2956-2960.

[31]. J. Ledy, H. Boeglen, A. Poussard, B. Hilt, R. Vauzelle, A Semi-deterministic channel model for VANETs simulations, *International Journal of Vehicular Technology*, 2012, pp. 1-8.

[32]. L. Shi, A. Fapojuwo, N. Viberg, W. Hoople, N. Chan, Methods for calculating bandwidth, delay, and packet loss metrics in Multi-Hop IEEE802.11 ad hoc networks, in *Proceedings of the IEEE Vehicular Technology Conference (VTC'08)*, May 2008, pp. 103-107.

[33]. Wireless LAN medium access control (MAC) and physical layer (PHY) specifications, *IEEE Standard 802.11*, Part 11, June 1999.

[34]. S. Su, Y. Su, J. Jung, A novel QoS admission control for ad hoc networks, in *Proceedings of the IEEE Wireless Communications and Networking Conference (WCNC'07)*, March 2007, pp. 4193-4197.

[35]. J. Weinmiller, H. Woesner, J.-P. Ebert, A. Wolisz, Analyzing the RTS/CTS mechanism in the DFWMAC media access protocol for wireless LANs, in *Proceedings of the TC6 Workshop for Personal Wireless Communication (IFIP'95)*, 1995.

[36]. I. Chlamtac, M. Conti, J. J.-N. Liu, Mobile ad hoc networking: imperatives and challenges, *Ad Hoc Networks*, Vol. 1, No. 1, 2003, pp. 13-64.

[37]. M. Zuniga, I. Irzynska, J. Hauer, T. Voigt, C. A. Boano, K. Roemer, Link quality ranking: getting the best out of unreliable links, in *Proceedings of the IEEE International Conference on Distributed Computing in Sensor Systems and Workshops (DCOSS'11)*, June 2011, pp. 1-8.

[38]. R. Fonseca, O. Gnawali, K. Jamieson, P. Levis, Four-bit wireless link estimation, in *Proceedings of the 6th Workshop on Hot Topics in Networks (HotNets VI)*, 2007.

[39]. G. Karbaschi, A. Fladenmuller, A link-quality and congestion-aware cross layer metric for multi-hop wireless routing, in *Proceedings of the IEEE International Conference on Mobile Adhoc and Sensor Systems Conference (MASS'05)*, November 2005.

[40]. B. Bakhshi, S. Khorsandi, Complexity and design of qos routing algorithms in wireless mesh networks, *Computer Communications*, Vol. 34, No. 14, 2011, pp. 1722-1737.

[41]. Z. Wang, J. Crowcroft, Quality-of-service routing for supporting multimedia applications, *IEEE Journal on Selected Areas in Communications*, Vol. 14, September 1996, pp. 1228-1234.

[42]. J. Ledy, H. Boeglen, B. Hilt, A. Abouaissa, R. Vauzelle, An enhanced AODV protocol for VANETs with realistic radio propagation model validation, in *Proceedings of the 9th International Conference on Intelligent Transport Systems Telecommunications (ITST'09)*, 2009, pp. 398-402.

[43]. O. Gnawali, M. D. Yarvis, J. Heidemann, R. Govindan, Interaction of retransmission, blacklisting, and routing metrics for reliability in sensor network routing, in *Proceedings of the IEEE Conference on Sensor and Ad-hoc Communication and Networks (SECON'04)*, Santa Clara, California, USA, May 2004, pp. 34-43.

[44]. K. Oudidi, A. Hajami, M. Koutbi, QoS routing using OLSR protocol, in *Proceedings of the 14th WSEAS International Conference on Communications (ICCOM'10)*, 2010, pp. 165-171.

[45]. T. Yélémou, P. Meseure, A.-M. Poussard, Using number of retransmissions to improve routing in ad-hoc networks, in *Proceedings of the 12th International Conference on Wireless and Mobile Communications (ICWMC'16)*, November 13-17, Barcelona, Spain, 2016, pp. 121-126.

Chapter 4
A Knowledge-Based Modeling and Simulation Approach for the Management of Sensor Networks

Mehdi Mekni

4.1. Introduction

Sensor Networks (SN) are distributed network systems composed of hundreds of sensor nodes [1, 2]. New capabilities such as micro-sensing and in-situ sensing as well as the wireless connection of these nodes open new possibilities for applications in various domains such as military, environment and disaster relief [3, 4]. The low per-node cost and the shrinking size of microprocessors in addition to the enhancement of their computation capacities, while decreasing their energy consumption, will allow dense distribution of these wireless networks of sensors and actuators [2]. SN can be thought of as a macro-instrument concept that allows for the spatio-temporal understanding of phenomena, which take place in geographic environments through the coordinated efforts of a large number of sensing nodes of different types [2]. However, once SN are designed, the deployment of such complex systems is a challenge because of the complexity and the large-scale of the geographic environment [4]. Sensor networks deployment is by nature a spatial problem since nodes are highly constrained by the geographic characteristics of the environment. Even if it is practical to evaluate research on the real hardware platform, it may not be practical to experiment in an appropriate environment. An example of this are sensor networks which operate on glaciers, remote wildlife habitats, volcanos, and other environments where in-situ sensing techniques are required and with which it is expensive or dangerous to experiment. Therefore, there is a need for an efficient modelling and simulation approach to address the issue of sensor networks management using actors representing sensor nodes evolving in and interacting with a digital representation of the physical geographic environment.

Mehdi Mekni
Department of Computer Science and Information Technology, St. Cloud State University, St. Cloud, MN, United Statess of America

In order to address the above-mentioned challenges, we propose a knowledge-based agent-oriented geo-simulation approach to support the simulation of SN deployments in Informed Virtual Geographic Environments (IVGE). A critical step towards the simulation of SN deployment is the creation of appropriate representations of the geographic space and of the sensors evolving in it. Such a representation should enable the sensors' spatio-temporal reasoning capabilities. Moreover, an IVGE should provide sensor agents with knowledge about the virtual environment in which they evolve and with which they interact. A number of challenges arise when creating and managing knowledge about the environment, among which we mention: 1) Building a geometrically-precise geometrically-precise and semantically-informed virtual geographic environments; 2) Representing knowledge about environments using a standard formalism; 3) Providing agents with tools and mechanisms to allow them acquire knowledge about the environment; and 4) Inferring on facts and drawing conclusions that characterise the geographic environment in order to support spatial agents' decision-making. This approach builds on previous works on Informed Virtual Geographic Environments [5-7], on spatially reasoning agents [5] and on qualitative reasoning about geo-simulation results [8].

The rest of the chapter is organised as follows. Section 4.2 presents a comprehensive related works on agent-based simulation tools for SN. Moreover, it provides a short survey of existing geographic environment models as well as on agents' spatial behaviours. It also introduces the affordance concept, presents the notion of knowledge about the environment, and outlines its importance for spatial agents to enable autonomous decision–making mechanisms that take into account the characteristics of the virtual geographic environment in which agents evolve. Section 4.3 briefly describes our IVGE computation model. Section 4.4 introduces the concept of Environment Knowledge (EK) and details our novel method to define it using Conceptual Graphs (CGs) [22]. It also presents the environment knowledge base along with the associated decision making process, which involves an inference engine. Section 4.5 provides a description of the proposed agent model and presents patterns of spatial behaviours. Section 4.6 presents our knowledge-based approach along with associated agent and action achetypes concepts. Section 4.7 illustrates and discusses, through a case study, our knowledge-based approach through a SN deployment scenario for weather monitoring purposes. Section 4.8 discusses the results and Section 4.9 concludes with our future works.

4.2. Related Work

4.2.1. Modelling and Simulation Tools for Sensor Networks

According to our literature review, architectures for the management of sensor webs involving the Geo-simulation paradigm do not exist. However, a few research projects have attempted to integrate the agent paradigm into sensor network architectures such as IrisNet [30], Abacus [31], Biswas and Phoha's architecture [32], and SWAP [33].

Most of these architectures identify the need for distributed data collection and processing, and propose layered architectures to achieve this. In Abacus, different agents in the processing layer detect and report alert conditions to a higher layer interacting with users [31]. IrisNet uses agents such as Sensor Agents (SA) and Sensor Organisers (SO) to collect and analyze data from sensors to answer specific classes of queries [30]. Biswas and Phoha's approach uses agents in the service layer to analyze data from sensors and transfer it to the application layer [32]. All these approaches deal with data collection by providing a distributed infrastructure for publishing, discovering and accessing sensor resources. They also address the challenge of data fusion, to some extent, and aim to provide end-users with the information they need. These approaches use the agent paradigm, which is the distribution of tasks. However, these applications do not take complete advantage of the multi-agent systems approach. Indeed, they use reactive agents which are efficient for alerting purposes, but are neither able to perform situated behaviors nor autonomous decision-making. On the one hand, situated behaviors include performing spatial reasoning and taking advantage of the virtual environment's description where sensor agents are located. On the other hand, autonomous decision-making includes managing sensor nodes in order to efficiently cover the area of interest while taking into account their limited capabilities as well as local spatial characteristics. We think that, in order to achieve intelligent and autonomous deployment of sensor webs, it is essential to use a multi-agent geo-simulation approach in which agents are endowed with advanced capabilities such as perception, navigation, memory, and knowledge management. The knowledge management process includes the following tasks: (1) Represent knowledge about geographic environments using standard formalisms; (2) Allow spatial agents to acquire knowledge about the environment; (3) Enable agents' reasoning and decision-making while taking into account knowledge about geographic environments.

As the above-mentioned tools do not address the challenging task of sensor network management, we propose the knowledge-based intelligent modelling and simulation approach for the management of sensor networks using informed virtual geographic environments.

4.2.2. Environment Representation

Virtual environments and spatial representations have been used in several application domains. For example, Thalmann et al. proposed a virtual scene for virtual humans representing a part of a city for graphic animation purposes [4]. Donikian et al. proposed a modelling system, which is able to produce a multi-level database of virtual urban environments devoted to driving simulations [25]. More recently, Shao et al. proposed a virtual environment representing the New York City's Pennsylvania Train Station populated by autonomous virtual pedestrians in order to simulate the movement of people [21]. However, since the focus of these approaches is computer animation and virtual reality, the virtual environment usually plays the role of a simple background scene in which agents mainly deal with geometric characteristics.

Despite the multiple designs and implementations of virtual environments frameworks and systems, the creation of geometrically precise and semantically enriched geographic content is still an open issue. Indeed, research has focused almost exclusively on the geometric and topologic characteristics of the virtual geographic environment. However, the structure of the virtual environment description, the optimization of this description to support large-scale and complex geographic environments, the meaning of the geographic features contained in the environment as well as the ways to interact with them have received less attention.

4.2.3. Spatial Behaviours and Knowledge Management

Research on spatial behaviours investigates the processes that take place when spatial agents representing people or other dynamic actors orient themselves and navigate through complex and large-scale virtual geographic environments [18]. In order to build agents that exhibit plausible spatial behaviours with respect to their capabilities and to the virtual environment characteristics in which they evolve, we need to analyse humans' spatial behaviours in the physical world [29]. We also need to determine how spatial agents can make decisions using knowledge provided by the virtual environment. In this section, we present several works related to spatial behaviours and affordances and outline the importance of knowledge about the environment for the support of agents' spatial behaviours.

Several theories in the field of human spatial behaviours have been proposed in order to explain how people navigate in the physical world, what people need to find their ways, and how people's visual abilities influence their decisions [5]. Actually, these theories point out the use of various spatial and cognitive abilities to apprehend the physical world in which people evolve and with which they interact [7]. Weisman identified four classes of environmental variables that influence spatial behaviours in physical worlds: visual access; architectural differentiation; signs to provide identification or directional information; and plan configuration [26]. Seidel's study at the Dallas/Fort Worth Airport showed that the spatial structure of the physical environment has a strong influence on people's spatial behaviours [20]. Arthur and Passini introduced the term environmental communication, arguing that the built environment and its parts should function as a communication device [1]. Information about the geographic environment along with the spatial and cognitive capabilities are fundamental inputs to the spatial decision-making process [7]. This knowledge include information collected using perception capabilities, memorised information resulting from past experiences, and information provided by the environment itself [8].

Knowledge is an important asset for agents because it allows them to reason about it and to autonomously make informed decisions [28]. By its very nature, knowledge is disparate and heterogeneous and can be represented in various ways (qualitatively and quantitatively), and can be either structured or unstructured. Knowledge usually includes information about the agent's characteristics, as well as about the description of the geographic environment in which it is situated. Thus, spatial agents require knowledge about their environment in order to reason about it, to infer facts, and to draw conclusions, which will guide them to make decisions and to act.

The main reason why virtual environments have received less interest from practitioners is that geographic environments may be complex, large-scale, and densely populated with a variety of geographic features. Consequently, formally representing knowledge about geographic environments is usually complex and time consuming [27]. Another issue, which needs to be addressed, is the way to allow spatial agents to acquire this knowledge to autonomously make decisions. There is a need for a knowledge management approach: (1) to represent knowledge about geographic environments using a standard formalism; (2) to allow spatial agents to acquire knowledge about the environment; (3) to allow agents to reason using knowledge about geographic environments.

4.3. Generation of Informed Virtual Geographic Environments

In this section, we briefly present our automated approach to compute the IVGE data using vector GIS data. This approach is based on four stages: input data selection, spatial decomposition, maps unification, and finally the generation of the informed topologic graph [14]. A detailed description of the spatial decomposition and layers integration techniques is provided in [12, 13], and in [17].

4.3.1. GIS Input Data Selection

The first step of our approach consists of selecting the different vector data sets that are used to build the IVGE. The input data can be organized into two categories. First, elevation layers contain geographical marks indicating absolute terrain elevations [14]. Second, semantic layers are used to qualify various types of data in space. Each layer indicates the physical or virtual limits of a given set of features with identical semantics in the geographic environment, such as roads or buildings. Fig. 4.1 illustrates various data provided by geographic information systems.

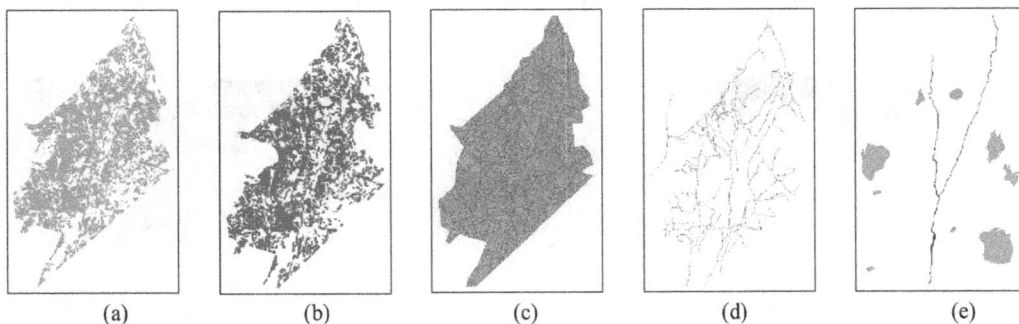

(a)   (b)   (c)   (d)   (e)

Fig. 4.1. Various geometric and semantic layers related to the Montmorency experimental forest, St. Lawrence Region, Quebec, Canada: (a) and (b) two types of vegetation characterizing the land cover; (b) water resources including rivers and lakes; and (b) road network.

4.3.2. Spatial Decomposition

The second step consists of obtaining an exact spatial decomposition of the input data into cells. First, an elevation map is computed using the Constrained Delaunay Triangulation (CDT) technique. All the elevation points of the layers are injected into a 2D triangulation, the elevation being considered as an attribute of each node. Second, a merged semantics map is computed, corresponding to a constrained triangulation of the semantic layers. Indeed, each segment of a semantic layer is injected as a constraint that keeps track of the original semantic data by using an additional attribute for each semantic layer.

4.3.3. Map Unification

The third step to obtain our IVGE consists of unifying the two maps previously obtained. This phase can be depicted as mapping the 2D merged semantic map onto the 2.5D elevation map in order to obtain the final 2.5D elevated merged semantics map. First, pre-processing is carried out on the merged semantics map in order to preserve the elevation precision inside the unified map. Indeed, all the points of the elevation map are injected into the merged semantics triangulation, creating new triangles. Then, a second process elevates the merged semantics map.

4.3.4. Informed Topologic Graph

The resulting unified map now contains all the semantic information of the input layers, along with the elevation information. This map is used as an Informed Topologic Graph (ITG), where each node corresponds to the map's triangles, and each arc corresponds to the adjacency relations between these triangles. Then, common graph algorithms are applied to this topological graph, and graph traversal algorithms in particular [13]. Fig. 4.2 depicts the IVGE representing the Montmorency experimental forest that will be used in the case study of the management of a sensor network for weather monitoring purposes presented in Section 4.7.

Fig. 4.2. IVGE representing the Montmorency experimental forest; (left) the 2.5D triangulated elevation map; (center) the semantic maps; and (right) the 3D unified map.

4.4. From Semantic Information to Environment Knowledge

In [15], we proposed a novel model along with a complete methodology for the automated generation of informed VGEs. Then, we presented our abstraction approach that enriches and structures the description of the IVGE, using geometric, topologic and semantic characteristics of the geographic environment. In order to represent semantic information that characterises our informed virtual environment model, we also proposed to use Conceptual Graphs (CGs) [22]. Our aim now is to evolve the semantic information to the level of knowledge and hence to build a knowledge-based virtual geographic environment in which spatial agents autonomously make informed decisions.

The process of making an informed decision has been modelled as a pyramid built on data [11] as shown on the left hand side of Fig. 4.3. Data corresponds to the transactional, incremental physical records [11]. In our IVGE model, this data corresponds to the geometric and geographic data provided by GIS. In and of itself this data is not sufficient to support spatial agents' decision-making. This data must be organized into information in order to be useful. Information is data that has been contextualized, categorized, often calculated (from initial data), corrected, and usually condensed [19]. In our IVGE model, information corresponds to the description of the IVGE resulting from the exact spatial decomposition of the geographic environment and enhanced with semantic information. Information often contains patterns within it and is sometimes useful for simple spatial behaviours such as motion planning. However, the context of these spatial behaviours can only be formed using some knowledge. Knowledge provides the next step of data organisation. For information to become knowledge, the context of the information needs to include predictive capabilities. Using predictive capabilities of knowledge, spatial agents can autonomously make informed decisions. The more complex and voluminous the underlying data sets are, the more effort is required to progressively organise it so that it becomes knowledge useful to the agents' decision-making. However, since our IVGE description is structured as a hierarchical topologic graph resulting from the geometric, topologic, and semantic abstraction processes, and since the semantic information is expressed using conceptual graphs, we are able to build knowledge about the environment to support agents' spatial behaviours.

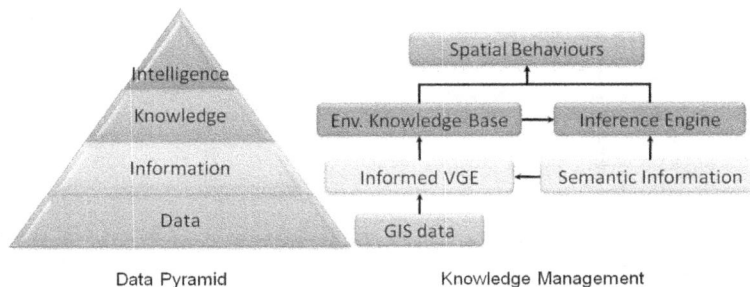

Data Pyramid   Knowledge Management

Fig. 4.3. The proposed knowledge management approach; on the left hand side, the pyramid data [11]; on the right hand side, the knowledge management approach relying on our IVGE model and involving a knowledge base coupled with an inference engine to support agents' spatial behaviours.

135

4.4.1. Environment Knowledge

We define the notion of Environment Knowledge (EK) as a specification of a conceptualization of the environment characteristics: the objects, agents, and other entities that are assumed to exist in the informed virtual geographic environment and the relationships that hold among them. Hence, EK is a description (like a formal specification of a program) of the spatial concepts (geographic features) and relationships (topologic, semantic) that may exist in a geographic environment. Users provide this description in order to enrich the qualification of the geographic features that characterise the environment. It is expressed using a standard formalism that is close to natural language and computer tractable.

Let us emphasize that enhancing agent-based geosimulation with EK, allows spatial agents to reason about the characteristics of the virtual geographic environment. Practically, EK is composed of spatial concepts (i.e., ask queries and make assertions) and spatial relationships (i.e., describe actions and behaviours). Our aim is to improve the perception-decision-action loop on which rely most agent models. Considering Newell's pyramid [16] that comprises the reactive, cognitive, rational and social levels of agent behaviours, we mainly focus on the knowledge acquisition process in order to support the decision-making capabilities of spatial agents. Fig. 4.4 illustrates two elements: (1) the knowledge acquisition process, and (2) the action archetype process, that we introduced in order to extend Newell's initial pyramid.

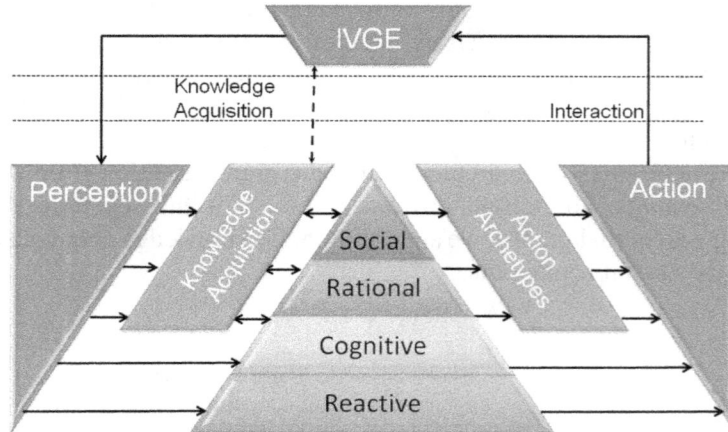

Fig. 4.4. The enhanced perception-decision-action loop.

The management of the environment knowledge is composed of two main parts: (1) an Environment Knowledge Base (EKB) that relies on spatial semantics represented using the CG formalism; and (2) an Inference Engine (IE) that allows to manipulate and to acquire environment knowledge in order to provide spatial agents with the capability of reasoning about it.

Conceptual graphs are widely used to represent knowledge [22, 23]. Actually, CGs enable us to formally represent spatial semantics characterizing our IVGE model and allow us to build a structured Environment Knowledge Base (EKB) based on a finite bipartite graph [24]. The EKB allows representing, using a standard formalism, the information characterizing the virtual environment as well as the objects and agents it contains. Moreover, the EKB enables us to explicitly specify affordances [6] in order to support the agents spatial interactions with the informed virtual geographic environment in which they evolve. The environment knowledge base, which is part of this process, relies on the notion of spatial semantics. Spatial Semantics (SS) consists of a structured, conceptualised, and organised representation of geographic features, agents, and objects that an informed virtual geographic environment may contain. Spatial semantics relies on two types of nodes: semantic concepts and semantic relations. Semantic concepts represent entities such as agents, objects and zones as well as attributes, states and events. Semantic relations represent the relationships that hold among semantic concepts.

The environment knowledge can be constructed by assembling percepts. In the process of assembly, semantic relations specify the role that each percept plays and semantic concepts represent the percepts themselves. Semantic concepts involve two types of functions; referent and type. The function referent maps semantic concepts to generic markers denoted by names starting with an asterisk "*" or individual markers usually denoted by numbers. For example, if the referent is just an asterisk, as in [HOUSE: ∗], the concept is called a generic semantic concept, which may be read as a house or some house. The function type maps concepts to a set of type labels. A semantic concept "sc" with type (sc) = t and reference (sc) = f is displayed as [t: f]. The function type can also be applied to relations. For example, if the referent is a number [HOUSE: #80972], the field to the left of the colon contains the type label HOUSE, the field to the right of the colon contains the referent #80972 which designates a particular house.

To sum up, the EKB contains knowledge about the informed virtual geographic environment that an agent may use. This knowledge is provided to enrich the qualification of the geographic features which characterise the IVGE. Finally, this knowledge is structured using semantic concepts and relations expressed using conceptual graphs.

4.4.2. Inference Engine

Now that we have defined the environment knowledge base as a structure that contains explicit descriptions of geographic features using CGs, let us describe the Inference Engine (IE) that is part of our knowledge management approach. The IE is a computer program that derives answers from our environment knowledge base. Therefore, the IE must be able to logically manipulate symbolic CGs using formulas in the first-order predicate calculus. In order to acquire knowledge about the virtual environment, agents use the IE and formulate queries using a semantic specification that is compatible with CGs. Agents interpret the answers provided by the IE and act on the environment. They can also enrich the EKB by adding new facts that result from their observation of the virtual environment (Fig. 4.5).

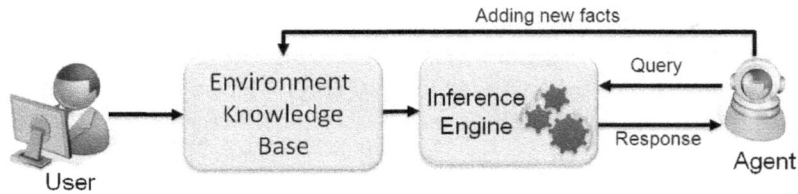

Fig. 4.5. The inference engine uses the EKB for the purpose to answer queries formulated by agents.

4.4.3. Mapping Knowledge Using Environment Knowledge and Inference Engine

In this sub-section, we first present how CGs allow us to map knowledge about the environment into first-order logic formulas. Then, we provide a short survey of existing tools that support the manipulation of CGs. We also discuss the capabilities of these tools to provide a programming language with CGs, related operations, and inference engine. Finally, we present the Amine platform [9], a platform to manipulate CGs using an inference engine embedded in PROLOG+CG language.

Conceptual graphs offer the opportunity to map knowledge about the environment into formulas in the first-order predicate calculus. Using formulas in the first-order predicate calculus, it is possible to build tools that allow spatial agents to manipulate knowledge about virtual environments represented using CGs. Moreover, it is possible to build tools that allow for logic and symbolic manipulations of environment knowledge and provide the opportunity to infer and to predict facts or assumptions about virtual environments. Several tools can be used to manipulate CGs (Amine, CGWorld, CoGITaNT, CPE, Notio, and WebKB). These tools can be classified under at least 8 categories of tools: CG editors, executable CG tools, algebraic tools (tools that provides CG operations), KB/ontology tools, ontology server tools, CG-based programming languages, IDE tools for CG applications and, agents/MAS tools. The category "CG-based programming language" concerns any CG tool that provides a programming language with CG, related operations, and inference engine. Only Amine belongs to this category, with its programming language: Prolog+CG. Therefore, we propose to use the Amine platform and Prolog+CG in order to logically manipulate symbolic CGs and to provide spatial agents with an inference engine that allows them to query the environment knowledge, to acquire environment knowledge and reason about it.

Using Amine platform, users can build an environment knowledge base (EKB) using CGs and query the Amine's inference engine (IE) to derive new knowledge from the content of the EKB using queries. The Amine platform provides a graphic user interface to support the manipulation of the EKB. Agents are able to send queries, during the simulation process, in order to acquire the knowledge they need to make a decision, using the Prolog+CG language that is supported by Amine platform. These queries are processed by the IE, which interrogates the EKB and sends back the response to agents.

In order to illustrate such a querying process, let us consider the following simple environment knowledge, composed of a set of two facts that provide an idea of the use of conceptual structures as a Prolog+CG data structure:

cg([Man:Mehdi]←agnt-[Study]-loc→[University]),

cg([Man:Mehdi]←agnt-[Play]-obj→[Soccer]).

And the following request: "Which actions are done by Man Mehdi ?",

?- cg([Man:Mehdi]←agnt-[x]).

The answer provided by the Amine platform using its Prolog+CG inference engine is:

x = Study; x = Play.

Now that we introduced the main parts of our environment knowledge management approach, namely EKB and IE, we detail in the following section the notions of agent and action archetypes and the way we use them to build spatial behaviours.

4.5. From Environment Knowledge to Spatial Behaviors

When dealing with arge number of spatial agents of various kinds, the specification of their attributes and associated spatial behaviours might be complex, time, and effort consuming. In order to characterise our spatial agents, we propose to specify: (1) the agent archetype, its super-types and sub-types according to the semantic type hierarchy; (2) the agent category (such as actor, object, and spatial area); and (3) the agent spatial behavioural capabilities, including moving within the IVGE content, perception of the IVGE and of other spatial agents. In the following subsections, we discuss these elements.

4.5.1. Agent Archetypes

In our environment knowledge management approach, the description of agents as well as objects and geographic features (spatial areas and zones) is enriched with semantic information. This means that these spatial agents belong to a semantic type hierarchy. Using the semantic type hierarchy allows us to take advantage of inheritance mechanisms. Hence, when modelling a large number of agents, we only need to specify the attributes that are associated with the highest-level types of agents that we call agent archetypes rather than repeatedly specifying them for each lower-level agents. Let us define the Prolog+CG rule used to build a semantic type-hierarchy as follows:

Supertype > Subtype1 > Subtype2 > ... > SubtypeN.

Below is an example of a portion of semantic type-lattice expressed in Prolog+CG whose graphical representation is provided in Fig. 4.6. Note how each line conforms to the rule given above:

Entity > Physical, Abstract,

Physical > Object, Process, Property. Object > Animate, Inanimate,

Animate > Human, Animal, Plant.

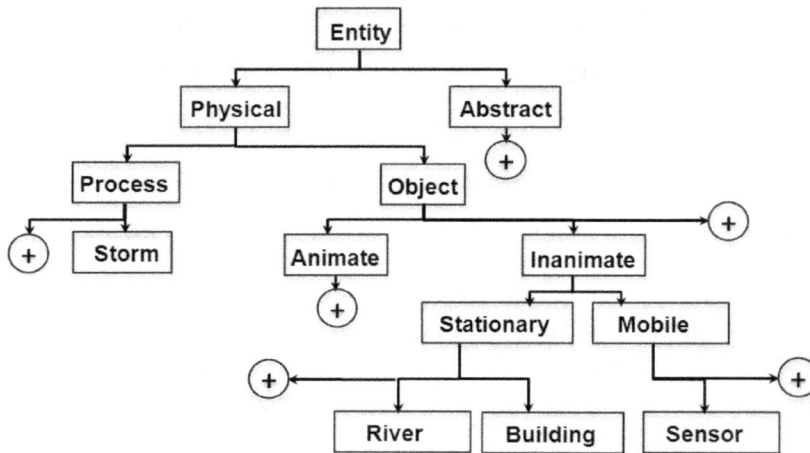

Fig. 4.6. A graph of Semantic Type Lattice with instances attached to agents archetypes (circle shapes).

We now explain this example. The example starts at the top of the lattice with Entity. This super-type is then declared to have two immediate sub-types: Physical, and Abstract. The Abstract node is not associated with any subtype, and so remains a leaf node. The Physical node is given three immediate subtypes: Object, Process, and Property, each of them being associated with subtypes. These subtypes may also have subtypes, and so on down the lattice.

Another important characteristic of agent archetypes is the multi-inheritance property which allows an agent type to belong to two (or several) different agent archetypes and hence to inherit from their characteristics. Let us consider the following example:

Adult > Woman, Man. Young > Girl, Boy,

Female > Woman, Girl. Male > Man, Boy.

Let us notice that Woman occurs at several places. This is allowed, as long as there is no circularity (i.e., as long as a type is not specified to be a subtype of itself) whether immediately or indirectly.

In Prolog+CG, we have two ways of saying that a type has an instance: (1) we can simply declare it as an individual in the referent of some CG; (2) we can declare it at the top of the program in a catalog of individuals. A catalog of individuals for a given type is written as follows:

Archeype = Instance1, Instance2, ..., InstanceN.

4.5.2. Action Archetypes

Since our research addresses the simulation of spatial behaviours, it has been influenced by some basic tenets of activity theory [2]. In particular, our approach to manage environment knowledge rests on the commitments in activity theory that: (1) activities are directed toward objects, zones, or actors [10]; (2) activities are hierarchically structured; and (3) activities capture some context-dependence of the meaning of information [2].

Theoretically, the common philosophy between our approach and activity theory is a view of the environment from the perspective of an agent interacting with it. Practically, we borrowed from activity theory two main ideas: (1) The semantics of activities and objects are inseparable [10]; and (2) Activities, objects as well as agents are hierarchically structured [2].We define an action archetype as a pattern of activities that are associated with agent archetypes. Hence, an action archetype describes a situation that involves one or several agent archetypes. We define an action archetype as a lattice of actions.

4.6. A Knowledge-Based Approach

The proposed approach (depicted in Fig. 4.7) relies on the agent paradigm in order to simulate the behaviour of a SN in a dynamic, complex, and large-scale informed virtual geographic environment. Sensors are modelled as intelligent agents embedded in a virtual space where dynamic phenomena can occur. Sensor agents have reasoning capabilities allowing them to reason about the virtual space and to react to its dynamic phenomena. Spatio-Temporal knowledge is used for two main purposes. First, it is used during the geo-simulation to support agents' reasoning capabilities. Second, it is used to analyze the results of the geo-simulation and to offer decision support to users. Finally, the results of the geo-simulation (which are inserted as facts in the Result Facts Base) are analyzed in order to offer decision support. In the following, we present these components.

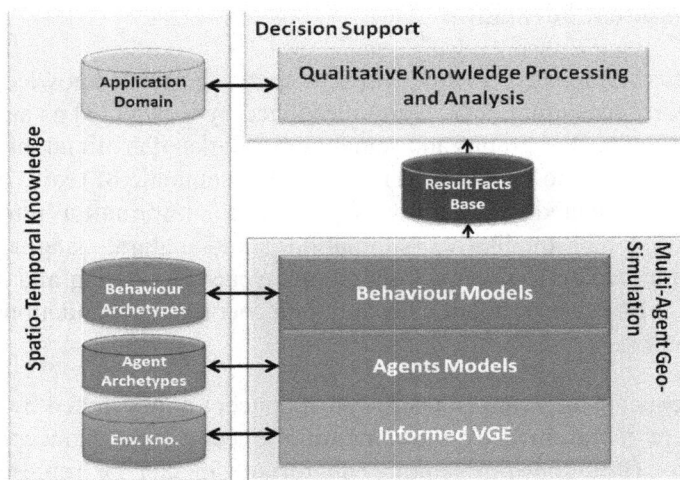

Fig. 4.7. The proposed knowledge-based geo-simulation model.

4.6.1. Agent-Based Simulation Using Informed Virtual Geographic Environments

The idea behind the agent-based simulation approach is to move the most intensive processing out of the Physical Sensor Network (PSN) into a parallel Virtual Sensor Network (VSN) operating on a base station or a remote server. The objective is to reproduce, in a realistic manner, the real world in a virtual environment. Indeed, in this virtual environment, which imposes no limits on data processing, energy consumption and communication capabilities, it is possible to create a system for the deployment of the physical sensor web. In order to faithfully mimic the physical sensor web deployed in the area of interest, we need to simulate, in a realistic way, the physical sensor nodes as well as the geographic environment where they are located. Physical sensor networks are represented in the virtual environment using software agents. An agent is a program with domain knowledge, goals and actions. An agent can observe and sense its environment as well as affect it. Agents' capabilities may include (quasi-) autonomy, perception, reasoning, assessing, understanding, learning, goal processing, and goal-directed knowledge processing. The reproduction of the geographic environment in which physical sensor nodes are deployed is based on reliable data obtained from Geographic Information Systems (GIS). The concept of agent-based geo-simulation evolves from such type of simulations involving software spatial agents immersed in a virtual geographic environment.

4.6.2. Spatio-Temporal Knowledge

As we mentioned so far, spatio-temporal knowledge is used in our approach 1) to support agents decision making during the geo-simulation and 2) to analyze the results of the geo-simulation in order to offer decision support. In the following we respectively present the representation formalism and the categories of spatio-temporal knowledge used in the proposed approach.

4.6.2.1. Representation Formalism

We use Conceptual Graphs (CGS) to represent spatio-temporal knowledge and to support spatio-temporal reasoning. CGs were introduced by Sowa [13] as a system of logic based on Peirce's existential graphs and semantic networks of artificial intelligence. They provide extensible means to capture and represent the semantic of real-world knowledge and have been implemented in a variety of projects for information retrieval, database design, expert systems, qualitative simulations, and natural language processing. However, their application to dynamic geographic spaces modeling and analyzing is an innovative issue. More details about CGs and their theoretical foundations can be found in [13], among others.

Syntactically, a conceptual graph is a network of concept nodes linked by relation nodes. Concept nodes are represented by the notation [Concept Type: Concept Instance] and relation nodes by (Relationship-Name). The formalism can be represented in either graphical or character-based notations. In the graphical notation, concepts are represented by rectangles, relations by circles and the links between concepts and relation nodes by arrows. The character-based notation (or linear form) is more compact than the graphical one and uses square brackets instead of boxes and parentheses instead of circles. Some examples are presented in the following sub-section.

4.6.2.2. Knowledge Categories

We distinguish three levels of spatio-temporal knowledge: 1) Knowledge about the environment, 2) Knowledge about actors and their behaviours and 3) Knowledge about the application domain (Fig. 4.8).

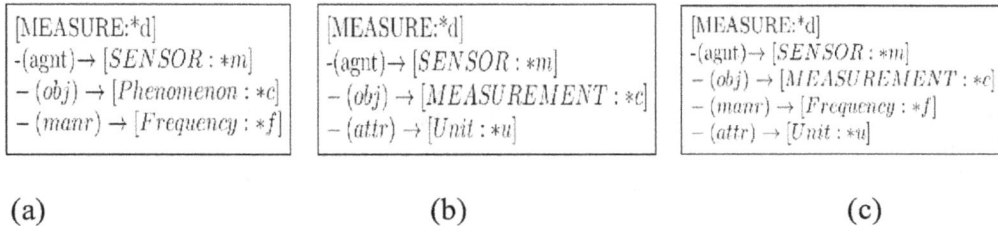

| (a) | (b) | (c) |

Fig. 4.8. Specification of rules using the Prolog+CG syntax.

4.6.2.2.1. Knowledge about the Environment

We define the notion of knowledge about the environment (Environment Knowledge (EK) for short) as a specification of a conceptualization of the environment characteristics. Hence, EK is a description of the spatial concepts (geographic features) and relationships (topologic, semantic) that may exist in a geographic environment. In multi-agent geo-simulation, EK is a specification used for enabling knowledge exploitation for spatial agents. Practically, EK is an agreement to use spatial concepts (i.e., ask queries and make assertions), spatial relationships (i.e., describe actions and behaviors), in a way that is consistent so we can share knowledge with and among spatial agents. Our aim is to improve the perception-decision-action loop on which relies most of the existing agents' models.

4.6.2.2.2. Knowledge about Actors and Behaviours Archetypes

Dealing with the specification of agents' attributes and associated spatial behaviors may be complex, time, and effort consuming. Agents' characterization aims to specify the agent archetype, its super-types and sub-types according to the semantic type hierarchy; and the behavior archetype that an agent archetype is allowed to perform within the informed VGE. Fig. 4.6 shows an example of a semantic type hierarchy of agent archetypes. Entity is an abstract node and Storm, River, Building and Sensor are instance nodes (leaves) of this agent archetype lattice. A key characteristic of agent archetype is inheritance. Agents belonging to one or several agent archetypes inherit the characteristics associates with these agent archetypes.

For example, let us consider two agent archetypes Temp-sensor and Press-sensor respectively sensing temperature and pressure. The Temp-sensor is characterised by a measurement frequency f. On the other hand, Press-sensor is characterised by a one meter circular sensing field. Consider now TP-sensor a multifunctional sensor that inherits from Temp-sensor and Press-sensor. Thanks to the inheritance property provided by agent archetypes, this agent performs measurements at a frequency f within a circular sensing area of one meter. Since our research addresses the simulation of spatial behaviors, it has been influenced by some basic tenets of active theory [14]. In particular, our approach to manage environment knowledge rests on the commitments in active theory that: (1) activities are directed toward objects, zones, or actors; (2) activities are hierarchically structure; and (3) activities capture some context-dependence of the meaning of information.

Theoretically, the common philosophy between our knowledge-based approach and activity theory is a view of the geographic environment from the perspective of an agent interacting with it [18]. Practically, the most important borrowings from activity theory are the idea that [15]: (1) the semantic of behaviors and objects are inseparable; and (2) behaviors, objects, as well as agents are hierarchically structured. Let us define the following behavior archetypes that we associate with the Sensor agent archetype as follows: See Fig. 4.8 (right) "an agent *m which is a sensor measures an object *c which is a phenomenon with a frequency *f "; Fig. 4.8 (center) "an agent *m which is a sensor measures an object *c which is a measurement of unit *u"). Since the above description is equal or more specific than the antecedent of the following behaviour, it can be inferred, by deduction, as shown in Fig. 4.9.

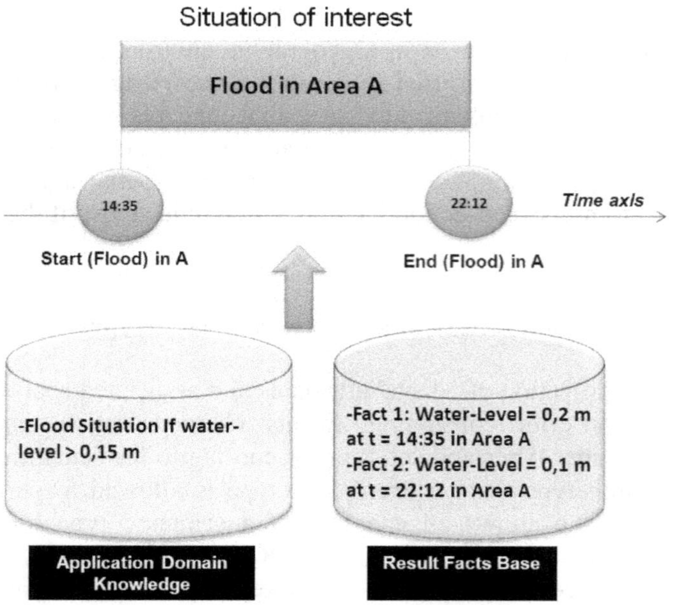

Fig. 4.9. An example of a simulation result analysis and processing.

4.6.2.2.3. Knowledge about the Application Domain

The above-mentioned levels of knowledge are used during the geo-simulation to support agents in their decision-making. In contrast, knowledge about the application domain is mainly used to qualitatively analyze the results of the geo-simulation and is thus more linked to decision support. In the context of SN deployment, nodes are aimed to collect measurements about phenomena of interest that vary according to the application domain (military, environmental surveillance, etc.). Knowledge about the application domain defines phenomena of interest in a particular application domain. In our approach, we use the concept of spatio-temporal situations [19] to model and reason about phenomena of interest.

A spatio-temporal situation represents a state, an event or a process situated in space and time and involving various objects of the world. Examples of spatio-temporal situations can be a sensor which brake down for a certain period of time at a certain spatial area (state), the start of rain in certain spatial area (punctual event) or a durative heavy rain in a given area (process). A spatio-temporal situation has a semantic type (rain, network breakdown, etc.), a start and end times and is located in space. Knowledge about the application domain defines spatio-temporal situations of interest according to their temporal (state, punctual event or durative process) and semantic characteristics. For example, the semantic punctual event "Start of rain" can be defined as the fact of water level exceeding a given threshold. Relationships between spatio-temporal situations (temporal and spatial) are also specified in the application domain knowledge, which enables defining complex phenomena. For example, a situation of storm can be defined as a situation of heavy rain followed by "/" accompanied with a situation of strong wind.

4.6.2.2.4. Decision Support

The decision support component analyzes the result of the geo-simulation using application domain knowledge in order to identify situations of interest to the user. This data analysis process is implemented using the approach proposed in [19]. Details of this approach are beyond the scope of this chapter. We only illustrate the principle using a simple example. In this example, the situation of interest is Flood. The application domain knowledge specifies that there is a flood situation if the water-level exceeds 0.15 meter. Otherwise, there is no flood situation. The decision support component uses this knowledge in order to analyze the facts collected by agents during the simulation (Result Facts Base). Particularly, the two facts are respectively interpreted as the start of flood in (Area A, t= 14:35) and the end of flood in (Area A, 22:12) (punctual events). Finally, start and end flood situations are used to identify the flood situation itself as a process located in Area A during the time interval [14:35, 22:12]. Obviously, detection of real complex situations requires taking into consideration other aspects (measurement errors, conflict of measurements between several sensors, etc.) that are beyond the scope of this chapter.

4.7. Case Study: Simulation of a Sensor Network for Weather Monitoring

In order to illustrate our knowledge-based geo-simulation framework, we propose to simulate a sensor web deployed in an IVGE representing the experimental forest of Montmorency (Quebec, Canada) for weather monitoring purposes. This scenario shows how agents adapt their spatial behaviors with respect to knowledge they acquire from the IVGE using our environment knowledge management along with their perception capabilities. The objective of the simulated sensor network is to identify and monitor a simulated storm evolving in the IVGE. In order to reason about knowledge, we used a platform to deal with CGs manipulation called Amine [20]. Amine platform provides a pattern matching and rule based programming paradigm embedded in Prolog+CG language that is an object-oriented and conceptual graphs-based extension of Prolog language.

A few agent archetypes representing different kinds of sensors are involved in this scenario. These sensors are first randomly deployed in the IVGE. Then, each sensor computes a path in order to reach its deployment position while taking into account the geographic environment characteristics. When sensors reach their final destinations, some of them stay active while other switch to idle in order to preserve the overall energy of the sensor web. Active sensors make measurements at a frequency f in order to monitor weather conditions. Active and idle sensors as well as the measurement frequency are specified in the simulation scenario.

The simulated storm appears after a period t from the beginning of the simulation. If an active sensor perceives the storm agent, it directly accesses to its properties and extracts the information that it monitors depending on its kind of sensor, i.e. temperature, pressure, wind speed and direction, or humidity. If a difference above a certain threshold Δ is observed, the sensor proceeds as follows: (1) it accelerates its measurements frequency, (2) it adds a new fact that keeps track of the event with its timestamp in the Result Facts Base; and (3) it sends a message to weak up idle all sensors of the same kind which are situated in a certain estimated distance. As the simulation time goes by and the storm agent evolves in the IVGE, most of initially idle sensors become active to sense the observed phenomenon. When the storm agent is out of the perception field of the sensor, this latter senses a new difference between the past and the current measurement. It notifies the Result Facts Base by adding a new fact that keeps track of the new event with its timestamp; Idle and active sensors switch states in order to preserve their energy.

In order to model the simulation described above, let us first consider the two agent archetypes ZONE and SENSOR. In contrast with the ZONE agent archetype, which represents a geographic area, the SENSOR agent archetype is associated with individual sensors deployed in the informed virtual geographic environment. Let us consider the two agent sub-types WEATHERZONE and STORMZONE. Agents of type WEATHERZONE are stationary and represent meteorological conditions within the geographic area they cover. Five instances of WEATHERZONE are created in order to approximately cover the monitored area (Fig. 4.10). In addition to their geometric characteristics, these agents encompass attributes which characterize the meteorological conditions such as temperature = 18° C, pressure = 1010 hPa, humidity = 30. When a

146

sensor detects a difference above the threshold, it adds a fact in the Results Facts Base using the CGs formalism. Consider the following example involving the sensortemp1 adding a fact describing an observed difference of temperature measurement of value 18 at cell 367 at 15:34:22 (See Fig. 4.10).

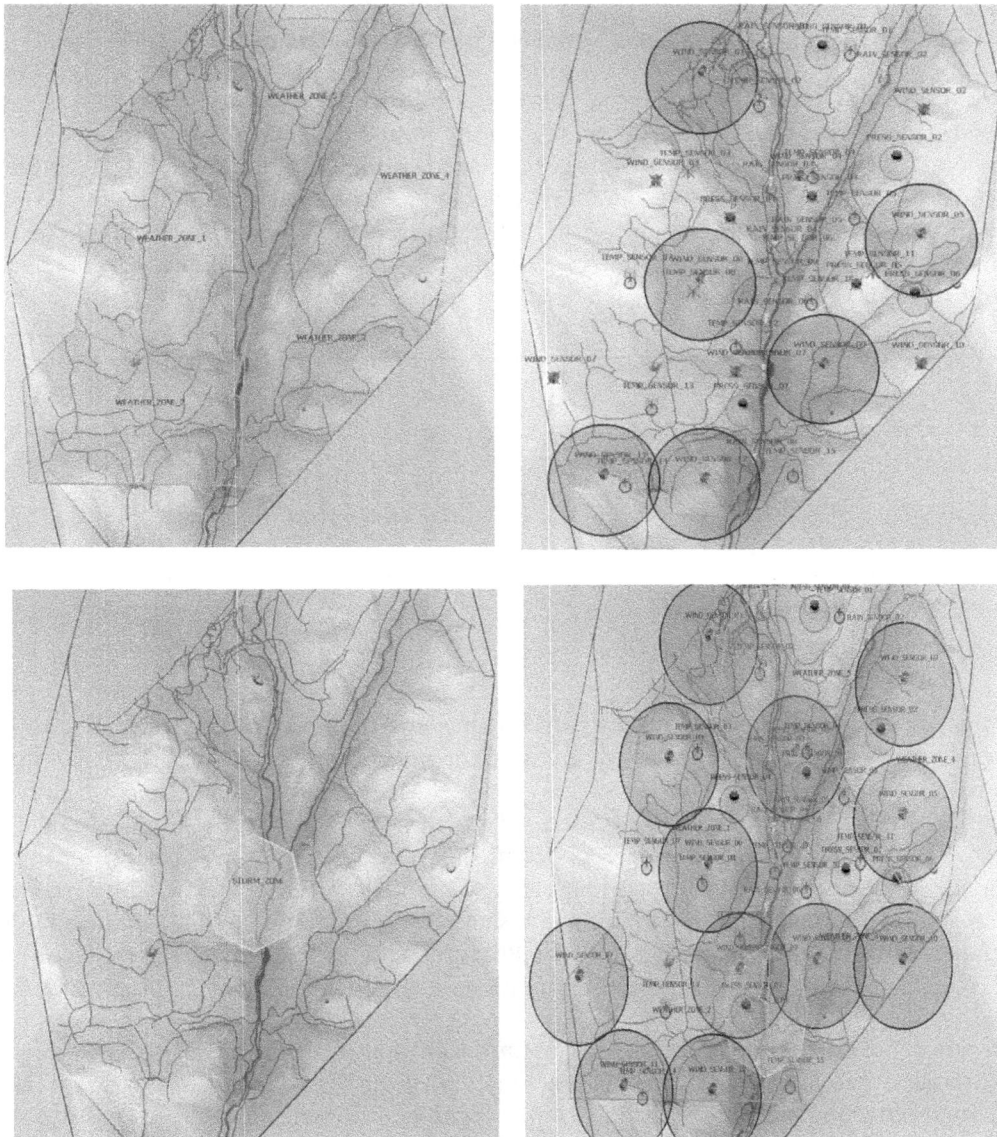

Fig. 4.10. Simulation results: (top left) WeatherZone agents covering the monitored area; (top right) the simulated sensors autonomously reaching their deployment position. The blue circle corresponds to the perception field for active sensors; (bottom left) the storm agent in yellow; and (bottom right) the simulation of the storm monitoring.

Let us now define the following sub-types of SENSOR archetype: TEMPSENSOR for temperature measurement, PRESSENSOR for atmospheric pressure measurement, WINDSENSOR for wind speed and orientation measurement, and HUMISENSOR for humidity measurement. In this scenario, the situation of interest is the storm. Let us suppose that we need to describe the evolution of the storm during the geo-simulation for decision support purposes. Knowledge about the application domain allows specifying how the presence of a storm phenomenon can be detected in a certain area.

For simplification, let us consider the following (Prolog+CG) rule specifying that a storm is detected at time T in area A if there are (in the results facts base) facts describing that temperature exceeds 20° C and wind speed exceeds 30 km/h at the same time T and the same area A (See Fig. 4.11). A storm detection event is triggered as soon as the conjunction of these two facts exists within the result facts base. The analysis of the simulation results conceptualize this event using the knowledge associated with the case study.

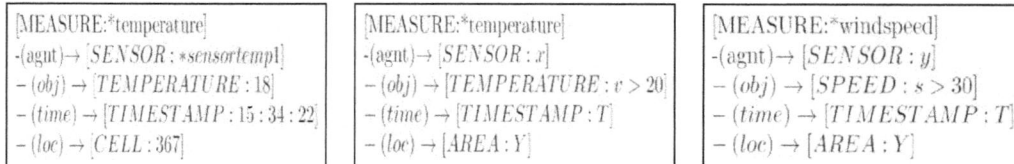

[MEASURE:*temperature]	[MEASURE:*temperature]	[MEASURE:*windspeed]
-(agnt)→ [SENSOR : *sensortemp1]	-(agnt)→ [SENSOR : x]	-(agnt)→ [SENSOR : y]
– (obj) → [TEMPERATURE : 18]	– (obj) → [TEMPERATURE : $v > 20$]	– (obj) → [SPEED : $s > 30$]
– (time) → [TIMESTAMP : 15 : 34 : 22]	– (time) → [TIMESTAMP : T]	– (time) → [TIMESTAMP : T]
– (loc) → [CELL : 367]	– (loc) → [AREA : Y]	– (loc) → [AREA : Y]

Fig. 4.11. Behaviour archetypes expressed using the linear notation of Conceptual Graphs (CGs).

4.8. Discussion

In this paper, we presented a knowledge management approach that aims to provide spatial agents with knowledge about the environment in order to support their autonomous decision making process. Our approach is influenced by some basic tenets of activity theory [2] as well as by the notion of affordance [6]. It is based on our IVGE model to represent complex and large-scale geographic environments. It uses the Conceptual Graphs formalism to represent knowledge about the environment (Environment Knowledge) structured as an Environment Knowledge Base (EKB). This approach also includes an inference engine that uses the Prolog+GC language to interrogate, infer and make deductions based on facts, cases, situations, and rules stored within the EKB.

Our environment knowledge management approach is original in various aspects. First, a multi-agent geo-simulation model that integrates an informed virtual geographic environment populated with spatial agents capable of acquiring and reasoning about environment knowledge did not exist. Second, a formal representation of knowledge about the environment using CGs that leverages a semantically enriched description of the virtual geographic environment has not yet been proposed. Third, providing agents with the capability to reason about a contextualised description of their virtual environment during the simulation is also an innovation that characterises our approach.

Nevertheless, some limits that characterise our environment knowledge management approach still need to be addressed. This approach, in its current version, is a proof of concept that demonstrates the capability of our IVGE model: 1) To integrate knowledge about the environment; 2) To allow agents to reason about it using an inference engine. Although the provided scenario is simplified, it illustrates the advantages of extending our IVGE model by: 1) Using a standard knowledge representation formalism (Conceptual Graphs) and; 2) Integrating an inference engine such as the Amine platform.

When the agent is acting, it uses the environment knowledge base, its observations of the virtual environment, and its goals and abilities to choose what to do and to update its own knowledge. Hence, the environment knowledge base corresponds to the agent's long-term memory, where it keeps the knowledge that is needed to act in the future. This knowledge comes from prior knowledge and is combined with what is learned from data and past experiences. The beliefs, intentions and desires of the agent correspond to its short-term memory. Although a clear distinction does not always exist between long-term memory and short-term memory, this issue might be addressed as part of the extension of our knowledge management approach. Moreover, there is feedback from the inference engine to the environment knowledge base, because observing and acting in the world provide more data from which the agents can learn. Evolving and allowing the agent model to learn from such data is another challenging task.

4.9. Conclusion and Future Perspectives

Our environment knowledge management approach is original at various aspects. First, a multi-agent geo-simulation model that integrates an informed virtual geo-graphic environment populated with spatial agents capable of acquiring and reasoning about environment knowledge does not exist. Second, a formal representation of knowledge about the environment using CGs that leverages a semantically enriched description of the virtual geographic environment has not yet been proposed. Third, providing agent with the capability to reason about a contextualized description of their virtual environment and phenomenon occurring within it during the simulation is also an innovation that characterizes our approach. We are currently working on the automated assessment of different simulation scenarios for sensor network management using our qualitative knowledge processing and analysis module. Indeed, users usually need to analyze and compare various scenarios in order to make informed decisions. This task may be complex and effort and time consuming. However, since our framework already supports the analysis of the multi-agent simulation results, it is easy to extend it to automatically assess scenarios.

Acknowledgements

This research was partially supported by the New Researcher Grant provided by the Office of Sponsored Program at St. Cloud State University, St. Cloud, Minnesota, United States. We thank our colleagues from the Department of Computer Science and Information

Technology who provided insight and expertise that greatly assisted the research, although they may not agree with all of the interpretations/conclusions of this chapter.

We would also like to show our gratitude to Dr. Adel Ali and Dr. Dale Buske, Dean and Associate Dean of the College of Science and Engineering (COSE) at St. Cloud State University for sharing their pearls of wisdom with us during the course of this research, and we thank the "anonymous" reviewers for their so-called insights. We are also immensely grateful to Cliff Moran for his comments on an earlier version of the manuscript, although any errors are our own and should not tarnish the reputations of this esteemed person.

References

[1]. P. Arthur, R. Passini, Wayfinding: People, Signs, and Architecture, *McGraw-Hill Companies*, 1998.
[2]. R. Bellamy, S. Bødker, E. Christiansen, Y. Engeström, V. Escalante, D. Holland, V. Kaptelinin, K. Kuutti, B. A. Nardi, A. Raeithel, J. Reeves, B. Velichkovksy, V. P. Zinchenko, Context and Consciousness: Activity Theory and Human-Computer Interaction, First edition, *The MIT Press*, November 1996.
[3]. I. Benenson, P. Torrens, Geosimulation: Automata-Based Modeling of Urban Phenomena, *John Wiley and Sons, Inc.*, 2004.
[4]. N. Farenc, R. Boulic, D. Thalmann, An informed environment dedicated to the simulation of virtual humans in urban context, *Proceedings of the Computer Graphics Forum (Eurographics'99)*, Vol. 18, 1999, pp. 309-318.
[5]. A. Frank, S. Bittner, M. Raubal, Spatial and cognitive simulation with multi-agent systems, in *Proceedings of the International Conference on Spatial Information Theory (COSIT'01)*, Morro Bay, CA, USA, September 2001, pp. 124-139.
[6]. J. J. Gibson, The Ecological Approach to Visual Perception, *Lawrence Erlbaum Associates*, London, 1979.
[7]. D. Guo, B. Ren, C. Wang, Integrated agent-based modeling with GIS for large scale emergency simulation, *Advances in Computation and Intelligence*, Vol. 5370, 2008, pp. 618-625.
[8]. P. Jansen-Osmann, M. Heil, The process of spatial knowledge acquisition in a square and a circular virtual environment, *Advances in Cognitive Psychology*, Vol. 3, Issue 3, 2007, pp. 389-397.
[9]. A. Kabbaj, Development of intelligent systems and multi-agents systems with Amine platform, in *Proceedings of the 14th International Conference on Conceptual Structures (ICCS'06)*, 2006, pp. 286-299.
[10]. V. Kaptelinin, B. Nardi, C. Macaulay, The activity checklist: a tool for representing the "space" of context, *Interactions Magazine*, Vol. 6, Issue 4, 1999, pp. 27-39.
[11]. D. Lenat, R. Guha, Building Large Knowledge-Based Systems; Representation and Inference in the Cyc Project, *Addison-Wesley Longman Publishing Co. Inc.*, Boston, MA, USA, 1989.
[12]. M. Mekni, Hierarchical path planning for situated agents in informed virtual geographic environments, in *Proceedings of the 3rd International Conference on Simulation Tools and Techniques (SIMUTools'10)*, Torremolinos, Malaga, Spain, March 2010, pp. 15-19.
[13]. M. Mekni, Abstraction of informed virtual geographic environments for the modeling of large-scale and complex geographic environments, in *Proceedings of the 3rd International Conference on Information, Process, and Knowledge Management (eKnow'11)*, 23-28 February, 2011, pp. 1-10.

[14]. M. Mekni, B. Moulin, Holonic modelling of large scale geographic environments, in *Proceedings of the 4th International Conference on Industrial Applications of Holonic and Multi-Agent Systems (HOLOMAS'09)*, Linz, Austria, 2009, pp. 266-275.

[15]. M. Mekni, B. Moulin, Hierarchical path planning for multi-agent systems situated in informed virtual geographic environments, in *Proceedings of the 2nd International Conference on Information, Process, and Knowledge Management (eKnow'10)*, Saint-Marteen, Netherlands Antilles, February 10-16, 2010. IEEE Computer Society, pp. 48-55.

[16]. A. Newell, Unified Theories of Cognition, *Harvard University Press*, Cambridge, Massachusetts, 1990.

[17]. S. Paris, M. Mekni, B. Moulin, Informed virtual geographic environments: an accurate topological approach, in *Proceedings of the* International Conference *on Advanced Geographic Information Systems & Web Services (GEOWS'09)*, 2009, pp. 1-6.

[18]. M. Raubal, Ontology and epistemology for agent-based wayfinding simulation, in *Proceedings of the Geographical Domain and Geographical Information Systems-EuroConference on Ontology and Epistemology for Spatial Data Standards*, France, September 2001, La Londe-les-Maures, pp. 85-87.

[19]. P. S. Sajja, Multi-agent system for knowledge-based access to distributed databases, *Interdisciplinary Journal of Information, Knowledge, and Management*, Vol. 3, 2008, pp. 1-9.

[20]. A. Seidel, Way-finding in public spaces: the dallas/fort worth, USA airport, in *Proceedings of the 20th International Congress of Applied Psychology*, Edinburgh, Scotland, 1982.

[21]. W. Shao, D. Terzopoulos, Environmental modeling for autonomous virtual pedestrians, in *Proceedings of the SAE Symposium on Digital Human Modeling for Design and Engineering*, 2005, pp. 1-8.

[22]. J. Sowa, Knowledge Representation: Logical, Philosophical, and Computational Foundations, *Course Technology*, August 1999.

[23]. J. Sowa, E. Way, Implementing a semantic interpreter using conceptual graphs, *IBM J. Res. Dev.*, Vol. 30, Issue 1, 1986, pp. 57-69.

[24]. J. F. Sowa, Conceptual Structures: Information Processing in Mind and Machine, *Addison-Wesley Longman Publishing Co. Inc.*, Boston, MA, USA, 1984.

[25]. R. Thomas, S. Donikian, A model of hierarchical cognitive map and human memory designed for reactive and planned navigation, in *Proceedings of the 4th International Space Syntax Symposium*, Londres, 2003, pp. 72-100.

[26]. J. Weisman, Evaluating architectural legibility: way-finding in the built environment, *Environment and Behavior*, Vol. 13, Issue 2, March 1981, pp. 189-204.

[27]. D. Weyns, K. Schelfthout, T. Holvoet, Exploiting a virtual environment in a real-world application, in *Proceedings of the International Workshop Environments for Multi-Agent Systems II (E4MAS'05)*, February 2005, pp. 218-234.

[28]. M. Wooldridge, N. Jennings, Intelligent agents: theory and practice, *Knowledge Engineering Review*, Vol. 10, Issue 2, 1995, pp. 115-152.

[29]. J. Zacharias, Exploratory spatial behaviour in real and virtual environments, *Landscape and Urban Planning*, Vol. 78, Issue 1-2, 2006, pp. 1-13.

[30]. P. Gibbons, B. Karp, Y. Ke, S. Nath, S. Srinivasan, Iris-net: an architecture for a worldwide sensor web, *Pervasive Computing*, Vol. 2, No. 4, 2003, pp. 22-33.

[31]. I. Athanasiadis, M. Milis, P. Mitkas, S. Michaelides, Abacus: A multi-agent system for meteorological radar data management and decision support, in *Proceedings of the 6th International Symposium on Environmental Software Systems (ISESS'05)*, Sesimbra, Portugal, 2005. https://issel.ee.auth.gr/wp-content/uploads/2016/02/Abacus-A-multi-agent-system-for-meteorological-radar-data-management-and-decision-support.pdf

151

[32]. P. Biswas, S. Phoha, A middleware-driven architecture for information dissemination in distributed sensor networks, in *Proceedings of the Sensor Networks and Information Processing Conference*, 2004, pp. 605-610.

[33]. D. Moodley, I. Simonis, New architecture for the sensor web: the swap framework, in *Proceedings of the 5th International Semantic Web Conference (ISWC'06)*, Athens, GA, USA, 2006, p. 17.

Security

Chapter 5
Reference Monitor-based Security Framework for Trust in Mobile Agent Computing

Sandhya Armoogum, Nawaz Mohamudally and Nimal Nissanke

5.1. Introduction

During the last decade, there have been major changes in distributed computing. Programs are no longer constrained to execute on one machine. Code can now be migrated to other hosts for execution. The most well-known example of mobile code is the use of Java applets and JavaScript code in a Web browser. This form of code mobility is referred to as code on demand. Code mobility has many uses, i.e., vendors can use mobile code to reconfigure software, Microsoft uses mobile code to distribute software patches, mobile code can also be used to manage distributed system by performing load balancing [1].

Mobile agents represent a more sophisticated and powerful form of code mobility. A mobile agent is a program that can move from one host to another according to its own internal logic. Mobile agent computing paradigm presents numerous advantages, compared to the traditional client-server based computing model, which include reduced network usage, better fault tolerance, adaptability to changes in the environment, and platform independence [2]. Thus, the mobile agent computing paradigm has become a natural and flexible way for implementing many applications on the network such as ecommerce and auctioning, network monitoring, real-time control systems and cloud computing. Recently, mobile agent computing proves to be very useful in the context of wireless sensor network (WSN) [3, 4]. As such, the mobile agent computing paradigm provides a very intuitive and flexible approach to solving old and new problems arising in many areas of computing (e.g. intrusion detection [5], web crawling [6], Big Data analysis [7], searching and communications in Internet of Things [8] and e-learning [9]), due to their mobility and autonomous nature, their ability to learn and adapt to changing

Sandhya Armoogum
Dept. Industrial Systems & Engineering, University of Technology, Mauritius

environments, their ability to communicate and collaborate with one another to perform some complex task, and their ability to clone themselves, if needed [10].

For mobile agent applications to be widely adopted, some security issues have to be addressed, as mobile agent applications, mainly deployed in open environments, may possibly be exposed to attacks. Currently, the lack of an integrated security solution is a main drawback, which has to be tackled before the mobile agent computing paradigm is widely accepted by industry. Four threat categories have been identified in [11] as follows: (1) mobile agent attacking an agent platform (AP), (2) AP attacking a mobile agent, (3) a mobile agent attacking another agent on the AP, and (4) other entities attacking the agent system. Some practical solutions exist for securing AP against malicious mobile agents. These include the use of Proof Carrying Code (PCC) [12], Model Carrying Code (MCC) [13, 14], Abstraction Carrying Code [15], Sandboxing [11, 16] and Code Signing [11].

However, securing the mobile agents against malicious AP is more challenging as the host has full control of the execution environment. Besides, the mobile agent computing model violates some of the fundamental assumptions of conventional security techniques, i.e., programs runs on hosts which are trusted. Other important assumptions are the identity and intention assumptions [17]. Usually, when a program attempts some action, the program acts on behalf of a known user and it is assumed that the person intends the action to be taken. When mobile agents execute on unknown hosts, then neither identity nor intention of the host is clear.

The fact that mobile agents may have to operate in such unknown and open environment requires the concept of trust of the hosting agent platforms despite that they may be previously unknown to the mobile agent's owner [18]. In this paper, we propose an approach for building trust in an AP by using the reference monitor. In this approach, the reference monitor entity is obtained from a trusted third party (TTP) and can provide or arbitrate several security services resulting in a system which supports secure mobile agent applications in an open system.

The rest of this paper is structured as follows. Section 5.2 discusses related work. Section 5.3 presents the proposed security framework based on trust provided by the Reference Monitor for providing a trusted computing environment on an AP. Section 5.4 discusses the integration and evaluation of the reference monitor entity in an agent platform. Section 5.5 presents the experimental setup used and results. Finally, Section 5.6 presents the conclusions drawn and outlines future work.

5.2. Background Study

Protecting mobile agents from malicious APs is thus essential. Typically, the malicious AP (MAP) can masquerade as another AP or as a trusted AP, to deceive a mobile agent and to extract sensitive information from the mobile agent. A MAP can cause a Denial of Service (DoS) attack on the mobile agent, for instance, by ignoring certain requests made by the agent, introducing critical delays for time-sensitive tasks, simply not executing the agent, or terminating an agent without notification. Similarly, a mobile agent can be livelocked by the MAP, where the agent is kept busy with requests and tasks such that it is prevented from achieving its own goal(s). The MAP can also prevent an agent from migrating to another AP, which is another example of a DoS. Furthermore, a MAP can perform eavesdropping whereby the MAP can intercept communications of the agent with other agents or its Home platform (HAP). Given that an agent can be represented as code, state and data that it carries, the MAP can also monitor the agent code, state and data. This can lead to cognition of the implementation strategy of the agent (decision making algorithm, other proprietary algorithms, trade secrets, negotiation strategies), as well as an understanding of the composition of code and data in memory of the AP (allowing for mobile code piracy). Understanding the agent code and data (code inspection) leads to knowledge about the agent structure and private data such as authentication credentials, private key, e-cash, and credit card details carried by the mobile agent. Similarly, the MAP may have knowledge of partial results or other data collected from previous executions of the mobile agent on other APs. These data may allow the MAP to cheat in its interaction with the mobile agent. In some cases, even if the sensitive data baggage of the mobile agent is protected (encrypted), information from the data flow and knowledge of the agent code allows the MAP to infer information such as the state of the agent or to infer other information about the agent's goals. Similarly, a MAP can manipulate the agent code, such as embedding some malware within the code such that the mobile agent becomes a transmission vector for the malware. Data and partial results carried by the agent can also be tampered with, i.e. the MAP can delete the itinerary such that the current AP is the last AP visited. Likewise, by changing the control flow of the agent, its behaviour can be changed.

Security mechanisms to secure mobile agents as they execute on remote AP exists, but most of these mechanisms mainly allows the detection of integrity attacks on the mobile agent. Agent code, data and/or state tampering can be detected by means of hash codes, message authentication codes, digital signatures as well as other techniques. However, inspection of the agent's code, data and/or state, replay attack and some types of DoS attack on the mobile agent cannot be detected when the agent is executing on a remote AP.

For securing mobile agents against possible MAPs, one proposed solution involves allowing mobile agents to migrate to known and trusted APs only [19]. In this case, agents are sent in encrypted form from one trusted AP to another, where they execute, often after authentication of the AP. However, such an approach seriously limits the agent's execution on a limited number of known and trusted APs. It is not suitable for some applications like search agents (searching for information) on the internet, and mobile ecommerce agents where the AP may not always be known; and is against the notion of an open multi-agent system where new APs can be dynamically added or removed from an agent's itinerary.

Another proposed solution for providing comprehensive security to mobile code advocates the use of trusted, tamper-proof hardware which is not controlled by the local system and which supports secure mobile agents' execution [20]. This secure trusted computing base on each AP, thus, provides the trusted environment for running critical code of the mobile agent. Here, mobile agents move from one trusted environment on an AP to another. Local resources on the system are accessed in a client-server mode. The system outside the trusted hardware has no access and thus cannot interfere with the execution of the mobile agent. The external system can only interact with the trusted tamper resistant hardware via some restricted interface. The main drawback of this approach is that every host AP has to be supplied with secure trusted hardware which is not a simple task as such hardware installation and maintenance may be expensive. Moreover, the tamper-resistant devices could also become a performance bottleneck in the execution of mobile agent especially in cases where smart cards are used as a cheap alternative for providing hardware trusted computing base [21].

Another approach for protecting mobile agents involves code obfuscation, whereby the agent's program is made illegible and data hidden, thus rendering it difficult to read and modify the agent code, data and partial results [22]. An obfuscated mobile agent is like a black-box entity; it only permits AP to provide inputs and read outputs from the mobile agent. Thus, even if the mobile code runs on unknown and untrusted APs, it can maintain confidentiality and integrity. However, this technique often only provides limited code confidentiality as given time, the malicious AP will be able to de-obfuscate the code. In [23], the authors show that complete obfuscation is impossible. Furthermore, the malicious host could still re-execute (replay) the mobile agent several times so as to observe its reaction and guess its decision-making strategy for example. Esparza, et al. [24] proposes to monitor the execution time on an AP. A longer than expected execution time on an AP is indication that the AP may be attempting to de-obfuscate the agent, modify the agent code, data and/or partial results or replay the mobile agent. The main drawbacks of the approach are that it requires the agent to return with the partial results from each host visited to the Home Agent Platform (HAP), i.e., HAP must be connected until the transaction is over. The application would then no longer support disconnected and asynchronous processing as promised by mobile agent computing paradigm, though it does provide security of mobile agent such as integrity, execution integrity, and detection of Denial of service (DoS) attack.

Finally, the use of mobile cryptography (also referred to as function hiding) which aims to offer provably strong protection to mobile agents against both modification and inspection attacks has been proposed. Mobile cryptography is such that a mobile agent program is encrypted into a ciphered executable program where it can execute on the untrusted AP while remaining in the ciphered form [25, 26]. The efficiency of this approach is unknown and to the best of our knowledge, there is no practical implementation. It is still unclear if such a scheme can be implemented for real-life applications.

Trust is an important component of mobile code security [27]. When agents are executing in open environments on unknown and untrusted APs, if they could have some guarantees of trust, this would provide a basis for better security solutions. In [28], the authors describe a trust management architecture (MobileTrust) that can be developed to manage security related trust relationships explicitly and to make trust decisions. In such a system, trust management brings an improvement in security as by leveraging the trust knowledge gained on the past behaviours of other execution hosts, the mobile agents itinerary can be composed such as to minimize security attacks from potential MAPs. Similar trust computation can allow the AP to evaluate the trustworthiness of a mobile agent from a specific agent owner. However, such a technique may not be suitable for all open environment applications. For example, in a mobile agent based e-commerce application, a mobile agent from an owner may not visit the same AP twice and thus the trust values of the current execution may never be useful. In [29], a trusted third party called a Clearing House is proposed to maintain trust information about APs. Before a mobile agent migrates to an AP, trust level associated with the AP may be found out. However, this solution heavily relies on the Clearing House which keep tracks of the trustworthiness of each AP. It may also not be suitable for applications where the mobile agent has a dynamic itinerary. Finally, such mechanisms of monitoring behaviour of AP may be unpractical since not all attacks can be detected and thus reported, especially breach in agent data and code confidentiality.

5.3. Reference Monitor Based Security Framework

In this research work, we propose a practical and pragmatic technique for providing mobile agent security based on the reference monitor (RM) concept for trust. Since its introduction, in the early 1970s, in the "Anderson Report" [30], the RM concept has been adopted for securing computer and network. This concept visualises a system component, called a *reference validation mechanism*, to be responsible for administering the system's security policy. It thus defines the requirements for implementing such a mechanism in a manner that ensures that malicious entities cannot circumvent policy enforcement [31]. The RM concept thus provides a trusted and verifiable security policy enforcement mechanism [32]. This is in line with the U.S Government's criteria for building secure systems, the Trusted Computer System Evaluation Criteria (the Orange book) [33], where the reference monitor is mentioned. The diagram in Fig. 5.1 depicts the logical structure of the RM.

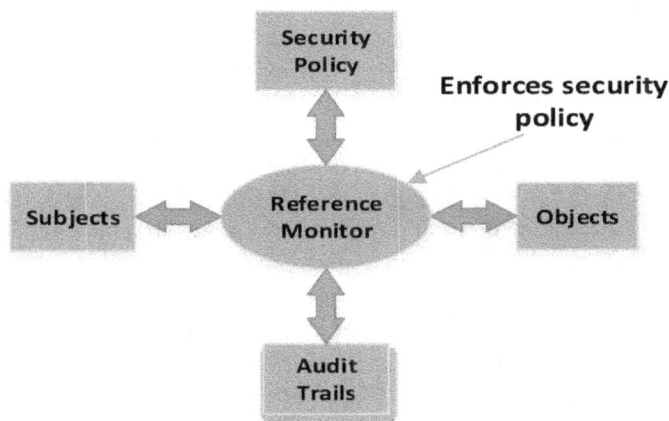

Fig. 5.1. Logical Structure of RM.

Implementations of the reference monitor concept usually make use of several traditional security measures, which apply to the agent environment. Such conventional techniques include cryptographic methods for authentication of AP/agent, encryption of data, integrity of data, and access control methods, thereby providing an integrated security solution. Consequently, the RM is used to enforce security by using the appropriate security mechanisms as per the requirements of the multi-agent system. An AP, which implements an RM entity, can be considered to be trustworthy. The mobile agent may then migrate to such an AP. In the case of mobile agent computing, we propose that the RM entity be in the form of an agent (RM-agent) which runs on each AP. We define the following fundamental assumptions about the trust association of all entities in the security framework, for providing mobile agent security. These trust associations comprise the trust model in our system.

- The Certificate Authority (CA) authenticates principals and issues certificates to entities such as the agent owner, AP administrator, RM-agent, trusted-third party (TTP) RM entity distributor. Users/agent owners and APs with valid certificates are considered trustworthy.

- Trusted Third Parties (TTPs) are principals with genuine certificates from CA and are trustworthy. These TTPs distribute standard RM-agents, which satisfy some design specifications, for use by any AP. Such TTPs can also act as subordinate CAs. The trusted CA delegates them the right to issue certificates to APs. Then, the TTP can also maintain a directory server where it stores information about each AP and their corresponding RM-agent.

- RM-agents on APs with genuine certificates from the TTP are trustworthy. These certificates are signed by the TTP who must have full knowledge of the RM agent's behaviour and capabilities and thus all the possible consequences of its operation. The system administrator of the AP can request for an RM-agent and its certificate from the TTP.

- Agents signed by trustworthy owners are trustworthy. The agent-owner is expected to have knowledge of the capability and behaviour of the agent and to take responsibility of the actions of the agent acting on behalf of the user/owner.

The underlying notion of our trust model is based on the usage of undeniable proofs like digital signatures, i.e., RM-agents are signed by the TTP, thus ensuring that the RM-agents are genuinely from the TTP. The digital signature of the RM-agent also allows to check the integrity of the RM-agent, i.e., indication of tampering, if any, on the RM-agent. This approach is used many systems whereby trusted entities are entrusted with correctly and reliably carrying out protocol steps that are decisive to guaranteeing certain targeted property. For example, trusted entities are relied upon for generating fresh keys in authentication protocols to guarantee authentication. In this work, we focus on the RMagent integration into the AP to establish trust, rather than on how the RM-agent actually enforces the security policy to provide the required security to mobile agents executing on the AP.

5.4. Integration and Evaluation of RM-Agent

The RM-Agent, which embodies the reference monitor concept, should be installed on an AP to establish trust. However, before a mobile agent can migrate and execute on the destination AP, the mobile agent must ensure that the RM-agent on the AP, and thus the AP, is trustworthy. In the next section, we describe how the RM-agent can be securely distributed and integrated into an AP, and how a mobile agent can assess the trustworthiness of the RM-agent before deciding to migrate to the remote AP.

5.4.1. Distribution and Integration of RM-Agent into an AP

The TTP is to act like a RM-agent distributor to provide RM-agents to APs upon request. Based on the described trust model, we detail, the different steps required for an AP to obtain a trusted RM-agent from a TTP. The presence of the RM-agent, thereafter, turns the AP into a secure trusted computing base. Fig. 5.2 illustrates the process.

Fig. 5.2. Distribution of RM-agent by TTP distributor.

Step 1: The AP registers with TTP and requests an RM-agent for enforcing security policy. The security to be provided by the RM-agent, may vary based on the security policy of the AP.

Step 2: The TTP generates a digital certificate ($RM_{TTPCert}$) for the RM-agent. This certificate contains information that would allow the receiving AP to verify the authenticity and integrity of the received RM-agent from TTP. Important information on the $RM_{TTPCert}$ certificate includes: the identity of AP to whom RM-agent is distributed; cryptographic hash value of RM-agent code; digital signature of the RM-agent, version no. of RM-agent, and security policy ID implemented by RMagent. The $RM_{TTPCert}$ certificate is itself signed by the TTP. Fig. 5.3 depicts how the signature is generated by the TTP.

Step 3: The TTP sends the RM-agent and the $RM_{TTPCert}$ certificate to the AP administrator/owner as shown in Fig. 5.4 below.

161

Fig. 5.3. RM-agent distributor generates digital signature of RM-agent code.

Fig. 5.4. RM-agent distributed to AP.

Step 4: Upon receipt of the RM-agent code and $RM_{TTPCert}$ certificate from the TTP, the AP administrator/owner can verify that the signature of the certificate is correct. The AP administrator/owner then uses the digital signature of the RM-agent, provided in the $RM_{TTPCert}$ to ensure that RM-agent was sent by the TTP (authenticity check). Finally, the AP administrator/owner, checks if the hash of the RM-agent file downloaded is the same as that on the certificate to ensure that the RM-agent has not been replaced or tampered with (integrity check) as depicted by Fig. 5.5. AP administrator/owner also checks other information on the certificate.

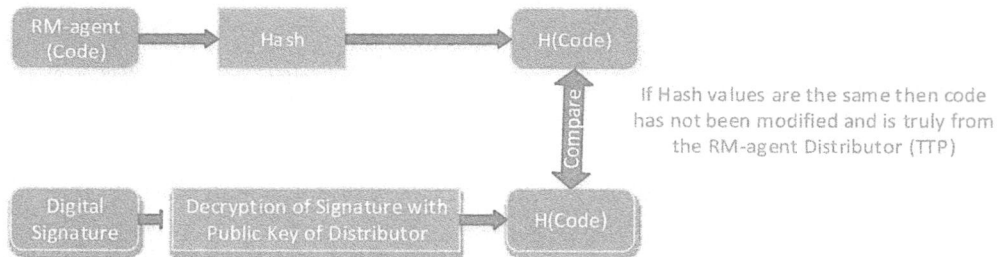

Fig. 5.5. AP checks integrity and authenticity of the RM-agent.

Step 5: Finally, the AP administrator integrates the RM-agent in the AP to establish the AP as a trusted computing base which provides security as per the security policy ID.

RM-agent can thus be easily and securely downloaded from a TTP by any AP in an open system and integrated into the agent execution environment to provide trust enhanced security.

5.4.2. Verifying Authenticity and Integrity of RM-Agent

One of the property of the RM is that it should be tamperproof and verifiable. Before a mobile agent migrates to the destination AP, the mobile agent has to ensure that the

162

RMagent running on the destination AP has not been modified and is truly from a TTP. This requirement is heightened by the fact that in an open system, the destination AP may not always be known by the mobile agent. Verifying the trustworthiness of the RM-agent on a remote AP is therefore necessary.

Verifying the certificate of the RM-agent ($RM_{TTPCert}$) allows to identify the TTP from which the RM-agent has been obtained, but the certificate is not enough to provide assurance that the RM-agent actually deployed on the AP is the same as that obtained from the TTP i.e. RM-agent has not been modified by the AP. However, the certificate provides such information as the security policy ID defining the security provided by the RM-agent; the hash value of the RM-agent as well as the digital signature of the RMagent.

A mobile agent on a source AP could determine the integrity and authenticity of the RMagent running on the destination AP prior to migration can be done by validating the signature of the RM-agent as follows. The mobile agent obtains the hash value of the RMagent from the certificate of the RM-agent ($RM_{TTPCert}$) and also requests the hash value of the RM-agent on the destination AP. If the two hashcodes match, then this would imply that the RM-agent is the same as that distributed by the TTP. But the hash value obtained from the destination AP, may not necessarily be that of the RM-agent running on the destination AP. A malicious AP may actually cache the hash value of the distributed RMagent prior to modifying the RM-agent and running the modified RM-agent on the platform. The mobile agent cannot compute the hash of the RM-agent, itself, as it is running on remote AP.

In such circumstances, the static validation approach fails to allow the mobile agent on a source AP to determine the integrity and authenticity of the RM-agent running on the destination AP prior to migration. Verifying the integrity and authenticity of code running at a remote destination is a challenging issue.

In [34-37], the authors describe the use of a Tag Generator, integrated in the code of the software to be evaluated, which continuously generates tags. At a different point in time, the newly generated tag is added to messages being communicated between the remote software and a Tag Checker software. The Tag Checker verifies the Tag. If the tag received is a tag, which is expected as per the Tag Generator software, then integrity of the software is established. Code obfuscation is used to hide the Tag generator code in the software to be evaluated. However, obfuscated code can be reverse engineered, to allow an attacker to modify the software but still be able to send the expected tags. Furthermore, this approach is not scalable beyond the client server configuration, and thus cannot be adopted in the context of agent computing. Agent computing being distributed, the interaction between agents on one platform with agents on another platform is similar to peer-to-peer interaction. Hence, we cannot distinguish "servers" among the AP. Hence, each AP/RM-agent would need to implement some Tag Checker software for every other platform it interacts with, which is not feasible in practice. In [35], it is proposed to inject additional computation (hashing code) into the software (host code) such as to generate the hash value of the execution trace of a portion of code, thereby, allowing to verify the run time behaviour of the software. However, adding hashing code in different parts of the software increases the code complexity and increases the overall execution time.

We identify the following requirements regarding the verification of the integrity and authenticity of the RM-agent running on the destination AP:

1. The source AP should be able to verify the integrity and authenticity of the RMagent on the destination AP by storing minimal information about the RM-agent.

2. The integrity and authenticity checking mechanism has to stay secure even if the destination AP is malicious.

3. The communication bandwidth used during the verification process should be minimal, i.e., it should not involve the transfer of large amount of data.

4. The verification mechanism should be efficient in terms of computation.

5. It should be possible to run the verification several times, if desired, to ensure that integrity and authenticity of the RM-agent is maintained at all times. Static validation actually fails to satisfy this requirement because the hash code can be stored and sent back by the destination AP whenever required.

We propose a dynamic validation mechanism that satisfies the above conditions based on the challenge response approach. The mobile agent on a source AP can thus effectively verify the integrity and authenticity (and thus the trustworthiness) of the RM-agent on the destination AP that it wishes to migrate to.

5.4.2.1. Authentication and Integrity Checking Protocol (AICP)

The task of checking the integrity and authenticity of the RM-agent on the destination AP in our security framework lies upon the RM-agent on the source AP, since it is a service that may be requested by any mobile agent. If the mobile agents themselves were to perform this task, it would add unnecessary burden on the mobile agent. With the aim to reduce bandwidth used by the mobile agents, the mobile agents remain lightweight and are programmed to only perform their tasks in the application.

Accordingly, when a mobile agent wishes to migrate from its current AP to another, it requests the RM-agent on the source AP to assess the security of the destination AP. Given that mobile agents can only interact with one another by communicating using an Agent Communication Language (i.e. an agent cannot invoke a method of another agent but it can request the destination agent for some processing), the proposed dynamic validation of authenticity and integrity of the RM-agent involves interaction which we refer to as the *Authentication and Integrity Checking protocol (AICP)*. Fig. 5.6 shows the sequence diagram of the AICP.

This interaction begins with the source RM-agent requesting the $RM_{TTPCert}$ certificate from the destination RM-agent. From the certificate, the source RM-agent learns about the TTP from which the destination RM-agent has been acquired and also obtains the hash value of the RM-agent. Alternatively, the source RM-agent can retrieve the distributed RMagent files *from the TTP* (same files/codes are running on the destination AP as the destination

RM-agent), so that the hash value of the RM-agent can be computed locally. It is assumed that the TTP makes available the RM-agent file for download for such verification process. It may be argued that downloading the RM-agent code for this purpose may require a large amount of bandwidth and storage space on the source AP. Nonetheless, our implementation of the RM-agent in JADE results in an RM-agent Java class file which is 17 KB in size. JADE is used as it is one of the most popular APs and it is Java-based; most of the existing APs are Java-based APs.

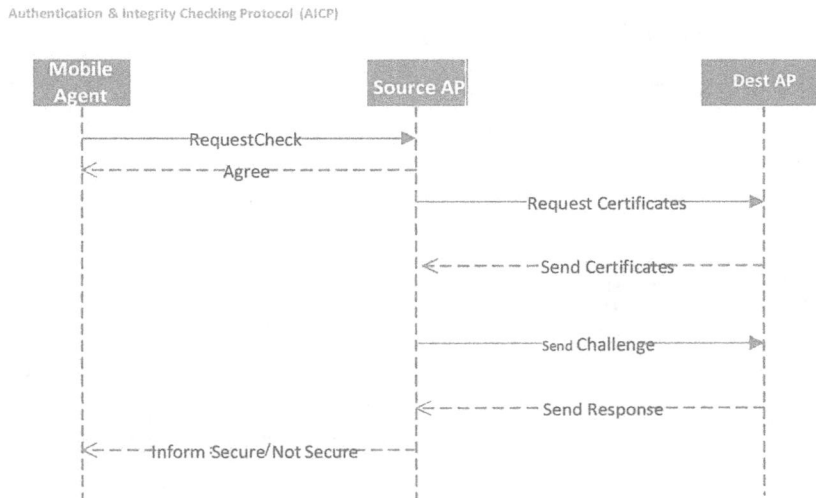

Fig. 5.6. Authenticity and Integrity Checking Protocol Sequence Diagram.

Given the increasing bandwidth capacity and high speed of today's network, retrieving and caching the RM-agent for a short period of time should not pose any problem. Assuming, there are ten different TTP sources with different implementations of the RMagent, and that each RM-agent is of the size 20 Kb, for an AP to temporarily cache all the RM-agents, only 200 Kb (0.2 Mb) of storage space is required. Thus the proposed protocol is storage efficient.

Next, the source RM-agent issues a challenge (random number -R) to the destination RMagent. This challenge (R) can be sent encrypted using the public key of the destination RM-agent.

The source RM-agent also calculates the expected response as follows and as depicted by Fig. 5.7.

- Challenge (R) is concatenated with the RM-agent class file (RMCF): R || RMCF;

- The concatenated output is hashed to obtain the hash value H (R || RCMF).

Similarly, the destination RM-agent uses its Private key to decrypt the encrypted Challenge sent by the source RM-agent. This step ensures that only the destination

165

RMagent has access to the Challenge (R). The destination RM-agent concatenates the Random Number, R, to the current actual RM-agent file and then calculates the hash value of the concatenated input as shown in Fig. 5.8.

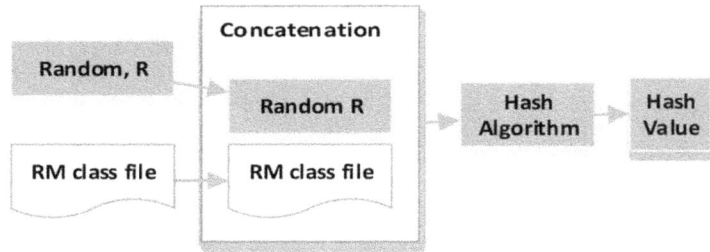

Fig. 5.7. Calculating the response of the challenge at source.

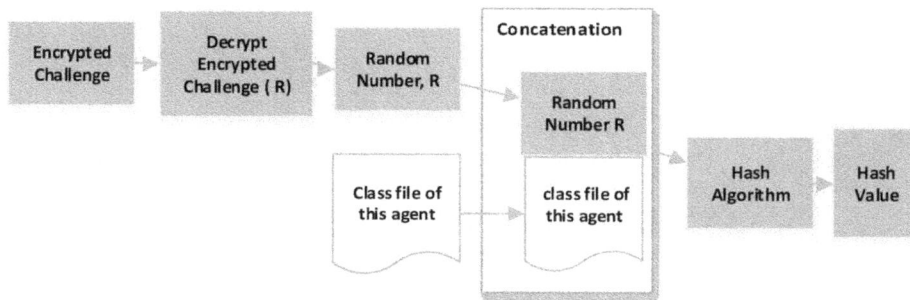

Fig. 5.8. Calculating the response of the challenge at destination.

The response from the destination RM-agent is compared with the expected response computed by the source RM-agent as shown in Fig. 5.9.

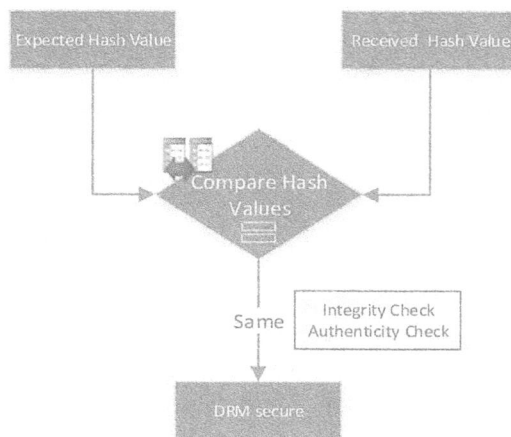

Fig. 5.9. Comparing the response of the challenge with the calculated response.

If they matches, then it is safe to assume that the destination RM-agent has not been tampered with (integrity). If the two hash values are the same, we can also assume authenticity of the destination RM-agent as the hash value has been computed using the RM-agent file from the TTP.

5.5. Experiments and Results

Experiments were designed with JADE agent development environment. JADE was chosen because it is FIPA compliant, thus mobile agent applications developed with JADE can interoperate with mobile agents from other applications. JADE is also continuously maintained and well supported by a large community of users.

The destination RM-agent uses its **own class file** for generating the Hash value. Using the this.getClass() method allows to find the agent's Java class filename. Once the class filename is known, the path of the file can be found. Because the RM-agent is running, the RM-agent java class file is read-only, i.e., cannot be modified. We read the file and copy it into another file, e.g., CopyFile. Concatenation is implemented by appending the Random number to the "CopyFile". The "CopyFile" is then hashed using SHA-256 to generate the hash value. Thus, it can be observed that the hash value generated truly correspond to the destination RM-agent. Similarly, on the source RM-agent, the RM-agent class file is read and copied into a new file, to which the random number is appended in the end (concatenation). This new file is passed as an argument to the hash function to generate the expected hash value. The same hash function is used by the source RM-agent and the destination RM-agent.

Given that in the context of agent computing all interaction takes place by means of ACL messaging between agents, the AICP is implemented as a series of messages exchanged between the source RM-agent and the destination RM-agent. The RM-agent is programmed by adding the following two behaviours: *InitiatorAICP* and *ResponseAICP* implemented as an OneShotBehaviour.

It was assumed that both the source AP and destination AP uses the same TTP source for the RM-agent. We run the AICP, with different versions of the destination RM-agent class file. It was observed that when the proper destination RM-agent was used at the destination AP, the response obtained was as expected. However, when the destination RM-agent code has been changed, the response obtained was different than the expected response. Thus, using the AICP, a source AP may successfully assess the authenticity and integrity of the destination RM-agent and consequently the trustworthiness of the destination AP.

5.6. Conclusions and Future Work

We introduced the RM-agent, as a fundamental component of our framework. The RMagent allows to implement the reference monitor concept, whereby a reference validation mechanism (implemented by the RM-agent), is responsible for enforcing the system's security policy. The RM-agent satisfies the requirements of the reference monitor concept in the sense that the mobile agent can verify that the RM-agent is from a trusted source (RM-agent is verifiable and trustworthy) and it is hasn't been tampered with. A novel dynamic verification approach was proposed for assessing authenticity and integrity of the remote RM-agent running on the destination AP.

RM-agents contribute to providing a secure computing environment to mobile agents. Altogether, with the security framework, it is possible for a mobile agent to migrate to any AP with the assurance of security. In this case, it is not important for a mobile agent to know every AP, as long as the AP deploys a RM-agent obtained from a TTP. Thus, our security framework is most suitable for open systems. Additionally, the security framework supports mobile agent computing with static as well as dynamic itineraries. Future works, revolves around equipping the RM-agent to ensure different security services by integrating different behaviours to the RM-agent. Then, the RM-agent will be able to enforce security such as code and data confidentiality, integrity, execution integrity.

References

[1]. R. R. Brooks, Mobile code paradigms and security issues, *IEEE Internet Computing*, Vol. 8, No. 3, May/June 2004, pp. 54-59.

[2]. D. B. Lange, M. Oshima, Seven good reasons for mobile agents, *Communications of ACM*, Vol. 42, Issue 3, March 1999, pp. 88-89.

[3]. R. K. Verma, S. Jangra, Significance of mobile agent in wireless sensor network, International *Journal of Advance Research in Computer Science and Management Studies*, Vol. 1, Issue 7, 2013, pp. 328-335.

[4]. R. Amine, K. Amine, B. Khalid, Z. Elhoussaine, O. Mohammed, Knowledge discovery in WSN using mobile agents, in *Proceedings of the Intelligent Systems and Computer Vision (ISCV'15)*, 2015, pp. 1-6.

[5]. T. T. Khose Patil, C. Banchhor, Distributed intrusion detection system using mobile agent in LAN environment, *International Journal of Advanced Research in Computer and Communication Engineering*, Vol. 2, Issue 4, 2013, pp. 1901-1903.

[6]. V. Upadhyay, J. Balwan, G. Shankar, Amritpal, A security approach for mobile agent based crawler, in *Proceedings of the 2nd International Conference on Computer Science, Engineering & Applications (ICCSEA'12)*, New Delhi, India, 2012, pp. 119-123.

[7]. Y. Essa, G. Attiya, A. El-Sayed, Mobile agent based new framework for improving big data analysis, in *Proceedings of the IEEE International Conference on Cloud Computing and Big Data (CloudCom-Asia'13)*, Fuzhou, 2013, pp. 381-286.

[8]. W. Godfrey, S. Jha, S. Nair, On a mobile agent framework for an internet of things, in *Proceedings of the IEEE International Conference on Communication Systems and Network Technologies (CSNT'13)*, 2013, pp. 345-350.

[9]. M. Higashino, Management of streaming multimedia content using mobile agent technology on pure P2P-based distributed e-learning system, In *Proceedings of the 27th IEEE International Conference on Advanced Information Networking and Applications (AINA'13)*, 2013, pp. 1041-1047.

[10]. S. S. Manvi, P. Venkataram, Applications of agent technology in communications: a review, *Computer Communications*, Vol. 27, pp. 1493-1508.

[11]. W. Jansen, T. Karygiannis, NIST Special Publication 800-19-Mobile Agent Security, Technical paper, Computer Security Division, *National Institute of Standards and Technology*, 2000.

[12]. G. Necula, P. Lee, Safe, untrusted agents using proof-carrying code, in Mobile Agents and Security (G. Vigna, Ed.), *Proceedings of the Lecture Notes in Computer Science*, Vol. 1419, Berlin, 1998, pp. 61-91.

[13]. R. Sekar, Model carrying code: a practical approach for safe execution of untrusted applications, *ACM SIGOPS Operating Systems Review*, Vol. 37, Issue 5, pp. 15-28.

[14]. S. Tuohimaa, M. Laine, V. Leppanen, Dynamic rights in model-carrying code, in *Proceedings of the International Conference on Computer Systems and Technologies, (CompSysTech'06)*, 2006.

[15]. E. Albert, G. Puebla, M. Hermenegildo, Abstraction-carrying code: a model for mobile code safety, *New Generation Computing*, Vol. 26, Issue 2, February 2008, pp. 171-204.

[16]. M. Alfalayleh, L. Brankovic, An overview of security issues and techniques in mobile agents, in *Proceedings of the 8th IFIP TC-6 TC-11 Conference on Communications and Multimedia Security*, Sept. 15–18, 2004, Windermere, The Lake District, United Kingdom, pp. 59-78.

[17]. C. Lin, V. Varadharajan, Trust enhanced security - a new philosophy for secure collaboration of mobile agents, in *Proceedings of the International Conference on Collaborative Computing: Networking, Applications and Worksharing (COLCOM'06)*, 2006, p. 76.

[18]. C. D. Jensen, The importance of trust in computer security, in *Proceedings of the Trust Management VIII: 8th IFIP WG 11.11 International Conference (IFIPTM'2014)*, 2014, pp. 1-12.

[19]. X. Guan, Y. Yang, J. You, POM-A mobile agent security model against malicious hosts, in *Proceedings of the IEEE 4th International Conference on High Performance Computing in Asia-Pacific Region*, Beijing, China, 2000, pp. 1165-1166.

[20]. S. Zhidong, T. Qiang, A security technology for mobile agent system improved by trusted computing platform, in *Proceedings of the 9th International Conference on Hybrid Intelligent Systems (HIS 2009)*, Shenyang, 2009, pp. 46-50.

[21]. S. Loureiro, R. Molva, Mobile code protection with smartcards, in *Proceedings of the 6th ECOOP Workshop on Mobile Object Systems: Operating System Support, Security and Programming Languages (ECOOP'2000)*, Cannes, France, 2000.

[22]. S. Armoogum, A. Caully, Obfuscation techniques for mobile agent code confidentiality, *Journal of Information & Systems Management*, Vol. 1, Issue 1, 2011, pp. 25-36.

[23]. B. Barak, On the (im)possibility of obfuscating programs, in *Proceedings of the 21st Annual International Cryptology Conference - Advances in Cryptology (CRYPTO'01)*, Santa Barbara, California, USA, 2001, pp 1-18.

[24]. O. Esparza, M. Soriano, J. L. Munoz, J. Forne, A protocol for detecting malicious hosts based on limiting the execution time of mobile agents, in *Proceedings of the IEEE Eight International Symposium on Computers and Communication (ISCC'03)*, Kemer-Antalya, Turkey, 2003, pp. 251-256.

[25]. T. Sander, C. F. Tschudin, Towards mobile cryptography, in *Proceedings of the IEEE Symposium on Security and Privacy*, Oakland, CA, 1998, pp. 215-224.

[26]. H. Lee, J. Alves-Foss, S. Harrison, The use of encrypted functions for mobile agent security, in *Proceedings of the IEEE 37th International Conference on System Sciences*, Hawaii, 2004, pp. 1-10.

[27]. U. G. Wilhelm, S. Staamann, L. Buttyn, On the problem of trust in mobile agent systems, In *Proceedings of the Network and Distributed Security Symposium (NDSS'98)*, San Diego, California, 1998, pp. 114-124.

[28]. C. Lin, V. Varadharajan, MobileTrust: a trust enhanced security architecture for mobile agent systems, *International Journal of Information Security*, Vol. 9, Issue 3, 2010, pp. 153-178.

[29]. D. Foster, V. Varadharajan, Security and trust enhanced mobile agent based system design, in *Proceedings of the 3rd International Conference on Information Technology and Applications (ICITA'05)*, Vol. 1, 2005, pp. 155-160.

[30]. J. Anderson, Computer Security Technology Planning Study, Technical Report ESD-TR-73-51, *Air Force Electronic Systems Division*, Hanscom AFB, Bedford, MA, 1972.

[31]. T. Jaeger, Reference Monitor, in Encyclopedia of Cryptography and Security, 2nd Ed., *Springer*, 2011, pp. 1038-1040.

[32]. C. E. Irvine, The reference monitor concept as a unifying principle in computer security education, in *Proceedings of the IFIP TC11 WG 11.8 1st World Conference on Information Security Education*, Kista, Sweden, 1999, pp. 1-11.

[33]. Trusted Computer System Evaluation Criteria, Orange Book, Library No. S225711, *Dept. of Defense CSC-STD-001-83*, 1983.

[34]. M. Baldi, Y. Ofek, M. Yung, Idiosyncratic signatures for authenticated execution of management code, in *Proceedings of the 14th IFIP/IEEE International Workshop on Distributed Systems: Operations and Management (DSOM'03)*, Heidelberg, Germany, 2003, pp. 204-206.

[35]. P. Falcarin, M. Baldi, D. Mazzacchi, Software tampering detection using AOP and mobile code*, in *Proceedings of the European Workshop on Software Architectures (EWSA'04)*, Saint-Andrews (UK), 2004.

[36]. P. Falcarin, R. Scandariato, M. Baldi, Y. Ofek, Integrity checking in remote computation, in *Proceedings of the Atti del XLIII Congresso Annuale (AICA'05)*, Udine, Italy, 2005.

[37]. P. Falcarin, R. Scandariato, M. Baldi, Remote trust with aspect-oriented programming, in *Proceedings of the 20th International Conference on Advanced Information Networking and Applications (AINA'06)*, Vienna, 2006, pp. 451-458.

Chapter 6
Risk Assessment Considering Configuration of Hybrid Cloud Computing

Shigeaki Tanimoto, Tsutomu Konosu, Motoi Iwashita, Shinsuke Matsui, Yoshiaki Seki, Hiroyuki Sato, Atsushi Kanai

6.1. Introduction

Recent years have seen great progress in Internet services and high-speed network environments. As a result, cloud computing has rapidly developed, with two main forms. The first is public clouds, which are operated by service providers, such as Google and Amazon. The second is private clouds, which are built and operated by individual enterprises for their own use. Generally, public cloud computing eliminates the cost of unnecessary facilities and offers rapid flexibility and scale. However, since a public cloud effectively has invisible features in a virtual configuration, enterprise users are uncertain about the cloud's security and aspects of practical use. In comparison, a private cloud offers the visualization of management since an enterprise operates its own facilities, and it guarantees security in accordance with the company's own policies. The drawbacks to a private cloud, however, include greater cost for maintenance and management of facilities [1].

As one example, an incident of missing and leaked data caused by a cloud operating company, called the 'big ripple', occurred in June 2012 in Japan [2]. When a cloud provider's management handles security poorly, serious risks occur only in the public cloud that it manages, so such incidents may become apparent to users. Thus, a hybrid cloud computing configuration, combining aspects of both public and private cloud computing, is now attracting attention.

Generally, in hybrid cloud computing, data requiring high security is handled within a private cloud, while data requiring easy operation at low cost is handled in a public cloud [3]. Thus, although the hybrid cloud computing configuration requires that two different cloud forms be maintained and managed, its operation also depends on the kind of data. Furthermore, various risk factors are involved, such as accidentally saving to a different

Shigeaki Tanimoto
Chiba Institute of Technology, Japan

cloud during data storage [4]. For these reasons, it is important to investigate risk management for this configuration.

In this paper, we first apply a risk assessment method for qualitative analysis from a comprehensive viewpoint. As a result, 21 risk factors in hybrid cloud computing are extracted, and countermeasures are proposed. However, this is only a qualitative study, meaning that a more practical quantitative evaluation still needs to be undertaken. Next, we describe a quantitative evaluation of the 21 risk factors obtained as a result of the qualitative analysis. Specifically, a risk value based on a formula is approximately calculated for each risk factor [5-7]. Then, on the basis of this value, the effect of the countermeasures on the risks can be quantitatively evaluated. It is shown that the countermeasures can reduce their corresponding risk factors by about 56 %. We believe that the results of this study will help to promote hybrid cloud computing services.

This paper is organized as follows. In Section 6.2, we review the hybrid cloud computing that has been studied so far. In Sections 6.3 and 6.4, we describe the risk assessment of the hybrid cloud computing configuration. Generally, risk assessment consists of three processes, that is, risk identification, risk analysis, and risk evaluation. In Section 6.3, risk identification and risk analysis are described, and we describe a qualitative risk analysis and proposed countermeasures for the risk factors of the configuration. In Section 6.4, we describe risk evaluation. Specifically, we describe a quantitative risk evaluation of the risk analysis result of Section 6.3. Section 6.5 is a conclusion and details future work.

6.2. Overview of Hybrid Cloud Computing Configuration

Cloud computing has now shifted to the practical use stage, and many cloud-related services increased in sales in 2011. Moreover, many user companies are verifying the possibility and practicality of cloud computing for introducing information and communications technology (ICT). Cloud computing analysis is thus recognized as a key stage in system configuration [8].

6.2.1. Reference Model of Cloud Computing

As shown in Fig. 6.1, software as a service (SaaS), platform as a service (PaaS), and infrastructure or hardware as a service (IaaS or HaaS) are classified as the main components of the present cloud computing model. Moreover, in terms of deployment models, cloud computing is classified into public, private, and hybrid or managed clouds. Finally, cloud computing includes the roles of cloud provider and cloud user [9].

6.2.2. Hybrid Cloud Computing

Although hybrid cloud computing appears in the reference model of Fig. 6.1, its concrete configuration combines public cloud computing and private cloud computing, as shown in Fig. 6.2. Usually, a company creates a hybrid cloud, and the company and a public cloud provider share executive responsibility. The hybrid cloud uses both public and private cloud computing services. Thus, when a company requires both public and private cloud computing services, hybrid cloud computing is optimal. In this case, the company can summarize its service targets and service requirements and then use public or private cloud computing services accordingly. Thus, service correspondence can be attained with hybrid cloud computing, not only for secure, mission-critical processes like employee salary processing but also for business information such as payment receipts from customers.

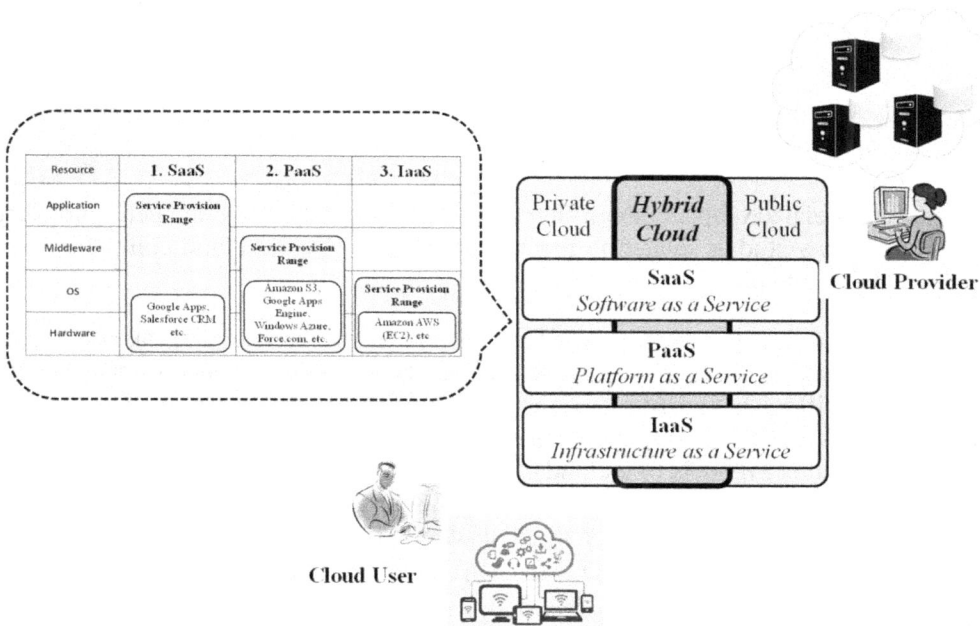

Fig. 6.1. Reference model of cloud computing [9].

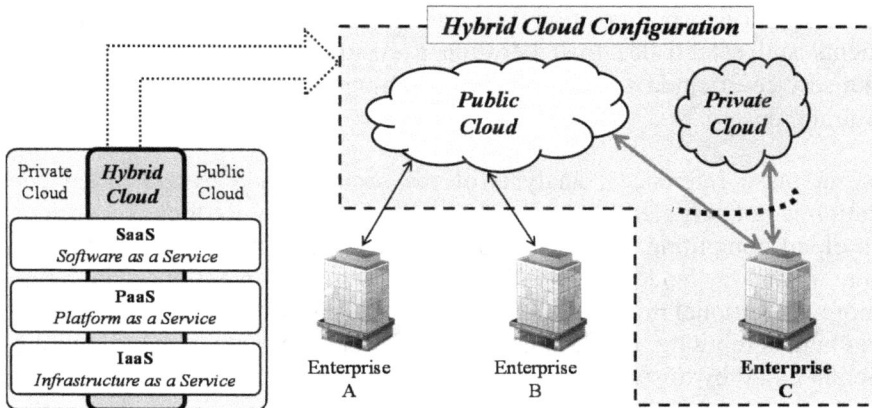

Fig. 6.2. Summary of hybrid cloud computing configuration.

However, the main problem with hybrid cloud computing is the difficulty of actually creating and managing such a solution. Public and private clouds must be provisioned as if they were one cloud, so implementation can become even more complicated. Therefore, since the hybrid cloud computing concept features a comparatively new architecture in cloud computing and best practices and tools have not yet been defined, companies are hesitant to adopt it in many cases [10].

Hence, in this paper, we examine the hybrid cloud computing configuration in terms of the risks that adoption entails. That is, from the viewpoints of both a cloud user and cloud provider, we consider what kinds of risks are assumed and develop a concrete risk management strategy.

6.2.3. Related Work

Many security-related papers about hybrid cloud computing have been published. Furthermore, many studies related to the security risks of hybrid cloud computing have also been conducted [11-18].

From the viewpoint of the security of cloud computing, reference [11] takes a holistic view of cloud computing security, spanning across the possible issues and vulnerabilities connected with virtualization infrastructure, from software platforms to legal compliance in the cloud. Reference [12] surveys security issues that have appeared due to the nature of the service delivery models of a cloud computing system. Furthermore, reference [13] also mentions general security issues in cloud computing. Another reference, [14], offers a clear view of the characteristics of a cloud, its deployment and delivery models, and the risks and challenges met in this increasingly important area.

From the viewpoint of hybrid cloud computing, reference [15] explores an efficient and secure mechanism for partitioning computations across public and private machines in a hybrid cloud setting. Moreover, reference [16] surveys recent research related to single and multi-cloud security and addresses possible solutions. Reference [17] aims to facilitate rapid service composition and agreements based on the necessary security requirements and establishes trust between a customer and provider. Reference [18] provides a service-oriented solution for provisioning secure storage services in a hybrid cloud environment.

However, in these references, analysis of the security in a hybrid cloud computing configuration is not sufficient. For example, these papers do not focus on various threats in hybrid cloud computing. Also, in hybrid cloud computing, there are threats such as operational mistakes made by cloud administrators, as mentioned in Section 6.1. Furthermore, operational mistakes made by users are also assumed to be a threat peculiar to hybrid cloud computing. For example, when a user saves data, the data could be saved on a different cloud by mistake.

6.3. Risk Identification and Analysis: Qualitative Risk Analysis of Hybrid Cloud Computing Configuration

Generally, risks are investigated in the following order: risk identification, risk analysis, and risk assessment [19]. In this section, we describe risk identification and risk analysis, shown in Fig. 6.3.

6.3.1. Risk Identification: Extraction of Risk Factors in Hybrid Cloud Computing Configuration

To extract the risk factors in the hybrid cloud computing configuration, we applied the risk breakdown structure (RBS) method, which is a typical method for risk management in project management [20]. Table 6.1 lists the extracted risk factors. As shown in the table, the hybrid cloud computing configuration was classified at the highest level in the system, operation, facility, and miscellaneous categories from a comprehensive viewpoint. A total of 21 risk factors were extracted [21].

Table 6.1. Risk factors extracted by risk breakdown structure method.

No	High level	Middle level	Low level	Risk factor
1	1. System	1.1 Software	1.1.1 Application	1.1.1.1 A risk of mistaken allocation of the program in hybrid Cloud
2				1.1.1.2 A risk of the mistaken allocation in the case of duplicate programs
3			1.1.2 Data	1.1.2.1 A risk of mistaken allocation of the data in hybrid Cloud
4				1.1.2.2 A risk of the mistaken allocation in the case of duplicate data
5		1.2 Hardware	1.2.1 Performance	1.2.1.1 A risk of the unexpected load for CPU throughput
6				1.2.1.2 A risk of unexpected use for memory size
7		1.3 Network	1.3.1 Performance	1.3.1.1 A risk of the access speed slowing during network congestion etc.
8	2. Operation	2.1 Public cloud		2.1.1 A risk when sharing resources with the other company in public Cloud
9				2.1.2 An operation risk of public Cloud's not being administrable by the company side
10				2.1.3 A risk of the service continuity by the side of public Cloud
11		2.2 Private cloud		2.2.1 A risk of cost exceeding estimation
12				2.2.2 A risk of the human resource development in private Cloud
13		2.3 Hybrid cloud		2.3.1 A risk of the data management mismatching between different Clouds
14	3. Facility	3.1 Public cloud		3.1.1 A facility risk of public Cloud's not being administrable by the company side
15				3.1.2 A risk of public Cloud's business continuity
16		3.2 Private cloud		3.2.1 A risk of an excess of facilities cost in private Cloud
17				3.2.2 A risk of the environmental construction in private Cloud
18				3.2.3 A risk of new business starting in private Cloud
19		3.3 Hybrid cloud		3.3.1 A risk of the optimal use ratio of public Cloud and private Cloud
20	4. Miscellaneous	4.1 Law		4.1.1 A risk of legal revision
21		4.2 Disasters		4.2.1 A risk of a disaster

6.3.2. Risk Analysis: Qualitative Analysis of Risk Factors in Hybrid Cloud Computing Configuration

6.3.2.1. Risk Analysis Method

Here, we apply risk analysis to address the risk factors of the hybrid cloud configuration listed in Table 6.1. Two typical risk analysis methods are the decision tree method and the risk matrix method, which are quantitative and qualitative methods, respectively [22-24]. In this paper, for the purpose of dealing with security issues, such as privacy protection in a hybrid cloud configuration, we adopt a qualitative viewpoint and apply the risk matrix method.

As shown in Fig. 6.3, this method can classify risks into four categories of risk management (risk avoidance, risk mitigation, risk acceptance, and risk transference) in accordance with their frequency and severity. These categories correspond with the risk management analysis in the following section. Fig. 6.4 shows an example of a template for deriving countermeasures after classifying risk factors by using the risk matrix method.

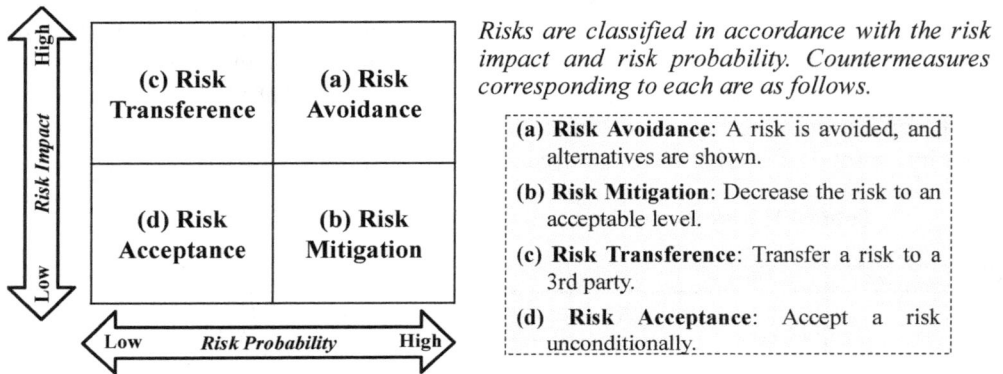

Risks are classified in accordance with the risk impact and risk probability. Countermeasures corresponding to each are as follows.

(a) **Risk Avoidance**: A risk is avoided, and alternatives are shown.

(b) **Risk Mitigation**: Decrease the risk to an acceptable level.

(c) **Risk Transference**: Transfer a risk to a 3rd party.

(d) **Risk Acceptance**: Accept a risk unconditionally.

Fig. 6.3. Risk matrix method.

6.3.2.2. Risk Analysis Result from Qualitative Viewpoint

Next, we conducted a detailed analysis of all 21 risk factors listed in Table 6.1, generating the results listed in Table 6.2. The risk matrix method was used to deduce these countermeasures [24]. As shown in Figs. 6.3 and 6.4, this method classified countermeasures into four kinds in accordance with their risk probability and risk impact, i.e., risk transference, risk mitigation, risk acceptance, and risk avoidance. Furthermore, it gives guidelines on drawing up countermeasures. Table 6.2 lists the risk matrix methods classified in correspondence with their proposed countermeasures.

6.3.2.3. Summary of Risk Analysis Results

Here, we summarize our conclusions concerning countermeasures for the risk factors in a hybrid cloud computing configuration, shown in Table 6.3.

As mentioned above, the choice to deploy multiple clouds and the precise knowledge of the data volume are important issues for an enterprise introducing a hybrid cloud. Third-party surveillance and assurances adapted for the cloud are also important. In particular, synthetic enhancement of not only the technical side but also the practical aspect of the facility side is important.

Risk Factor	2.1.3 Continuity of service

In this place, the risk factor extracted by the RBS method shown in Table 1 is filled in.

Classification of Risk Countermeasure

Risk Transference

Transfer a risk to a 3rd party.

(c) Risk Transference	(a) Risk Avoidance
(d) Risk Acceptance	(b) Risk Mitigation

Risk Impact — High / Low

Low   *Risk Probability*   High

In this place, the result classified according to the risk matrix method is filled in.

Detail Explanation of Risk Factor

The service continuity plan by the cloud provider is either nonexistent or insufficient. Therefore, the cloud user is uncertain about the possibility of stable, continuous service.

Detailed explanation of a risk factor is filled in in this place.

Cause of Risk Factor

The cloud provider has unstable management, leading to problems in service operation..

Main causes of a risk factor are filled in in this place.

Countermeasures of Risk Management

Select multiple cloud providers and prepare a backup plan.

Countermeasures of a risk factor are filled in in this place.

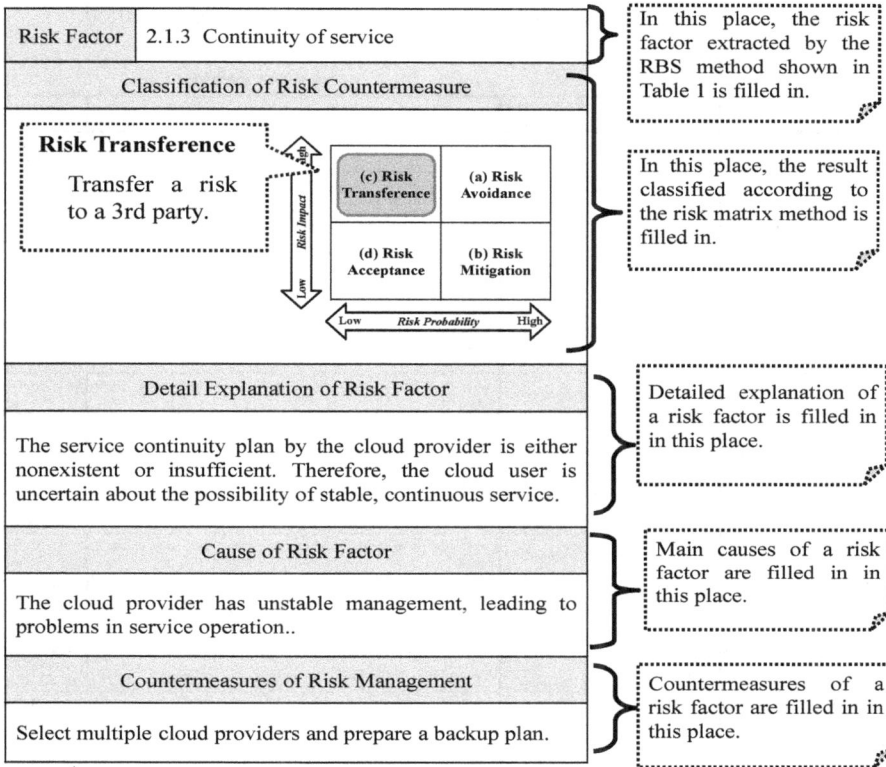

Fig. 6.4. Example of risk analysis result.

6.4. Risk Evaluation: Quantitative Risk Evaluation of Hybrid Cloud Computing Configuration

In the previous section, risk identification and risk analysis were described. In this section, risk assessment is described. Specifically, the validity of a countermeasure is evaluated through the quantification of the risk factors shown in Table 6.2. First, a risk formula used in the field of information security management systems (ISMS) is shown [5-7]. Next, an approximation is described for calculating a risk value on the basis of our previous qualitative results [25-26]. Finally, a risk value for hybrid cloud computing services is deduced by using the formula and approximation [27].

6.4.1. Risk Formula

Each risk value is quantified by using Eq. (6.1), which is used in the field of ISMS 5-7.

$$\text{Risk value} = \text{value of asset} * \text{value of threat} * \text{value of vulnerability.} \quad (6.1)$$

177

Table 6.2. Risk factors extracted by risk breakdown structure method.

No	Level 3: Risk Factors	Risk Impact	Risk Probability	Proposed Countermeasures		
				Classification of Risk Matrix	Main Countermeasure Contents	Classification of Overview for Countermeasure
1	1.1.1.1 A risk of mistaken allocation of the program in hybrid Cloud	High	Low	Risk transference	Strengthen the management system upon deploying data and programs.	Design reinforcement of the Cloud construction
2	1.1.1.2 A risk of the mistaken allocation in the case of duplicate programs	High	Low	Risk transference	Even if the cloud is used mainly on active standby, prepare an additional cloud on cold standby to enable program exchange through manual operation.	Design reinforcement of the Cloud construction
3	1.1.2.1 A risk of mistaken allocation of the data in hybrid Cloud	Low	High	Risk mitigation	Prepare a data management manual. Upon cloud introduction, educate and train employees.	Design reinforcement of the Cloud construction
4	1.1.2.2 A risk of the mistaken allocation in the case of duplicate data	High	Low	Risk transference	Even if the cloud is used mainly on active standby, prepare an additional cloud on cold standby to enable program exchange through manual operation.	Design reinforcement of the Cloud construction
5	1.2.1.1 A risk of the unexpected load for CPU throughput	Low	High	Risk mitigation	During cloud design, include a significant performance margin to enable efficient cloud usage even when system utilization exceeds estimates.	Design reinforcement of the Cloud construction
6	1.2.1.2 A risk of unexpected use for memory size	Low	High	Risk mitigation	Guarantee sufficient storage capacity to handle cases of excessive system utilization.	Design reinforcement of the Cloud construction
7	1.3.1.1 A risk of the access speed slowing during network congestion etc.	Low	High	Risk mitigation	During cloud design, properly consider scale, cost, enterprise usage pattern, and so forth.	Design reinforcement of the Cloud construction
8	2.1.1 A risk when sharing resources with the other company in public Cloud	High	High	Risk avoidance	Do not use the public cloud but protect the company by using the private cloud.	Mainly using private cloud
9	2.1.2 An operation risk of public Cloud's not being administrable by the company side	High	High	Risk avoidance	If public cloud operation is unsuitable, switch to private cloud operation, and vice versa.	Mainly using private cloud
10	2.1.3 A risk of the service continuity by the side of public Cloud	High	Low	Risk transference	Select multiple cloud providers and organize backups and other processes in other public clouds.	Utilization of multiple public clouds
11	2.2.1 A risk of cost exceeding estimation	Low	High	Risk mitigation	Reduce cost by educating employees so that the cloud's operation can be corresponded as much as possible in its company.	Design reinforcement of the Cloud construction
12	2.2.2 A risk of the human resource development in private Cloud	High	Low	Risk transference	The training for the Cloud operation is held regularly in an enterprise. Accordingly, when a security incident occurs, the system which can correspond promptly is built.	Strengthen employee cloud management skills
13	2.3.1 A risk of the data management mismatching between different Clouds	Low	High	Risk mitigation	During cloud construction, fully investigate security so as to unify the security control methods of both the private and public clouds.	Design reinforcement of the Cloud construction
14	3.1.1 A facility risk of public Cloud's not being administrable by the company side	High	Low	Risk transference	Deploy multiple public clouds.	Utilization of multiple public clouds
15	3.1.2 A risk of public Cloud's business continuity	High	Low	Risk transference	Take out an insurance policy upon public cloud utilization. In addition, request third-party evaluation and survey the cloud provider.	Third-party surveillance, and cloud insurance
16	3.2.1 A risk of an excess of facilities cost in private Cloud	Low	Low	Risk acceptance	Investigate the cost of private cloud construction sufficiently, and ensure that cloud facilities are used efficiently, such as through diversion.	Design reinforcement of the Cloud construction
17	3.2.2 A risk of the environmental construction in private Cloud	Low	Low	Risk acceptance	If a particular situation is judged necessary for the enterprise, approve it in order to develop the business.	Design reinforcement of the Cloud construction
18	3.2.3 A risk of new business starting in private Cloud	Low	High	Risk mitigation	Private Cloud's operation is made to permeate as an enterprise rule beforehand.	Design reinforcement of the Cloud construction
19	3.3.1 A risk of the optimal use ratio of public Cloud and private Cloud	Low	High	Risk mitigation	Determine a utilization policy for data handling.	Design reinforcement of the Cloud construction
20	4.1.1 A risk of legal revision	Low	Low	Risk acceptance	Respond flexibly to changes in law.	Flexible response to regulations
21	4.2.1 A risk of a disaster	High	Low	Risk transference	Prepare multiple, separate backups for the private and public clouds.	Utilization of multiple public clouds

Table 6.3. Summary of risk analysis results based on risk matrix method.

Risk Classification	Number of Risk Factors	Main Features of Countermeasure
Risk avoidance	2	Practical use of a public cloud entails a data sharing problem with other companies. If such data sharing is unacceptable, the use of a private cloud is optimal, thus demonstrating one benefit of a hybrid cloud configuration.
Risk transference	8	Many countermeasures for risk factors in this category involve the public cloud as a general trend. Specifically, these include the deployment of multiple public clouds, third-party surveillance, and use of assurances adapted for the cloud.
Risk mitigation	8	Many risk factors in this category require advanced countermeasures, such as the choice of cloud, precise estimation of data volume, and granting of access permissions.
Risk acceptance	3	Direct problems, such as the scale of an enterprise and its facilities, follow enterprise-level policies. Indirect problems, such as legal revision, require more flexible responses.

178

Generally, all elements of the right-hand side of (6.1) are very difficult to calculate. In this paper, the following approximation is used to simplify these elements [25-27]. It is explained in detail with Fig. 6.5.

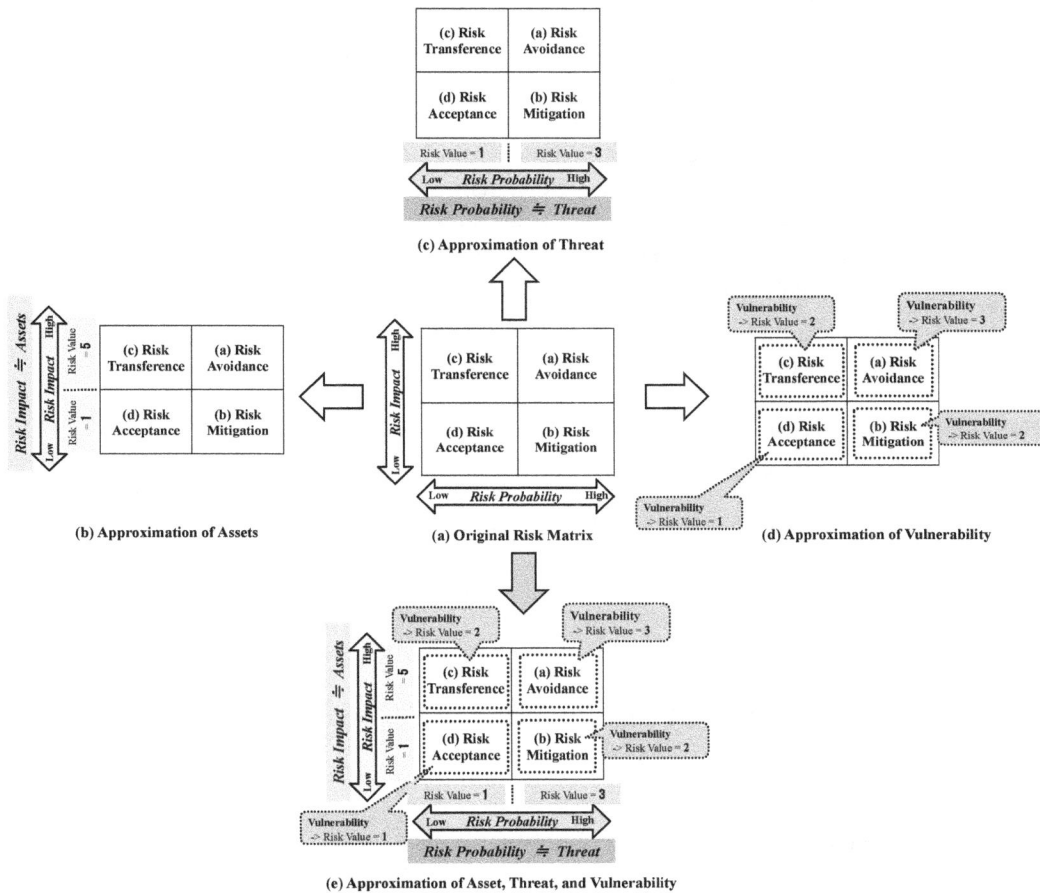

Fig. 6.5. Approximation of risk value in risk matrix.

6.4.1.1. Approximation of Asset Value

Here, the asset value of Eq. (6.1) is approximated in terms of the risk impact in a risk matrix, shown in Fig. 6.5 (b). This approximation was created for the following reasons. The amount of damage is considered for assets. As a further approximation, it was decided that the amount of damage was the risk impact. Additionally, references [5-7] rate the risk impact as 1 (low) to 5 (high).

As a further approximation, these values are mapped in terms of the risk impact to a risk matrix [25-27]. As shown in Fig. 6.5 (b), the risk impact of the risk matrix is divided in two. For the sake of simplicity, the higher of the two divisions is approximated to the

179

maximum risk impact (risk value ≒ 5). Similarly, the lower of the two divisions is approximated to the minimum risk impact (risk value ≒ 1).

6.4.1.2. Approximation of Threat Value

The threat value of Eq. (6.1) is approximated in terms of the risk probability in the risk matrix, as shown in Fig. 6.5 (c). This approximation was created for the following reasons. It was supposed that a threat is strongly dependent on the risk probability. From references [5-7], the risk probability is defined in a range from 1 (low) to 3 (high). These values are mapped to the risk probability of the risk matrix in Fig. 6.5 (c) as well as the above-mentioned risk impact approximation. That is, the higher of the two divisions is approximated to the maximum risk probability (risk value ≒ 3), and the lower of the two divisions is approximated to the minimum risk probability (risk value ≒ 1).

6.4.1.3. Approximation of Value of Vulnerability

The evaluation of vulnerability is defined in references [5-7] as well. It is defined on a three-level scale: 3 (high), 2 (medium), and 1 (low). These levels are approximated in accordance with the classification of the risk matrix in Fig. 6.5 (d). Here, the four domains of the figure are classified into three categories in accordance with the risk probability and risk impact as follows.

- Risk Avoidance: both the risk probability and risk impact are high. This category approximately corresponds to the highest risk classification.

- Risk Transference and Risk Mitigation: either the risk probability or the risk impact is high. These categories approximately correspond to the second highest risk classification.

- Risk Acceptance: both the risk probability and risk impact are low. This category approximately corresponds to the lowest risk classification.

In the above-mentioned classification, risk avoidance cases are approximated to 3 (high), risk transference and risk mitigation cases to 2 (medium), and risk acceptance cases to 1 (low).

As mentioned above, the asset value, threat value, and value of vulnerability with (6.1) is approximated, as shown in Fig. 6.5 (d). Accordingly, (6.1) is approximated as (6.2) below. In addition, the approximate value of each parameter of (6.2) becomes as shown in Tables 6.4 and 6.5.

Risk value = value of risk impact * value of risk probability * value of vulnerability. (6.2)

Table 6.4. Approximate value of risk impact and risk probability of (6.2).

	Risk Impact	Risk Probability
High	5	3
Low	1	1

Table 6.5. Approximate value of vulnerability of (6.2).

	Vulnerability
Risk Avoidance	3
Risk Transference	2
Risk Mitigation	2
Risk Acceptance	1

6.4.2. Calculation of Risk Value

The risk values before applying countermeasures against risks are calculated using Eq. (6.2) and shown in Table 6.6.

Next, the risk values after applying the countermeasures shown in Table 6.2 were calculated. Table 6.7 (After Countermeasures) shows the resulting risk values when performing the countermeasures.

Here, supposing an ideal case, vulnerability was assumed to be 0 as a result of using the proposed countermeasures. Moreover, supposing an actual case, these countermeasures are not always perfect. Thus, the vulnerability of an actual case is approximated to 1 (the minimum level).

Table 6.8 summarizes the results shown in Tables 6.6 and 6.7.

6.4.3. Discussion

As shown in Table 6.8, it turned out that the risk reduction rate when the countermeasures were taken was about 56 percent. These results also show that a detailed numerical expression can treat a risk more specifically by quantifying it and the prospective countermeasure. For example, on the basis of this result, it can be expected that the management layer will also review the risk countermeasures of hybrid cloud computing more specifically.

As an aside, as shown in Table 6.7, it is assumed that the vulnerability value after a countermeasure against a risk is 1 (low). Originally, from the viewpoint of an ideal result, the vulnerability value after a countermeasure should be 0. Therefore, this evaluation means not an ideal condition but a real-world situation.

However, the problem of cost might also affect these countermeasures. Generally speaking, these countermeasures can become expensive because design reinforcement, the utilization of multiple public clouds, third-party surveillance, cloud insurance, etc. are needed. In the future, we will have to devise a verification considering such cost.

Table 6.6. Risk values before countermeasures.

No.	Level 3: Risk Factors	Asset ≒ Risk Impact	Threat ≒ Risk Probability	Vulnerability	Value of Risk
1	1.1.1.1 A risk of mistaken allocation of the program in hybrid Cloud	5	1	2	10
2	1.1.1.2 A risk of the mistaken allocation in the case of duplicate programs	5	1	2	10
3	1.1.2.1 A risk of mistaken allocation of the data in hybrid Cloud	1	3	2	6
4	1.1.2.2 A risk of the mistaken allocation in the case of duplicate data	5	1	2	10
5	1.2.1.1 A risk of the unexpected load for CPU throughput	1	3	2	6
6	1.2.1.2 A risk of unexpected use for memory size	1	3	2	6
7	1.3.1.1 A risk of the access speed fall in network congestion etc.	1	3	2	6
8	2.1.1 A resource share risk with the other company in public Cloud	5	3	3	45
9	2.1.2 An operation risk of public Cloud's not being administrable by the company side	5	3	3	45
10	2.1.3 A risk of the service continuity by the side of public Cloud	5	1	2	10
11	2.2.1 A risk of an excess of cost more than estimation	1	3	2	6
12	2.2.2 A risk of the human resource development in private Cloud	5	1	2	10
13	2.3.1 A risk of the data management mismatching between different Cloud	1	3	2	6
14	3.1.1 A facility risk of public Cloud's not being administrable by the company side	5	1	2	10
15	3.1.2 A risk of public Cloud's business continuity	5	1	2	10
16	3.2.1 A risk of an excess of facilities cost in private Cloud	1	1	1	1
17	3.2.2 A risk of the environmental construction in private Cloud	1	1	1	1
18	3.2.3 A risk of new business starting in private Cloud	1	3	2	6
19	3.3.1 A risk of the optimal use ratio of public Cloud and private Cloud	1	3	2	6
20	4.1.1 A risk of legal revision	1	1	1	1
21	4.2.1 A risk of a disaster	5	1	2	10
	Total				221

Table 6.7. Risk values after countermeasures.

No	Level 3: Risk Factors	Classification of Overview for Countermeasure	Asset ≒ Risk Impact	Threat ≒ Risk Proablitiy	Vulnerability	Valu of Risk
1	1.1.1.1 A risk of mistaken allocation of the program in hybrid Cloud	Design reinforcement of the Cloud construction	5	1	1	5
2	1.1.1.2 A risk of the mistaken allocation in the case of duplicate programs	Design reinforcement of the Cloud construction	5	1	1	5
3	1.1.2.1 A risk of mistaken allocation of the data in hybrid Cloud	Design reinforcement of the Cloud construction	1	3	1	3
4	1.1.2.2 A risk of the mistaken allocation in the case of duplicate data	Design reinforcement of the Cloud construction	5	1	1	5
5	1.2.1.1 A risk of the unexpected load for CPU throughput	Design reinforcement of the Cloud construction	1	3	1	3
6	1.2.1.2 A risk of unexpected use for memory size	Design reinforcement of the Cloud construction	1	3	1	3
7	1.3.1.1 A risk of the access speed slowing during network congestion etc.	Design reinforcement of the Cloud construction	1	3	1	3
8	2.1.1 A risk when sharing resources with the other company in public Cloud	Mainly using private cloud	5	3	1	15
9	2.1.2 An operation risk of public Cloud's not being administrable by the company side	Mainly using private cloud	5	3	1	15
10	2.1.3 A risk of the service continuity by the side of public Cloud	Utilization of multiple public clouds	5	1	1	5
11	2.2.1 A risk of cost exceeding estimation	Design reinforcement of the Cloud construction	1	3	1	3
12	2.2.2 A risk of the human resource development in private Cloud	Strengthen employee cloud management skills	5	1	1	5
13	2.3.1 A risk of the data management mismatching between different Clouds	Design reinforcement of the Cloud construction	1	3	1	3
14	3.1.1 A facility risk of public Cloud's not being administrable by the company side	Utilization of multiple public clouds	5	1	1	5
15	3.1.2 A risk of public Cloud's business continuity	Third-party surveillance, and cloud insurance	5	1	1	5
16	3.2.1 A risk of an excess of facilities cost in private Cloud	Design reinforcement of the Cloud construction	1	1	1	1
17	3.2.2 A risk of the environmental construction in private Cloud	Design reinforcement of the Cloud construction	1	1	1	1
18	3.2.3 A risk of new business starting in private Cloud	Design reinforcement of the Cloud construction	1	3	1	3
19	3.3.1 A risk of the optimal use ratio of public Cloud and private Cloud	Design reinforcement of the Cloud construction	1	3	1	3
20	4.1.1 A risk of legal revision	Flexible response to regulations	1	1	1	1
21	4.2.1 A risk of a disaster	Utilization of multiple public clouds	5	1	1	5
	Total					97

Table 6.8. Evaluation results: summary of risk values before and after countermeasures.

	Before countermeasures against risk factors (①)	After countermeasures against risk factors (②)
Total risk value	221	97
Risk reduction rate = (① - ②)/①	-	0.56

6.5. Conclusion and Future Work

We are interested in promoting hybrid cloud computing services as a next-generation digitized infrastructure by assessing their risks and proposing countermeasures. First, we qualitatively analyzed a hybrid cloud computing configuration. We extracted 21 risk factors for the configuration comprehensively and proposed countermeasures for these factors.

Next, we performed a quantitative evaluation that used a risk value. It was found that the countermeasures could reduce their corresponding risk factors by about 56 %. This result means that the effect of these countermeasures can be more specifically evaluated by introducing a risk value.

In the future, we will further improve the countermeasures and verify their cost effectiveness.

Acknowledgements

This work was supported by the Japan Society for the Promotion of Science (JSPS, KAKENHI Grant Number 15H02783).

References

[1]. S. Nakahara, N. Fujiki, S. Ushijima, Cloud traceability (CBoC TRX), *NTT Technical Journal*, Vol. 10, 2011, pp. 31-35 (in Japanese).
[2]. O. Inoue, First server failure, *Nihon Keizai Shimbun*, Vol. 6, 2012 (in Japanese).
[3]. Shift to hybrid cloud computing (in Japanese), http://www.microsoft.com/ja-jp/opinionleaders/economyict/100701_2.aspx
[4]. S. Tanimoto, M. Hiramoto, M. Iwashita, H. Sato, A. Kanai, Risk management on the security problem in cloud computing , in *Proceedings of the First ACIS/JNU International Conference on Computers, Networks, Systems and Industrial Engineering (ACIS CNSI'11)*, Korea, 2011, pp. 147-152.
[5]. M. S. Toosarvandani, N. Modiri, M. Afzali, The risk assessment and treatment approach in order to provide LAN security based on ISMS standard, *International Journal in Foundations of Computer Science & Technology*, Vol. 2, No. 6, Nov. 2012, pp. 15-36.
[6]. H. Sato, T. Kasamatsu, T. Tamura, Y. Kobayashi, Information Security Infrastructure, *Kyoritsu Shuppan Co. Ltd.*, 2010 (in Japanese).
[7]. ISMS Risk Assessment Manual v1.4, https://ru.scribd.com/document/202271054/ISMS-Risk-Assessment-Manual-v1-4
[8]. A. Goto, T. Nishihara, The concept for the cloud computing technology CboC, *NTT Technical Journal*, Vol. 9, 2009, pp. 64-69 (in Japanese).
[9]. N. Uramoto, Security and compliance issues in cloud computing, *IPSJ Magazine*, Vol. 50, Issue 11, 2009, pp. 1099-1105.
[10]. T. Dillon, C. Wu and E. Chang, Cloud Computing: Issues and Challenges, in *Proceedings of the 24th IEEE International Conference on Advanced Information Networking and Applications*, 2010, pp. 27-33.

[11]. S. Sengupta, V. Kaulgud, V. S. Sharma, Cloud computing security - trends and research directions, in *Proceedings of the IEEE World Congress on Service*, 2011, pp. 524-531.

[12]. S. Subashini, V. Kavitha, A Survey on security issues in service delivery models of cloud computing , *Journal of Network and Computer Applications*, Vol. 34, 2011, pp. 1-11.

[13]. M. K. Srinivasan, K. Sarukesi, P. Rodrigues, M. S. Manoj, P. Revathy, State-of-the-art cloud computing security taxonomies - A classification of security challenges in the present cloud computing environment, in *Proceedings of the International Conference on Advances in Computing, Communications and Informatics (ICACCI'12)*, 2012, pp. 470-476.

[14]. L. Savu, Cloud computing - deployment models, delivery models, risks and research challenges, http://ieeexplore.ieee.org/stamp/stamp.jsp?arnumber=5778816

[15]. K. Y. Oktay, Risk-aware workload distribution in hybrid cloud computing s, in *Proceedings of the IEEE Fifth International Conference on Cloud Computing (CLOUD'12)*, 2012, pp. 229-236.

[16]. M. A. AlZain, E. Pardede, B. Soh, J. A. Thom, Cloud computing security: from single to multi-clouds, in *Proceedings of the 45th Hawaii International Conference on System Sciences (HICSS'12)*, 2012, pp. 5490-5499.

[17]. K. Bernsmed, M. G. Jaatun, P. H. Meland, A. Undheim, Security SLAs for federated cloud services, in *Proceedings of the 6th International Conference on Availability, Reliability and Security (ARES'11)*, 2011, pp. 202-209.

[18]. S. Nepal, C. Friedrich, L. Henry, S. Chen, A secure storage service in the hybrid cloud computing , in *Proceedings of the 4th IEEE International Conference on Utility and Cloud Computing (UCC'11)*, 2011, pp. 334-335.

[19]. K. Noguchi, Risk Management - Management Technology which Supports Goal Achievement, in Japanese, *Japanese Standards Association*, 2009.

[20]. Risk Breakdown Structure, http://www.justgetpmp.com/2011/12/risk-breakdown-structure-rbs.html

[21]. S. Tanimoto, A study of risk management in hybrid cloud computing configuration, *Computer and Information Science*, Vol. 493, 2013, pp. 247-257.

[22]. IPA, Information security management and PDCA cycle: http://www.ipa.go.jp/security/manager/protect/pdca/risk.html

[23]. Risk Management Guide for Information Technology Systems, http://csrc.nist.gov/publications/nistpubs/800-30/sp800-30.pdf

[24]. Cox's Risk Matrix Theorem and Its Implications for Project Risk Management, http://eight2late.wordpress.com/2009/07/01/cox%E2%80%99s-risk-matrix-theorem-and-its-implications-for-project-risk-management/

[25]. S. Tanimoto, A study of risk assessment quantification in cloud computing , in *Proceedings of the 8th International Workshop on Advanced Distributed and Parallel Network Applications (ADPNA'14)*, 2014, pp. 426-431.

[26]. S. Tanimoto, H. Sato, A. Kanai, Risk assessment quantification of ambient service, in *Proceedings of the 9th International Conference on Digital Society (ICDS'15)*, Lisbon, Feb. 2015, pp. 70-75.

[27]. S. Tanimoto, T. Konosu, M. Iwashita, H. Sato, A. Kanai, Risk assessment quantification in hybrid cloud configuration, in *Proceedings of the 9th International Conference on Emerging Security Information, Systems and Technologies (SECURWARE'15)*, 2015, pp. 1-6.

Chapter 7
The Mathematical Modeling of Road Transport in Context of Critical Infrastructure Protection

Jan Mrazek, Lucia Duricova and Martin Hromada

7.1. Introduction

Critical infrastructure has an important role in modern society. Critical infrastructure elements are national priorities all over the world. Not only their reliability, safety, continuous operation, maintenance, but also their protection is important for each state. All sectors of critical infrastructure are interconnected through links. When an emergency occurs in one sector, there is a disruption or limitation of its functionality. This occurrence can even cause a paralyzing of one or more elements of critical infrastructure.

Critical infrastructure's most important sub-sectors include road transport, which is also the most vulnerable sector. Currently many major international projects deal with this issue, but none of them deals with issues of dynamic modelling of the road transport system, which minimizes impacts on society and other critical infrastructure sectors.

Several categories are included in crisis areas, but the key ones are natural disasters. Predicting their course or their duration, or the time of occurrence, is very complicated. When comparing with the area of traffic accidents, the possibility of minimizing risks is incomparably more demanding.

On the facts designated as "Review on modeling and simulation of interdependent critical infrastructure systems" [17], a fundamental knowledge basis for the delimination of dynamic system state space has been created - and is presented in detail in the following chapters.

Jan Mrazek
Tomas Bata University in Zlin, Faculty of applied informatics
Nad Stranemi 4511, Zlín, 76001 Czech Republic

7.2. Critical Infrastructure

Critical infrastructure is mostly known as an element or a system whose functional degradation leads to a significant impact on national security. National security considers the basic needs of the population of a country, its health and economy.

Individual elements of critical infrastructure have linkages to each other, which guarantee the correct functioning of individual elements. These linkages ensure the interdependence of its various sectors and elements.

The elements we call critical infrastructure in particular. These elements are determined by the cross-cutting and sector specific criteria, if the element of critical infrastructure part of the European Critical Infrastructure. In this case, it is considered as an element of European critical infrastructure [14].

7.2.1. The Cross-Cutting Criteria

In order to assess the severity of the impact of disruption to critical infrastructure element, we consider the elements from the following perspectives:

a) Casualties criterion (assessed in terms of the potential number of fatalities or injuries);

b) Economic effects criterion (assessed in terms of the significance of economic loss and/or degradation of products or services; including potential environmental effects);

c) Public effects criterion (assessed in terms of the impact on public confidence, physical suffering and disruption of daily life; including the loss of essential services).

7.2.2. The Sectoral Criteria

The sectoral criteria to identify critical infrastructure elements are set by Government Decree no. 432/2010 Coll. [14], on criteria for determining the elements of critical infrastructure.

The basic classification of sector specific criteria to identify critical infrastructure element are [14]:

- Energy;
- Water management;
- Food and agriculture;
- Health;

- Transport;
- Communication and information systems;
- Financial market and currency;
- Emergency services;
- Public administration [14].

According to the process management, we can examine the results. According to the results, we can define the corrective action, which can be implemented into the inputs. This whole process of function steps can improve the entire process. This principle was used to the proposed solution. According to the inputs, we can identify the points, where we need integrate corrective actions. The inputs define to us the points, where we need to integrate the corrective actions. According to this principle we know, that we need to know the whole process of function each of the buildings and objects. And the last, but not least, we are aware, that we need to improve whole process no only one unit.

7.3. Critical Road Infrastructure

The individual elements of the critical road infrastructure are important for the state in terms of not only passenger and material. Whether it's rail, water, air or road transport. The individual elements are also in risks that affect any linkage occurs to the crisis or emergency. These situations can impair or limit the functionality of this state. In case of limited functionality, there is a disruption of its security. In our case, we focus on the road critical infrastructure, which could be described as the most important element in the transport [6].

The current state of road infrastructure can be analyzed from two perspectives. The first view is within road safety. The second view shows us linkage with other elements of critical infrastructure.

Traffic safety is influenced by two inputs. The first entry is homeland security, which is aimed at reducing accidents and creating traffic rules. External security expresses anti-terrorism, vandalism, the elements etc.

We can conclude that road transport is the most vulnerable and the most significant element of the critical infrastructure. Shows us the risks that may arise during minor traffic accidents.

When looking at Fig. 7.1, we can see that the riskiest places in the Czech Republic are the main arteries of motorways and Class I roads. The biggest risk for the Czech Republic is without any doubt the D1 motorway. Looking at Fig. 7.1. It is possible to draw attention to the utilization of major roads that is overloaded by a large number of vehicles not only personal but also from the reason to save the finances of the carriers to the toll. This issue would deserve its own article.

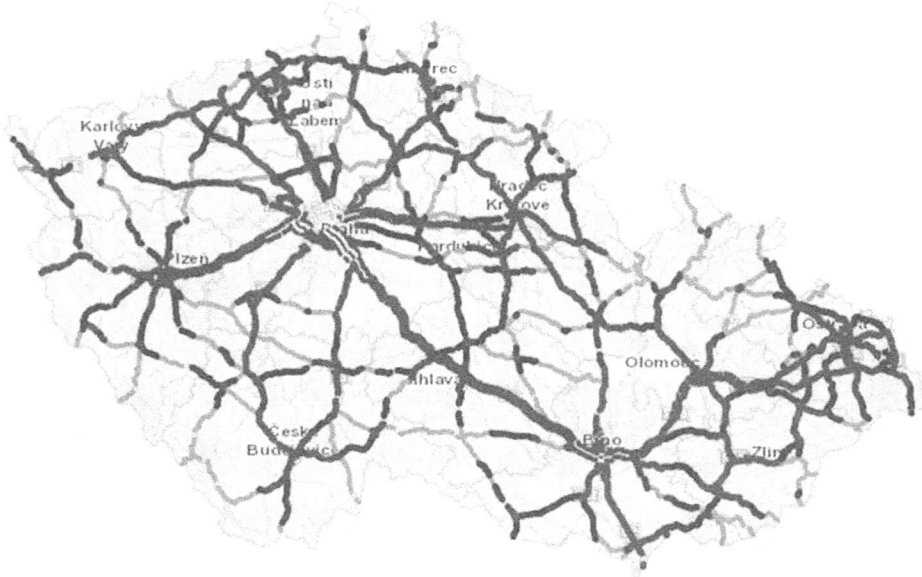

Fig. 7.1. Intensity of Transport in Czech Republic 2015 [13].

7.4. Mathematical Models of Road Transport

The goal of mathematical modeling in road transport is what the most trusted model vehicle movements and their interaction in traffic. One of the criteria is the failure of random factors and variables. These models can then be divided into stochastic and deterministic. The difference between these two models is substantial. Stochastic models operate with the probability of certain events and take into account the random effects, while deterministic models with random effects are not counted. Deterministic model is act upon strict mathematical, statistical and logical relationships. These relations predetermine the behavior [9, 10].

A very important role in modeling road transport is the request for range of investigated network. This fact is closely related to the requirement for input data. When modeling large transport networks, we can generalize some detail or possibly neglected. For small transport networks, it cannot be afforded such a procedure, it is necessary to give weight to each transport movement. In this model we divide the macroscopic and microscopic or their combination. When you merge macroscopic and microscopic models, there is a new model that we call mesoscopic. With this model it is possible to meet only rarely. [9, 10]. The next table defines the parameters, which have to be defined in the dynamic part of the assessment. In our solution, the proposed tool don't examine the features of the building in the continuous process. The dynamic solution is based on the linkages between outside and inside. The features of the soft targets cannot be monitored without delay, or not all features cannot be monitored without delay. We can monitor the situation in the object immediately, but not features. One of the features is how the object controls the situation

190

in the object. According to the methodical principle, we can define the interval for the repeated assessment.

7.4.1. Macroscopic Model

The most commonly are used to simulate large-scale communication networks. These models are mainly used for prognostic purposes.

$$q = v \times k, \tag{7.1}$$

where q is the intensity of transmission services [ks·h-1], v is the current speed of vehicle [km·h-1], k is the density of vehicles in the stream [ks·km-1].

Greenshields model is among the simplest and oldest linear model. It is based on measuring the speed and intensity when the help of these data we can calculate the density. Fundamental assumption is the linear dependency of velocity on the density. The following formula expresses this dependency [5]:

$$v(k) = v_{max} \times \left(1 - \frac{k}{k_{max}}\right), \tag{7.2}$$

where v_{max} is the maximum speed [km·h^{-1}]; k_{max} is the density congestion [ks·km^{-1}].

In areas with a low density, this model acts as unrealistic. This behavior is caused by insufficient speed and density is then symmetric parabolic dependence on the density of intensity. At lower densities, it leads to maximum intensity.

7.4.2. Macroscopic Model

They are based on modeling of individual vehicles driving along the road when there is a consideration as the communication parameters and the vehicle and the driver behavior. During a traffic simulation, we meet mostly these models.

Input parameters are achieved vehicles vehicle speed, engine power, size, acceleration and deceleration. Other essential parameters for input features are network users and their interaction. The input data are never accurate, mainly due to the uniqueness of each participant and the vehicle. The determination of acceleration depending on the environmental conditions is essential. Generally, we can express the acceleration in microscopic models as follows:

$$a = f(v, \Delta v, \Delta x), \tag{7.3}$$

where a is the acceleration [m·s^{-2}]; v is the speed [m·s^{-1}]; Δv is the relative speed relative to the preceding vehicle [m·s^{-1}]; Δx is the distance from the preceding vehicle [m].

Consequently, the examination of the traffic flow, we filter out two basic ways of influencing the vehicle between them. The first approach is developed based on observance of a safe distance between vehicles, and when changing the speed of the vehicle prior. The second approach is based on the distance between vehicles [12, 4].

The specific and direct relationships between these traffic flow status variables are the subject of long-term systematic research and thus cannot currently be described as universally valid (see Fig. 7.2).

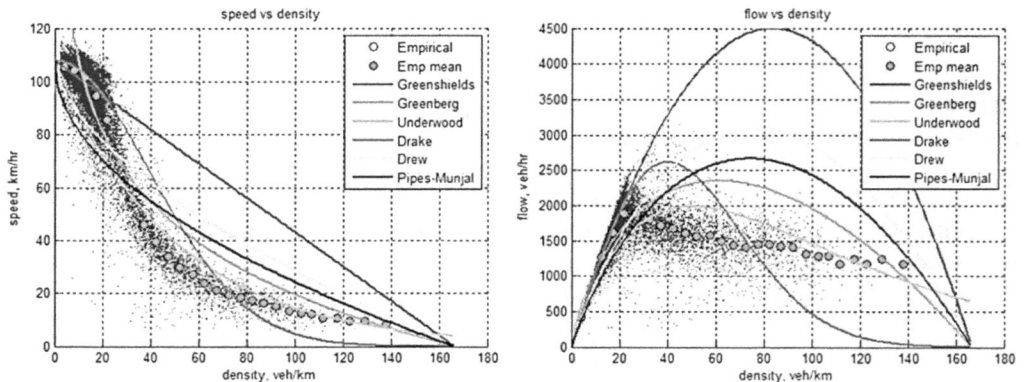

Fig. 7.2. Basic diagrams comparing macroscopic models [16].

The oldest, and simplest, macroscopic model is Greenshields´ Linear Model - based on the relationship between speed and intensity, and which also allows one to calculate jam density. [15]

Weidman's model – The model is a reaction of the driver, which is carried out at a certain distance and the difference in speed between the vehicles. This difference estimated driver only relatively. Using the model, we are looking threshold level drivers that are running in its decisions. These limits are divided into four categories:

- Free movement;
- Approximation;
- Monitor;
- Emergency braking.

These four categories are shown in Fig. 7.3, which you can see below.

In the free movement, the drivers try to or do reach maximum speed. In cases where an increase of density on the road is, there occurs a state where the vehicle is traveling at a higher speed than the preceding vehicle for this vehicle approaching. Once there is the approaching to the vehicle prior to the vehicle driver tries to adjust not only the speed of the preceding vehicle but also it' driving style. This leads to copying of previous vehicle, not only in terms of speed but also in view of maintaining a constant distance between vehicles. At this point, the most important element is the reaction time of the driver, which can lead to two different states. The first condition can be described as the best possible, and that is the driver can react quickly and reduce the existing distance between vehicles. The second condition leads to a bad reaction that could cause a collision.

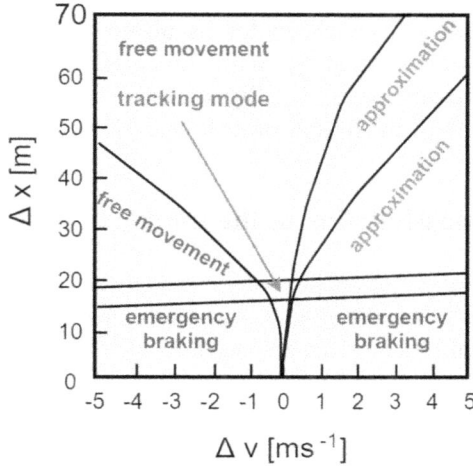

Fig. 7.3. Wiedemann Model [11].

Gipps model – This model can be also known as non- crash. It is based on the speed limit, when as the result of this action is no crash. This model was the first realistic model. The positive aspect is its ability to reproduce the characteristics of real traffic flow without necessity introducing parameters. These properties are not related to the driver. The driver is limited to maximum accelerations and decelerations that are in full speed distance between vehicles and the relative speed. Therefore, the vehicle will never exceed the maximum speed and acceleration should in the free flow of traffic drop to zero. Based on the vehicle speed in the traffic flow calibration has been experimental data expressed by the following formula:

$$v_n(t + \tau) \leq v_n(t) + 2.5a_n\tau \times (1 - \frac{v_n(t)}{V_n}) \times \sqrt{(0.025 + \frac{v_n(t)}{V_n})}, \qquad (7.4)$$

where a_n is the maximum vehicle acceleration [m·s^{-2}], τ is the reaction time of the driver [s], V_n is the target vehicle speed corresponding free traffic without restrictions [m·s^{-1}], $v_n(t)$ is the vehicle speed at time t[m·s^{-1}] [11].

IDM model – The biggest advantage of the above-mentioned model is the number of input parameters. These source parameters are intuitive character. From this perspective, it saves a great deal of time because they do not have to work on long analysis to obtain the input data. The basic equation of this model is expressed dependence acceleration:

$$a_i(s_i, v_i, \Delta v_i) = a_{i0}[1 - (\frac{v}{v_{i0}})^\delta - (\frac{s_{opt}(v_i, \Delta v_i)}{s_i})^2], \qquad (7.5)$$

where a_i is the acceleration of the vehicle [m·s^{-2}], a_{i0} is the comfortable acceleration [m·s^{-2}], v_{i0} is the target vehicle speed [m·s^{-1}], v is the vehicle speed [m·s^{-1}], s_{opt} is the optimum distance vehicles [m], s_i is the immediate distance from the preceding vehicle [m], δ is the acceleration factor of realism of the reference vehicle [11].

Usefulness of the models has in every sector its pros and cons. In the community of mathematicians and physicists, dealing with this issue it has become the most recognized IDM model also termed as "Intelligent Driver Model". Gipps and Wiedemann models and are most prevalent in simulation software [7, 8].

7.5. The Use of Dynamic Modeling of the Impacts of Road Transport

Macroscopic and microscopic models have their uses for modeling in road transport. The output element of these models is the evaluation of vehicle movement and their interaction in the road. We can say that it is modern approaches. On this basis, we developed a group of authors on the subject of dynamic system modeling in road transport. This system is based on a deterministic approach. The basis of this system is the analysis section where it disposes critical element and detour routes based on the macroscopic model. The brain of the system is the algorithm that evaluates the performance parameters of the sections, the degree of intensity transmittance pass routes in the transport and effects of operating parameters on alternate roads.

In the Fig. 7.4 below we can see the progress of the algorithm for dynamic system modeling the impact of road transport.

The whole process takes five steps, which are further divided into three sectors. The first sector is the identification that takes place in steps one and three. Another sector is the analysis that takes place in the second and the fourth step. The last sector is the evaluation that is dealt with in the last step, the fifth step.

After an evaluation, this process does not end due to a proper control. After the evaluation, there is the fourth step, so called Controls, which is shown in Fig. 7.4. Here we go back to step two and perform this analysis along with step four. Last of all, we evaluate the process.

Now we will get acquainted with the individual steps and take a closer look at each of them:

Step 1. Identifying critical element in road infrastructure - In case of rejection of a critical element of road transport is the first step no doubt his identification. The identification is based on a directive of the European Union, which leads to their identification and subsequent designation by sectoral and cross-cutting criteria. To be able to fulfil the cross- cutting criteria, you need to model the results of the impact that these values be able to compare the values of crosscutting. This hypothesis is the initial step due to confirm or refute criticism.

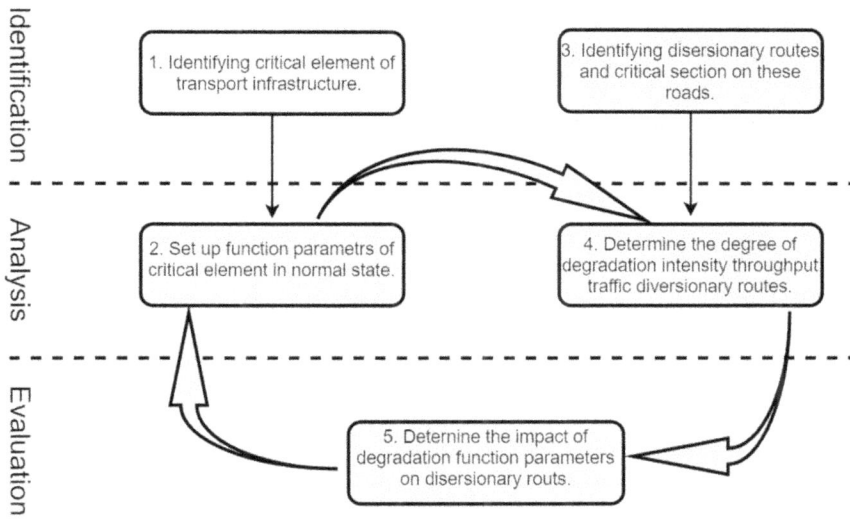

Fig. 7.4. Algorithm of Dynamic Modeling of the Impacts of Road Transport [1].

Subsequently, these operational parameters need to be determined for the whole route on which this critical element is located. For better demonstration of the system: this route - located between Points A and B, will be marked as Route Z (see Fig. 7.5).

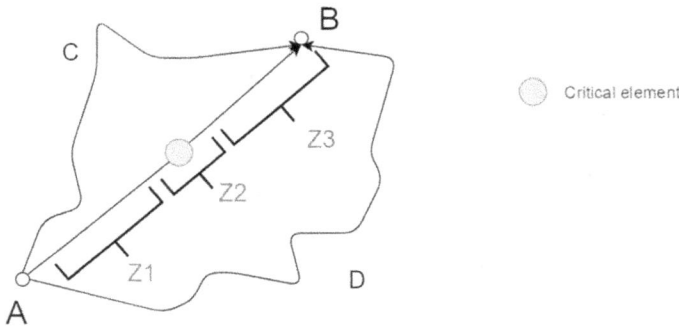

Fig. 7.5. Graphic depiction of the route with identified critical element.

Step 2. Determination of critical element functional parameters in a normal state - After identifying follows assessment of functional parameters critical element in a normal state. This means that a certain intensity and traffic throughput as the latest phase in this step set these parameters along the entire route, where this is a critical element.

Step 3. Finding detour routes and critical sections of these routes - At a time when there is loss of a critical element we need to be prepared for alternative roundabout routes. Certainly, we should not forget the possible risk sections also on alternative routes so that we are ready for the emergence of other events.

Step 4. Determine the degree of intensity throughput traffic on the detour routs. Determination of functional parameters should be done not only on the main road but also on alternative routes. We can see this situation in Fig. 7.6. Functional parameters means to determine the intensity of transmission traffic across alternative route.

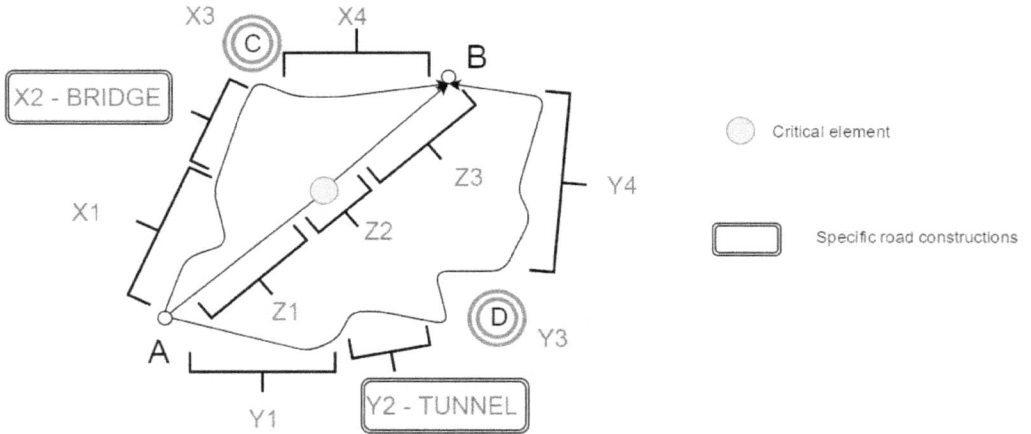

Fig. 7.6. Graphic depiction of detour roads divided into given segments.

Route X consists of several sections, with different permeability intensities (X_1, X_2... X_n), which are the key to the calculation of Route X's intensity (q_x) and are the arithmetical average of the permeability intensities of the given sections (Formula 7.6):

$$q_X = \frac{\sum_{i=1}^{n} q_{X_i}}{n}. \tag{7.6}$$

Route Y consists of several sections, with different permeability intensities (Y_1, Y_2... Y_n), which are the key to the calculation of Route Y's traffic intensity (q_y), which is the arithmetical average of the permeability intensities of the given sections (Formula 7.7):

$$q_Y = \frac{\sum_{i=1}^{n} q_{Y_i}}{n}. \tag{7.7}$$

Now, it is possible to continue with the determination of the degradation level (i.e. the percentage decline) of traffic permeability intensity on the detour roads - in comparison with the original Route Z. The degradation level of traffic permeability intensity (d) for Route X is calculated according to Formula (7.8) and for Route Y according to Formula (7.9):

$$d_X = \frac{q_Z - q_X}{q_Z}, \tag{7.8}$$

$$d_Y = \frac{q_Z - q_Y}{q_Z}. \tag{7.9}$$

Step 5. Determination of the impact on functional parameters on alternate roads. Finally, to represent a number of casualties and economic loss. The final statement gives additional operating costs and we lost gross domestic product (GDP).

The length of the routes can easily be obtained from maps; however, the time needs to be defined in relation to the density and permeability intensity of the given routes. Therefore, it is sufficient to use for the calculation of the permeability intensity and - by its alteration, to obtain information about the time needed to drive through the given routes:

$$q = v \cdot k = \frac{s \cdot k}{t} \rightarrow t = \frac{s \cdot k}{q}, \qquad (7.10)$$

where q is the traffic permeability intensity [pcs·h^{-1}], v is the vehicle speed in flow [km·h^{-1}], k is the vehicle density in flow [pcs·km^{-1}], s is the length of the evaluated route [km], t is the time needed to drive through the given route [h].

Afterwards, the economic loss due to operational costs caused by taking detour Route X as a consequence of the dysfunctionality of Route Z can be calculated (Formula 7.11):

$$E_{OC(X)} = (s_X - s_Z) \cdot q_Z \cdot e_{OC}, \qquad (7.11)$$

where $E_{OC(X)}$ is the economic loss due to operational costs caused by taking detour Route X as a consequence of the disfunctionality of Route Z [\$·h^{-1}], s_X is the detour Route X length [km], s_Z is the original Route Z length [km], q_Z is the traffic permeability intensity on Route Z [pcs·h^{-1}], e_{OC} is the operational costs [\$·km^{-1}].

The economic loss to GDP caused by taking detour Route X as a consequence of the dysfunctionality of Route Z can be calculated using Formula (7.12):

$$E_{GDP(X)} = (t_X - t_Z) \cdot p \cdot e_{GDP}, \qquad (7.12)$$

where $E_{GDP(X)}$ is the economic loss to GDP caused by taking detour Route X as a consequence of the dysfunctionality of Route Z [\$·h^{-1}], t_X is the time needed to drive through Route X [h], t_Z is the time needed to drive through the original Route Z [h], p is the number of persons being transported [pcs·h^{-1}], e_{GDP} is the GDP per capita [\$·pcs^{-1}].

The resulting numbers of affected persons (p) and economic losses ($E_{OC} + E_{GDP}$) must be averaged at a minimum for at least 24 hours afterwards so that they can be compared with the numbers determined by the cross-cutting criteria.

7.6. Conclusion

Road transport is a very specific sub-sector of critical infrastructure. It belongs among its important parts and its disruption or failure can have a serious impact on other sectors dependent on transport. As a result, there is an increase in the negative impact on society, not only in terms of safety.

In transport, that is as an element of critical infrastructure, is decisive to minimize risks but also to respond to emergencies.

One of the suitable solutions can be a dynamic modelling based on the impact of road transport as described in the article. The system was created using existing model simulation in transport, especially in cross-sectional criteria.

The benefit of this article is the objective selection of preventive measures to minimize the impact of threats to selected critical infrastructure sub-sectors. Especially in terms of safety, security and crisis preparedness.

Based on dynamic modelling by assessing the critical infrastructure of this model, the procedure will be progressively modified so that a list of critical road and rail infrastructure can be identified. The model also allows the modelling of the individual parts' defects and their impact on the surroundings.

The model further suggests safety measures of technical supervision. The outputs will serve as a basis for identifying elements of critical road transport infrastructure and for developing of modelling methodology.

The dynamic impact modeling is not only suitable for the analysis of impacts spread over sectors combined with a passage of time; but also - for a better understanding the functioning of road transport systems.

Acknowledgements

This project is realized as the research with a doctoral student, and it is the primary input for next research, which we will develop in next term. It was realized with the support of the university. This work was supported by Internal Grant Agency of Tomas Bata University under the project No. IGA/FAI/2017/016.

References

[1]. M. Hromada, L. Lukas, Methodology for selected critical infrastructure elements and elements system resilience evaluation, in *Proceedings of the IEEE Symposium on Technologies for Homeland Security (HST'15)*, Waltham, USA, 2015.
[2]. L. Lukáš, M. Hromada, Simulation and modelling in critical infrastructure protection, *International Journal of Mathematics and Computers in Simulation*, 2011, pp. 386-394.
[3]. S. Atta, Markov Chains and Dynamical Systems: The Open System Point of View, http://math.univ-lyon1.fr/~attal/Mesarticles/cosa_attal.pdf
[4]. D. Ni, Equilibrium Traffic Flow Models, in book Traffic Flow Theory, Characteristics, Experimental Methods, and Numerical Techniques, *Elsevier*, 2016, pp. 51-71.
[5]. R. Bogo, L. Gramani, E. Kaviski, Modelling the flow of vehicles by the macroscopic theory, *Revista Brasileira de Ensino de Física*, Vol. 37, Issue 1, 2015, pp. 1-8.
[6]. G. Oliva, S. Panzieri, R. Setola, Agent-based input– output interdependency model, *International Journal of Critical Infrastructure Protection*, Vol. 3, 2010, pp. 79-82.

[7]. G. Upreti, P. V. Rao, R. S. Sawhney, I. Atuahene, R. Dhingra, Increasing transport efficiency using simulation modeling in a dynamic modeling approach, *Journal of Cleaner Production*, Vol. 85, 2014, pp. 433-441.

[8]. W. Young, A. Sobhani, M. G. Lenné, M. Sarvi, Simulation of safety: A review of the state of the art in road safety simulation modelling, *Accident Analysis & Prevention*, Vol. 66, 2014, pp. 89-103.

[9]. T. Apeltauer, Microscopic Traffic Models in Work Zones*, University of Technology*, Brno, 2013 (in Czech).

[10]. T. Apeltauer, P. Holcner, J. Macur, Verification of Some Models of Traffic Flow, Článek v Silnice a Železnice, *KONSTRUKCE Media*, Ostrava, 2009 (in Czech).

[11]. T. Apeltauer, P. Holcner, J. Macur, Validation of microscopic traffic models based on GPS precise measurement of the vehicle dynamics, *Promet - Traffic&Transportation*, Vol. 25, 2013, pp. 157-167.

[12]. T. Apeltauer, Romodis – Rozvoj Moderních Dopravních Inteligentních Systémů (modul 8 – Využití Simulačních Modelů v Dopravě), http://www.romodis.cz/files/169118867.pdf (in Czech).

[13]. Nation Traffic Census (in Czech), http://scitani2010.rsd.cz/pages/map/default.aspx

[14]. Act N. 432/2010 Coll. Government Regulation on Criteria for Determining Critical Infrastructure Element (in Czech), https://www.zakonyprolidi.cz/cs/2010-432

[15]. R. Bogo, L. Gramani, E. Kaviski, Modelling the flow of vehicles by the macroscopic theory, *Revista Brasileira de Ensino de Física*, Vol. 37, pp. 1-8.

[16]. D. Ni, Traffic Flow Theory: Characteristics, Experimental Methods and Numerical Techniques, *Elsevier*, Oxford, 2015.

[17]. M. Ouyang, Review on modeling and simulation of interdependent critical infrastructure systems, *Reliability Engineering & System Safety*, Vol. 121, pp. 43-60.

[18]. J. Mrazek, L. Duricova, M. Hromada, The mathematical modeling of road transport in context of critical infrastructure protection, in *Proceedings of the 10th IARIA International Conference on Emerging Security Information, System and Technologies (SECURWARE'16)*, July 24-28, 2016, pp. 95-99.

Chapter 8
The Assessment of the Soft Targets

Lucia Duricova and Martin Hromada

8.1. Introduction

The soft targets and crowded places are closely related to a risk of the attacks. Soft targets are specified as the objects with a large number of the visitors in a closed place at the same time; for example, the buildings, which do not have integrated the special security and safety measures into the management of the processes. The special secure and safe measures are the measures, which react to the process of an incident. These soft targets can be divided into two groups. The first group includes the buildings and the second group includes the events; both groups can be called as a soft target. For example, cinemas, schools, universities, shopping center, football match and other buildings could be identified as such [1, 10].

These objects do not currently integrate the special secure measures into the management of the processes. It is linked to a lack of special law's requirements. Currently, the law manages Occupational Health and Safety and Fire Protection in Czech and Slovak Republic. However, the system security is not required by statute. In other groups of the objects (critical infrastructure) the state (The Czech Republic, The Slovak Republic, and others) directs the security in these kinds of the objects by the law's requirements. In the commercial objects (the production companies), the situation is a little bit different. The commercial objects are managed by the owners and by the customer requirements. For example, this kind of companies needs to integrate ISO management into the processes. ISO management can be the tool for achieving the higher level of the security in the companies [3].

This analytical software was proposed to an evaluation of the object, which could help to define the numerical parameters as a risk, a probability and so on. Currently, these objects are evaluated by the analytical methods or some analytical tools, which are focused on specified problems. For example the environment issues, the simulation of the leakage of dangerous substances, and others. Each of them has some advantages and some

Lucia Duricova
Tomas Bata University in Zlin, Faculty of applied informatics
Nad Stranemi 4511, Zlín, 76001 Czech Republic

disadvantages. This proposed analytical tool is more general as the tools, which work separately. However, it caused that the proposed solution can be more applicable. The parameters of this tool represent a situation in an object by a numerical value. The evaluation is based on the multi-criteria solving. The next analytical step is a comparison, which compares the object's coefficient with the incident's coefficient. The analytical solution is a part of the proposed software solution. This paper defines the primary basis of the software tool.

According to ISO/IEC 31010 the relevant terms and definitions are defined:

- RISK is an effect of uncertainty on objectives.

- RISK MANAGEMENT- is coordinated activities to direct and control an organization with regard to risk.

- RISK ASSESSMENT is an overall process of risk identification, risk analysis, and risk evaluation. Risk identification is the process of finding, recognizing and describing risks. Risk identification involves identifying risk sources, events, their causes and their potential consequences. Risk identification can involve expert opinions and stakeholder's needs.

- RISK SOURCE is an element which alone or in combination has the intrinsic potential to give rise to risk.

- EVENT is occurrence or change of a particular set of circumstances.

- LIKELIHOOD is a chance of something happening.

- RISK CRITERIA – terms of reference against which the significance of a risk is evaluated.

- RISK EVALUATION is the process of comparing the results of risk analysis with risk criteria to determine whether the risk and/or its magnitude is acceptable or tolerable.

- RISK TREATMENT is the process to modify risk [8].

Also, we can choose more definitions of risk and other terms. We chose the definitions according to ISO/IEC 31010 because these terms are applied in a whole range of the commercial objects. However, if we define risk in the safety processes, we should use the definition according to British Standard OHSAS 18001:2007 (Occupational Health and Safety). International Standardization Organization is respected worldwide.

The current situation reflects the requirements of security and safety in these kinds of objects. The commercial organizations apply ISO standardization into the management processes. This standardization they apply because the certification represents the quality of their products and services. We can conclude, that these standards are applicable to practice and therefore we need to apply this approach to the proposal of the solving.

Furthermore, the proposed solution should be applicable and useful in the commercial organizations. The secure and safe requirements are divided into a lot of groups. The main is a security of the building, a safety of visitors and employees, security of information technologies and the security of the processes. In many states in the world, law's requirements of these objects (soft targets) are divided into two groups. Fire Safety and Occupational Health and Safety requirements. However, according to the mentioned standards the organizations ensure the safe situation in other categories of security. For example ISO/ IEC 27001- Information security management systems.

We propose the software, which main aim is the analysis of the object, the analysis of an incident and finally to propose the efficient solutions. These solutions can support the decision making processes in case of the incidents. On the other hand, this software can be used to the definition of preventive actions in the building.

Along with standards, the analytical methods was used in the software. The proposed analytical part analyses the object according to more analytical methods, and we chose them according to a usability in practical using. For example: Failure Mode and Effect Analysis (FMEA), Failure tree analysis (FTA) and analysis according to risk criteria.

8.2. Methodology of the Solving

We can say, that whole methodology is supported by a whole range of the standards, as we wrote in the first part of this chapter. Besides that, we used the principal of PDCA model, which is used in the whole range of standards too. We used ISO 3100, ISO/ IEC 27001, ISO 18001, ISO 14001 and ISO 9001 too. These standards are used in a lot of companies around the world and according to this applicability, we can say, that this principal is functional.

8.2.1. Deming's Circle Methodology

Deming circle is used in a lot of variety of the standards. The using is certificated according to a lot of ISO standard too. In our proposal, we define the PDCA model (Fig. 8.1) as one of the main principal, which must be implemented into the management and assessment of the security in the object. The process of the evaluation should be interactive and based on continuity of the activities.

The security and safety are two kinds of very familiar themes. Model PDCA can be applicate into the whole range of companies, facilities, systems and objects. We can say, that this process can ensure the continual improvement. However, if we need to define the effective solutions, we need to monitor current states in the objects.

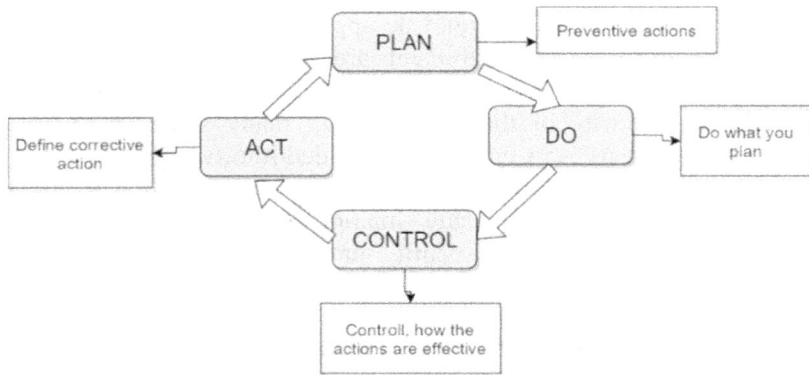

Fig. 8.1. The proposed application of PDCA model [8].

8.2.2. Processes Management

We proposed the management of this solution according to the processes management. The main reason, why we agree and propose this methodology is, that processes management is currently used in a whole range of organizations around the world. The main reason is that these processes can be simply described and the main advantage is that these processes can be simply managed. The principle of management is described in Fig. 8.2.

We can say, that every process is described according to the inputs. These inputs are known as 6M (man, material, machine, mother nature, measurement, methods). The process can be defined as sequents of the events, which has defined the exact order. The process transforms inputs to the outputs. Outputs are characterized according to the requirements to the outputs. We can set the requirements, and we can measure the outputs. According to the results of the measurement, we can examine the state of the object or system, which is evaluated.

According to the process management, we can examine the results. According to the results, we can define the corrective action, which can be implemented into the inputs. This whole process of function steps can improve the entire process. This principle was used to the proposed solution. According to the inputs, we can identify the points, where we need integrate corrective actions. The inputs define to us the points, where we need to integrate the corrective actions. According to this principle we know, that we need to know the whole process of function each of the buildings and objects. And the last, but not least, we are aware, that we need to improve whole process no only one unit.

8.2.3. Crisis Escalation

The process of crisis escalation is based on two types of the principles. The first is a preventive part. This part can be called as risk management. The second part is called as managing crises. The first part examines the features of the system before incidents. This

204

examination consists of the study the conditions according to the indicators and to measure them. According to the measure of the indicators, we can say, how is process establish.

In Fig. 8.3, we can see the whole process of the crises escalation.

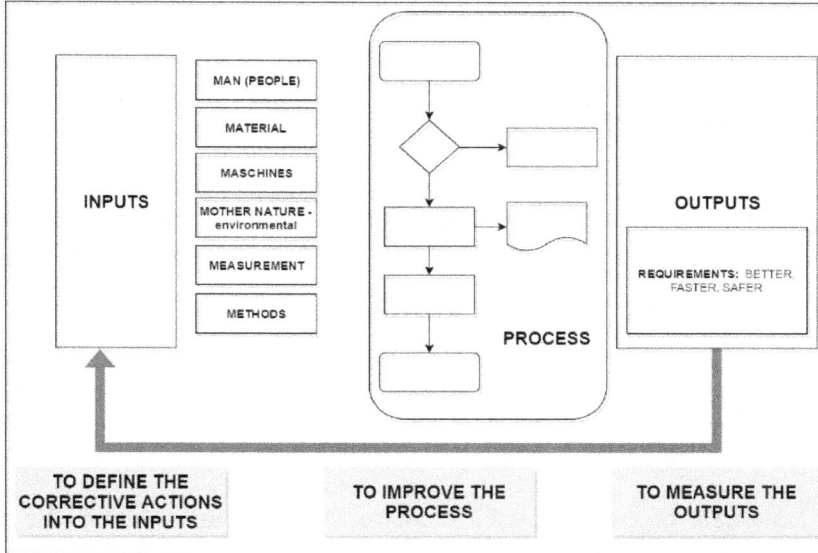

Fig. 8.2. Process approach.

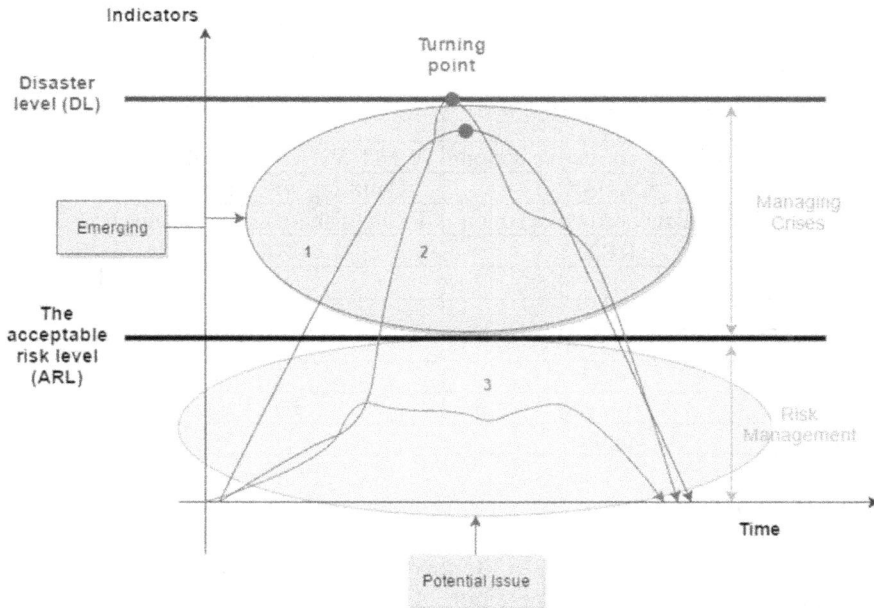

Fig. 8.3. The principle of crisis escalation.

In the first phase, we need to define the indicators, which we will monitor. According to the results, we need to define the acceptable risk level. This acceptable risk level is crucial for the operator in the object and for integrated rescue system too. According to this value, we can say, if the preventive action is recommended or its application is required. In Fig. 8.2, we can see how we can work with the object's features, and how much influence can we have on the buildings. Failure Mode and Effect Analysis is closely related to this part of the methodical principle.

8.2.4. Failure Mode and Effect Analysis

FMEA is used to estimate severity, occurrence, and detection of a failure mode based on expert's evaluation as well as to establish the priorities of accident cause removal methods regarding the risk priority number (RPN), which is the product of these three factors.

$$RPN = Severity * Occurence * Detection. \qquad (8.1)$$

The main advantage, why this analysis is well for this proposed solution is that this analysis needs to define an acceptable level of RPN. That means, we need to set the level of RPN, which can tell us, if we need to determine the corrective actions or not. In the next table, we can see the ways, which can be used.

In Table 8.1, we can see the proposed FMEA worksheet for our solution. We define the acceptable level of RPN to the value of 150.

Table 8.1. The proposal of the FMEA worksheet.

New number of problem	OK
RPN	126
D	3
S	7
O	6
Permanent corrective action	WHO WHAT DEADLINE
CAUSE	Handling with open fire
Immediate corrective action	Extinguishers in each room
RPN	216
D	4
S	9
O	6
Process	2
Problem	Fire
Number	1.1

Many organization and companies define this value as crucial according to the examination whole range of values. In the next part of the research, we can change this value according to the results of the research. A crucial fact is that we need to identify the definition of the each parameter (Occurrence, Severity, and Detection). According to this definition, we can calculate the each of these parameters with a similar process. The immediate and permanent corrective actions are actions which are defined as preventive solutions. However, the main difference between them is, that immediate corrective action is effective only limited time. On the other hand, the permanent corrective action is applied to the point of the causes. The permanent corrective action can increase the probability of occurrence of the undesired events in the next period (future) [8].

8.3. The Proposed Static Assessment

The proposed software is divided into three parts. The first part analyses the current conditions and the state of the object. After that, the system can derive the parameters, which define the secure and safe state in the object. These objects are categorized into the groups, which will be implemented into the general analysis.

The categorization has an impact on an evaluation of the object in a static part of the software. According to the category and the localization, an operator can determine the general secure coefficient of the object. The conditions and features of the object can influence the probability of the risk state. These conditions are found out in the first static analysis. We can identify these object's conditions or features by the analysis; for example: by the static part of the analysis. Firstly, the object must be evaluated, and then the object can be included in the main assessment.

The proposed assessment is based on the evaluated criteria, which are assessed in the range between 0 and 10. These criteria examine the whole range of capabilities of the object. Each of these objects is characterized according to the inputs parameters, which were described in previous part [3].

In Fig. 8.4 can be seen two parts of the assessment. The first is general and the second is a specific part of the assessment. This general part is based on the quick evaluation process, which can examine the general information. According to the general information, we can say the general statement. The general statement defines the general coefficient of security. This type of the assessment is not so specific, and an operator can not say, where we need integrate the preventive actions. On the other hand, an operator can immediately tell, in which category of threatening is his object.

As you can see in Fig. 8.5, the object is defined by the primary input's parameters in the first step – general assessment. According to this parameters we can assign the basic parameters to the objects. These parameters are general and described in the Table 8.2.

Table 8.2. The definition of the object's criteria [3].

OBJECT	Coefficient of the locality
	Coefficient of the category
	Capacity

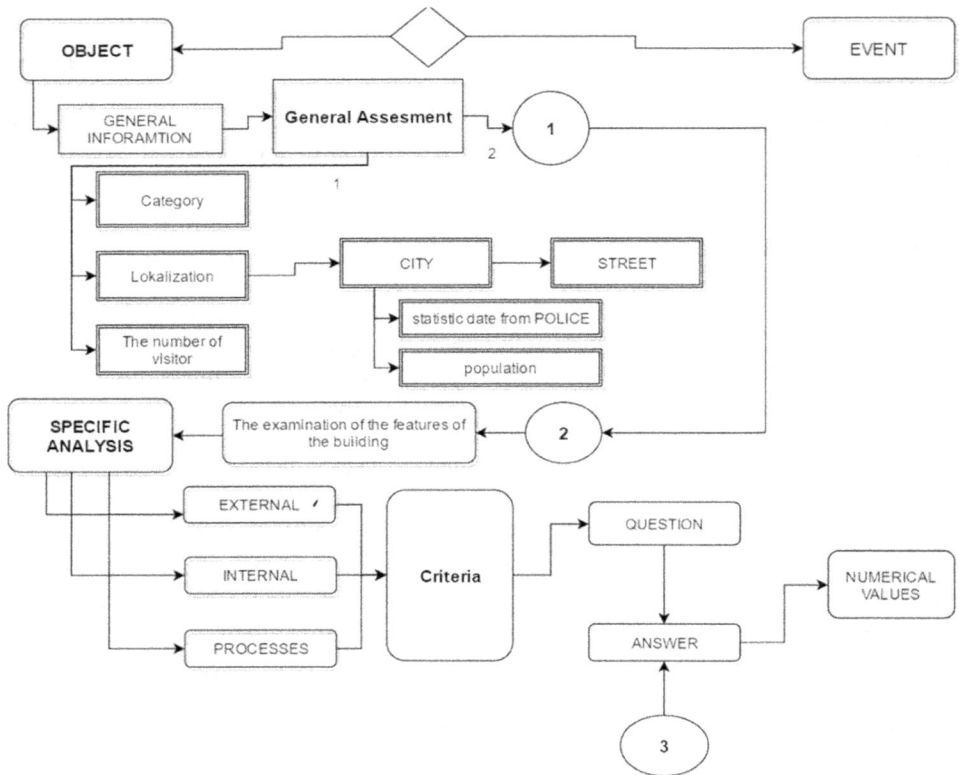

Fig. 8.4. The process of the static assessment.

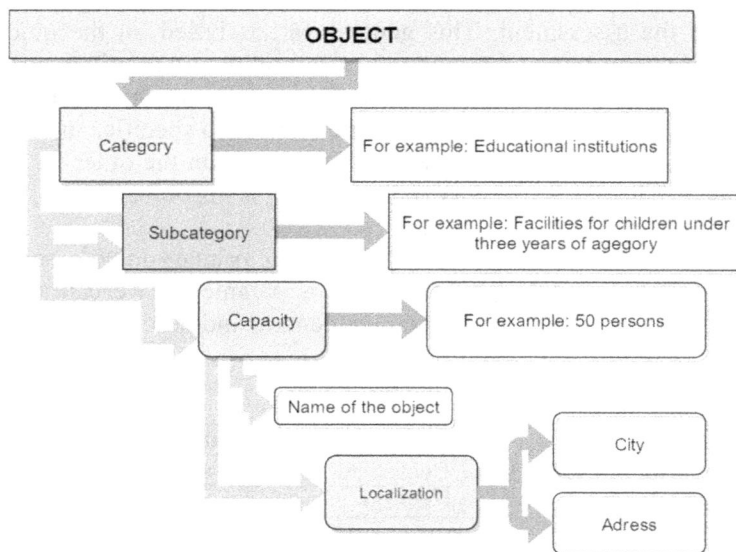

Fig. 8.5. The general information about object.

The criteria of the capacity can have a significant impact on the assessment of the object. In the plan of the proposed solution, we want to define the dependency between criminality and capacity of the object. We suggested that this dependency can be represented by the coefficient of the capacity (CC). This coefficient can increase the security situation in dependence on the capacity.

$$C_L = \frac{S_{KD}}{P_O} \times KKD, \tag{8.2}$$

where C_L is the coefficient of the locality, S_{KD} is the statistic part of criminality according to the kind, and KKD is the coefficient criminality according to kind.

$$K_K = \frac{\sum_{i=1}^{n} CAL_i}{n}, \tag{8.3}$$

where KK is the coefficient of the category, CAL_i is the coefficient of the whole security of i-element.

According to these input data, we can identify the primary general coefficient of the object or building. The general coefficient can define the general state of the security in the building. According to this general state, we can not say, what we need to do or which corrective action we need to integrate into the process. However, according to this general coefficient, we can identify the primary risks, which can threaten the building or object.

The second kind of the analysis is more specifically. The specific analysis is based on the evaluation according to the criteria. These criteria can be divided into three groups according to the aim of the properties. The crucial factor can define according to the risk. That means each of this security coefficients can be changed according to the analyzed risk.

Table 8.3. Three types of the analysis [3].

Exterior	Interior	Process
Exterior analysis	Interior analysis	Analysis of the processes
Examine the surrounding of the building	Examine the aim and function of the interior settings	Examine the processes of the function – aimed at security

In the Table 8.3 we can see the three kinds of the analysis, which are based on the static assessment. That means the system calculate only static coefficients, no dynamic in the time.

8.4. The Proposed Dynamic Assessment

The dynamic assessment is based on the static analytical tool. This part of the software tool examines the process of the improvement of the object's security according to the outside effects. We can say, that outside effects can have the significant impact on the security situation in the object. This effect can be caused by the higher number of the visitors per day, with the higher population in the close area and with the higher unsecured and dangerous situation in the narrow area.

The Table 8.4 defines the parameters, which have to be defined in the dynamic part of the assessment. In our solution, the proposed tool don't examine the features of the building in the continuous process. The dynamic solution is based on the linkages between outside and inside. The features of the soft targets can not be monitored without delay, or not all features can not be monitored without delay. We can monitor the situation in the object without delay, but not features. One of the features is, how the object controls the situation in the object. According to the methodical principle, we can define the interval for the repeated assessment.

Table 8.4. The dynamic part of the tool.

Calendar FROM - TO	Type of the events	Area of the localization	Coefficient of the impact
Date of the events.	Which category of event is it?	Identification the place, where the vent is and the radius of the event.	How is value of the impact to the other soft targets?

According to Table 8.4, we need to define the category of the events. These events can be categorized according to the features (for who are defined, how is a security risk of the event, which types of incidents can occurrences). For example, we can say, that football match can be defined as a sporting event. On the other hand, the Olympic game has a different type of threats as the football match.

This dynamic part can be used to Police and integrated rescue system too. Police can plan the events, and can simply see how can the event threats other groups of buildings or soft targets.

8.5. The Proposal of the Preventive Actions

In the next part of the paper, we identify where is the main linkage between the assessment of the features of the building and the proposal of the corrective actions. The preventive action can be proposed as the sequents of actions, which are proposed for the maximum effectivity. This proposal utilizes FMEA and in the next part can be used Fault Tree Analysis (FTA) too. This two parts of analyses are organized in preventive part of the proposal.

In Fig. 8.6, can be seen the whole process of the proposed tool. In this figure, we can see each of the parts of the tool. The first part is the assessment, which is based on the evaluation of the features of the buildings. The second part is the dynamic part, which can use the time coefficients and time linkages according to the calendar's data. In the last part, the tool can support the decision-making process and define the corrective action into the buildings. Each of this date needs to use the localization according to maps or geographical background.

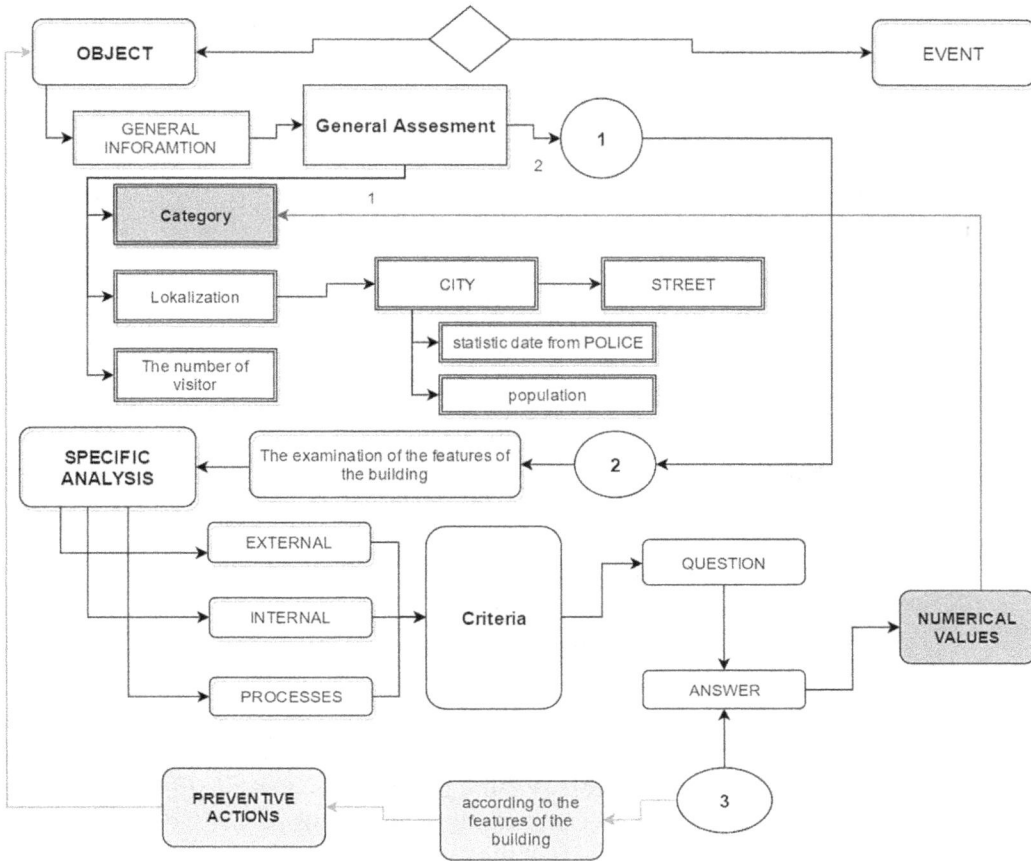

Fig. 8.6. The whole process of the evaluation the proposed tool.

In Fig. 8.7, can be seen, the process of the application of the preventive actions to the buildings. We can see the main linkage between the answer to the question (criteria) and the proposed solution. According to the types of the current security solution, we can define the proposed corrective actions, and how an operator can achieve the higher secure situation in the object.

8.6. Conclusions

Effective protection is not only about definition the corrective action, but also we need to define the linkages between these measures and examine the influences between them. The proposed solution is aimed to the evaluation the features of the buildings. According to the specification of the criteria, the software tool can support preventive solution in the building. We are aware that this system must be examined with case studies and according to the results we can say if this proposal is well proposed. However, criteria are defined for the operator and simultaneously very specific for the system. The system needs to know the expert's statements and according to them can evaluate the current state of the object.

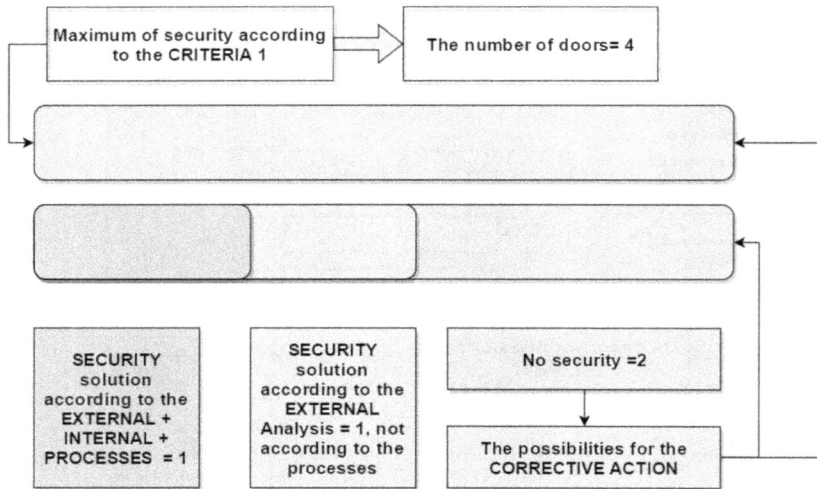

Fig. 8.7. The relationship between criteria and preventive action.

The next part of the research will be focused on the definition the security coefficient for the locality in the Czech Republic. According to the coefficients of locality, we can start with the part of the confirmation the proposed tool in the practical using. In this phase, we confirmed this tool only in one city, because we didn't have this part of the proposal. We can say, if each of the examined objects is in the same city than the coefficient of locality is negligible.

We apply into this tool the currently used approaches and methods, what can have the significant impact to the effectivity of the proposed preventive action. We used process management, FMEA analysis, and crisis escalation and so on. These methods are now used in the whole range of organization, and this fact can influence the practical uses in commercial business. In the end, this tool is proposed for these types of the organizations.

Acknowledgements

This project is realized as the research with a doctoral student, and it is the primary input for next research, which we will develop in next term. It was realized with the support of the university. This work was supported by Internal Grant Agency of Tomas Bata University under the project No. IGA/FAI/2017/017 and by the research project VI20172019073 "Identification and methods of protection of Czech soft targets against violent acts with elaboration of a warning system", supported by the Ministry of the Interior of the Czech Republic in the years 2017-2019 and by the research project VI20172019054 'An analytical software module for the real-time resilience evaluation from point of the converged security', supported by the Ministry of the Interior of the Czech Republic in the years 2017-2019.

References

[1]. S. Crockett, et al., Protecting Soft Targets: A Case Study and Evaluation of High School Security, https://courses.cs.washington.edu/courses/csep590/05au/whitepaper_turnin/soft-targets-white-paper.pdf

[2]. C. K. Chae, J. W. Ko., FTA-FMEA-Based Validity Verification Techniques for Safety Standards, *Korean Journal of Chemical Engineering*, 2017, pp. 1-9.

[3]. L. Duricova, M. Hromada, J. Mrazek, Security and safety requirements for soft targets in Czech Republic, in *Proceedings of the 10th IARIA International Conference on Emerging Security Information, System and Technologies (SECURWARE'16)*, July 24-28, 2016, pp. 271-275.

[4]. L. Duricova, M. Hromada, J. Mrazek, Security and safety processes in Czech Republic Universities, in *Proceedings of the 10th IARIA International Conference on Emerging Security Information, System and Technologies (SECURWARE'16)*, July 24-28, 2016, pp. 105-110.

[5]. L. Duricova, J. Mrazek, M. Hromada, The proposal of security and safety management system with fuzzy logic support, in *Proceedings of the 6th International Defense and Homeland Security Simulation Workshop (DHSS'16)*, 2016, pp. 10-14.

[6]. A. P. H. de Gusmão, L. Henriques, L. C. Silva, M. M. Silva, T. Poleto, A. P. C. S. Costa, Information security risk analysis model using fuzzy decision theory, *International Journal of Information Management,* Vol. 36, Issue 1, 2016, pp. 25-34.

[7]. F. Alexander, Safety and security target levels: opportunities and challenges for risk management and risk communication, *International Journal of Disaster Risk Reduction,* Vol. 2, 2012, pp. 67-76.

[8]. Risk Management – Principles and Guidelines, *ISO/IEC 31000:2009.*

[9]. C. Liu, C.-K. Tan, Y.-S. Fang, T.-S. Lok, The security risk assessment methodology, *Procedia Engineering,* Vol. 43, 2012, pp. 600-609.

[10]. C. Raspotnig, A. Opdahl, Comparing risk identification techniques for safety and security requirements, *Journal of Systems and Software*, Vol. 86, Issue 4, 2013, pp. 1124-1151.

[11]. T. S. Parsana, M. T. Patel, A case study: a process FMEA tool to enhance quality and efficiency of manufacturing industry, *Bonfring International Journal of Industrstrial Engineering and Management Science,* Vol. 4, 2014, pp. 145-152.

Communications

Chapter 9

Performance Analysis of Quadrature Amplitude Modulation Schemes in Amplify-and-Forward Relay Networks over Rayleigh Fading Channels

Nagendra Kumar, Vimal Bhatia and Dharmendra Dixit

9.1. Introduction

In recent years, there has been a lot of interest in cooperative communication systems since it provides high data rate, enhanced transmission reliability, enlarged system capacity, extended network coverage and improved quality of service in wireless communication systems. The basic idea about cooperative communication is discussed in details in [1-3]. Relaying schemes has been recognized in the standardization process of next-generation mobile broadband communication systems such as third generation partnership project (3GPP) long-term evolution (LTE)-Advanced, IEEE 802.16j and IEEE 802.16m [4]. The main objective behind use of the relaying scheme is to assign a dedicated relay between source and destination to provide reliable communication. There are two most common relaying schemes such as amplify-and-forward (AF) and decode-and-forward (DF) which are broadly used at relay node. In the DF relaying scheme, source's transmitted signal is first decoded and then re-encoded at the relays. After re-encoding, the received signal is sent towards the destination. DF scheme is generally used to provide better performance; however, complexity at the relay node is increased considerably. Details of the DF scheme are given in [5-8]. In AF relaying, the source's transmitted signal is amplified at the relay and then this amplified signal is retransmitted to the destination. The AF relaying scheme is preferred than DF because of low complexity and simplicity in implementation, and it also provides longer battery life than other schemes. Detailed study about AF is given in [9-11].

The AF relaying scheme can further be classified on the availability of channel state information (CSI) as: - 1) fixed-gain AF relaying and 2) variable-gain AF relaying. Fixed-

Nagendra Kumar
Discipline of Electrical Engineering, Indian Institute of Technology Indore, India

gain AF relaying requires partial CSI and is more attractive from the practical implementation perspective due to its lower complexity. However, variable-gain AF relaying requires full CSI which amplifies the received signal using the instantaneous CSI of the previous hop. Variable-gain is a higher complexity scheme, but gives the performance advantages over fixed-gain AF relaying. The details about fixed and variable-gain AF schemes are given in [12]. Since variable-gain provides better performance, the scope of the work is limited to variable-gain AF relaying. At the receiver multiple copies of the signal are combined using diversity combining scheme to achieve better signal quality, reduced transmission power, better coverage and higher capacity. Various diversity combining schemes are available in the literature such as maximal ratio combining (MRC), selection combining (SC) etc., which are used to improve reliability and strength of the links at the destination. Conventional MRC is optimum combining scheme which combines all the signals coming from the different links at the destination and maximizes the end-to-end SNR [13, 14]. However, SC is very attractive which is considered as the least complex scheme among others and has ability to maintain the full diversity order without utilizing full amount of required resources [11, 15, 16].

On the other hand, quadrature amplitude modulation (QAM) is considered as promising modulation scheme to be used in high-rate data transmission over wireless communication links. The family of QAM schemes such as square QAM (SQAM), rectangular QAM (RQAM) and cross QAM (XQAM) is preferred modulation schemes in digital communication systems due to their high efficiency in power and bandwidth. RQAM is also known as generic modulation scheme which includes SQAM, binary phase-shift keying (BPSK), orthogonal binary frequency-shift keying, quadrature phase-shift keying (QPSK) and multilevel amplitude-shift keying as special cases [17]. RQAM is preferred due to its practicality and bandwidth efficiency in wireless communication systems and is frequently used to implement rate-adaptive transmission strategies designated to cope with the needs of current and future multimedia wireless applications [18]. Further, XQAM is considered as an optimal QAM constellation for transmission of odd number of bits per symbol as it has low peak and average power than RQAM [19, 20]. The practical applications of XQAM with constellation from 5 bits to 15 bits have been used in asymmetric digital subscriber lines (ADSL) and very high speed digital subscriber lines (VDSL). The 32 and 128-XQAMs are adopted in digital video broadcasting-cable (DVB-C) [21]. Recently, XQAM has been found to be useful in blind equalization [22] and adaptive modulation schemes wherein the constellation size is adjusted depending on the channel quality [23, 24].

From literature review, many research works have been found on average symbol error rate (ASER) performance over several modulation schemes. In [2], a closed-form expression of MGF is derived to analyze the outage probability and ASER performance over Weibull fading channels for cooperative AF relaying network using BRS scheme. The BER performance of several modulation schemes such as M-PSK, M-QAM, etc., for multi-hop DF relaying system over log-normal fading channels has been analyzed in [6]. In [8], theoretical expressions for average BER, spectral efficiency and outage probability have been investigated in DF relaying system using M-ary QAM scheme. In [10], theoretical analysis of outage probability and symbol error rate (SER) have been derived using fixed-gain AF relaying with orthogonal space time block code over Nakagami-m fading channels. Closed-form expression of outage probability and bit error rate (BER) expression for BPSK have been derived using MRC scheme in [14]. The closed-form expressions for the exact SEP and BEP for XQAM over i.n.i.d. generalized η-μ channels with MRC diversity scheme have been derived in [21]. Theoretical expression of exact symbol error probability (SEP) for XQAM over AWGN channel and different fading channels are derived in [25]. In [26], the exact expression of SEP for arbitrary XQAM over Rayleigh flat fading channel has been derived in terms of elemental function. In [27], exact average SEP expression of XQAM signal for multi-branch MRC receiver is derived over independent but not necessarily identically distributed (i.n.i.d.) fading channels including Rayleigh, Nakagami-m, Nakagami-n (Rice) and Nakagami-q (Hoyt) channels. Analytical expressions for probability density function (PDF), cumulative density function (CDF), MGF, and ASER performance for SQAM with BRS scheme for AF cooperative communication network with N relays and a direct link over Rayleigh fading channels are presented in [28]. The work of [28] is extended in [29] to analyze ASER performance for RQAM with BRS scheme using MGF approach.

In this chapter, the lower-bound expressions of ASER are derived by using MGF-based approach in variable-gain AF dual-hop cooperative relaying network with N relays and a direct link over Rayleigh fading channels for M-ary RQAM and M-ary XQAM using MRC scheme. Further, the lower-bound expression of ASER for M-ary XQAM using BRS scheme is also derived. Although, there has been lot of research on performance analysis over Rayleigh fading channel, however, the important aspects covered in this work are so far not considered in the literature.

Rest of the chapter is structured as follows. Section 9.2 describes the system model. Derivation of ASER expressions for ASER is presented in Section 9.3. Numerical and simulation results are presented in Section 9.4. Section 9.5 concludes the overall work.

9.2. System Model

This section provides description of system model in which we consider a cooperative AF dual-hop relaying network with N relays, one source and one destination as shown in Fig. 9.1. The source node 'S' communicates with a destination node 'D' through a direct link represented as $S \rightarrow D$ and N indirect links represented as $S \rightarrow R_i \rightarrow D$ where $R_i (i = 1, 2, 3, ..., N)$ represent the relay nodes, over independent and identically distributed (i.i.d.) Rayleigh fading channels. All the nodes have one transmitting and one receiving antenna, and they communicate with each other in half-duplex mode. It is assumed that variable-gain AF relaying is used at relay node, thus, full CSI is known at the destination. The system model shown in Fig. 9.1 completes the communication in two time phases. In the first phase, source transmits signal to the destination and N relays. The signals received by the destination and the ith relay are represented, respectively, as

$$y_{SD} = h_{SD}\sqrt{E_s}x + n_{SD},$$ (9.1)

$$y_{SR_i} = h_{SR_i}\sqrt{E_s}x + n_{SR_i},$$ (9.2)

219

where h_{SD} and h_{SR_i} represent the channel gain coefficients of $S \to D$ and $S \to R_i \to D$ links, respectively, n_{SD} and n_{SR_i} denote the additive white Gaussian noise (AWGN) components of $S \to D$ and $S \to R_i \to D$ links, respectively, x represents the transmitted signal from the source and E_s represents average energy per symbol. During the second phase, i^{th} relay amplifies the signal received from the source and retransmits it towards the destination. Thus, the received signal by the destination is represented as

$$y_{R_iD} = h_{R_iD}\sqrt{E_s}x + n_{R_iD},$$ (9.3)

where h_{R_iD} and n_{R_iD} are the channel gain coefficient and AWGN component, respectively, of $R_i \to D$ link and G_i represents the variable-gain factor at the ith relay, which is given as

$$G_i = \sqrt{\frac{E_s}{E_s|h_{SR_i}|^2 + N_0}},$$ (9.4)

where N_0 is the noise variance of AWGN, which is assumed to be same for all the links. By using MRC scheme, the instantaneous end-to-end SNR at the destination is given as [28]

$$\gamma_t = \gamma_{SD} + \sum_{i=1}^{N}\frac{\gamma_{SR_i}\gamma_{R_iD}}{1+\gamma_{SR_i}+\gamma_{R_iD}}.$$ (9.5)

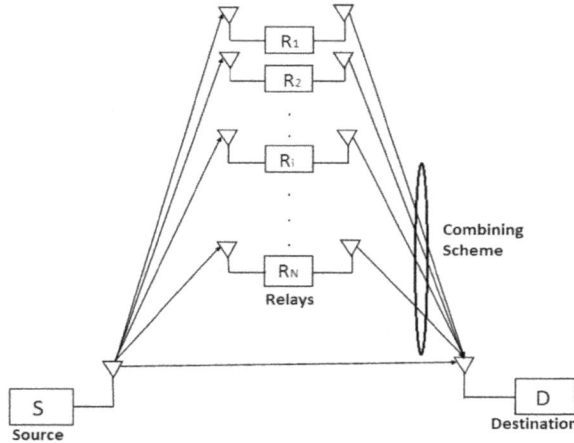

Fig. 9.1. System model.

In order to simplify the analysis, the upper bound for the equivalent SNR can be written as

$$\gamma_t \leq \gamma_{SD} + \sum_{i=1}^{N}\gamma_i,$$ (9.6)

where $\gamma_i = \min(\gamma_{SR_i}, \gamma_{R_iD}) \geq \frac{\gamma_{SR_i}\gamma_{R_iD}}{1+\gamma_{SR_i}+\gamma_{R_iD}}$. Further, by using BRS scheme at the destination, the instantaneous end-to-end SNR can be given as [28]

$$\gamma_t \leq \gamma_{SD} + \max_{i \in R}\{\min(\gamma_{SR_i}, \gamma_{R_iD})\}, \tag{9.7}$$

where $\gamma_{SD} = |h_{SD}|^2 E_s/N_0$, $\gamma_{SR_i} = |h_{SR_i}|^2 E_s/N_0$ and $\gamma_{R_iD} = |h_{R_iD}|^2 E_s/N_0$ represent the instantaneous SNRs of the $S \to D$, $S \to R_i$ and $R_i \to D$ links, respectively. The upper-bound values of SNR given in (9.6) and (9.7) are quite accurate at medium and high SNR values [30].

9.3. ASER Performance Analysis

In this section, ASER expressions of RQAM and XQAM scheme for cooperative relay network over i.i.d. Rayleigh fading channels with MRC and BRS schemes are discussed. The expression for ASER for any modulation scheme can be evaluated as

$$P_s(e) = \int_0^\infty P_s(e|\gamma)f_{\gamma_t}(\gamma)d\gamma, \tag{9.8}$$

where $f_{\gamma_t}(\gamma)$ is the pdf of the end-to-end SNR, γ_t at the receiver, and $P_s(e|\gamma)$ represent the conditional symbol error rate (SER) for AWGN channels.

9.3.1. ASER of General Order RQAM for MRC Scheme

The conditional symbol error rate (SER) for M-ary RQAM modulation scheme is presented in [31, (9.9)] as

$$P_s(e|\gamma) = 2pQ_z\left(a\sqrt{\gamma}, \frac{\pi}{2}\right) + 2qQ_z\left(b\sqrt{\gamma}, \frac{\pi}{2}\right) - 2pq[Q_z\left(a\sqrt{\gamma}, \frac{\pi}{2} - \arctan\left(\frac{b}{a}\right)\right) +$$
$$Q_z(b\sqrt{\gamma}, \arctan(b/a))], \tag{9.9}$$

where $M = M_I \times M_Q$, wherein M_I and M_Q are the number of in-phase and quadrature-phase constellation points, respectively. Further, $p = 1 - \frac{1}{M_I}$, $q = 1 - \frac{1}{M_Q}$, $a = \sqrt{\frac{6}{(M_I^2-1)+(M_Q^2-1)\beta^2}}$, $b = \sqrt{\frac{6\beta^2}{(M_I^2-1)+(M_Q^2-1)\beta^2}}$ and $\beta = d_Q/d_I$, in which d_I and d_Q are the in-phase and the quadrature-phase decision distances, respectively. Further, the integral value of $Q_z(x, \varphi)$ which is the alternative representation of 1-D and 2-D Gaussian Q-function is given in [17] as

$$Q_z(x, \varphi) = \frac{1}{\pi}\int_0^\varphi \exp\left(-\frac{x^2}{2\sin^2\theta}\right)d\theta, x \geq 0.$$

Substituting (9.9) into (9.8) and after some mathematical manipulations, ASER expression for RQAM with MRC scheme can be obtained as

$$P_s^{RQAM,MRC}(e) = 2pI\left(a, \frac{\pi}{2}\right) + 2qI\left(b, \frac{\pi}{2}\right) - 2pq[I\left(a, \frac{\pi}{2} - \arctan\left(\frac{b}{a}\right)\right) +$$
$$I\left(b, \arctan\left(\frac{b}{a}\right)\right)], \tag{9.10}$$

where $I(.,.)$ is defined in [31] as

$$I(x, \varphi) = \int_0^\infty Q_z(x\sqrt{\gamma}, \varphi) f_{\gamma_t}(\gamma) \, d\gamma$$
$$= \frac{1}{\pi} \int_0^\varphi \int_0^\infty \exp\left(-\frac{x^2\gamma}{2\sin^2\theta}\right) f_{\gamma_t}(\gamma) \, d\gamma \, d\theta$$
$$= \frac{1}{\pi} \int_0^\varphi M_{\gamma_t}\left(\frac{x^2}{2\sin^2\theta}\right) d\theta, \tag{9.11}$$

where $M_{\gamma_t}(s)$ represents the MGF of the end-to-end SNR, γ_t. Further, to evaluate the value of $I(x, \varphi)$ in (9.11), we need an expression of $M_{\gamma_t}(s)$ for MRC scheme which is given in [28] as

$$M_{\gamma_t}(s) = \frac{1}{(1+\bar{\gamma}_{SD}s)} \prod_{i=1}^N \frac{1}{(1+\bar{\gamma}_C s)}, \tag{9.12}$$

Assuming all links as i.i.d, thus, (9.12) can also be written as

$$M_{\gamma_t}(s) = (1 + \bar{\gamma}_{SD}s)^{-1}(1 + \bar{\gamma}_C s)^{-N}, \tag{9.13}$$

where $\bar{\gamma}_{SD} = \bar{\gamma}_0 = E_s/N_0$ and $\bar{\gamma}_C = \frac{\bar{\gamma}_{SR_i}\bar{\gamma}_{R_iD}}{\bar{\gamma}_{SR_i} + \bar{\gamma}_{R_iD}} = \frac{\bar{\gamma}_0}{2}$. By using partial fraction expansion approach, (9.13) can be re-written as

$$M_{\gamma_t}(s) = \frac{2^N}{(1+\bar{\gamma}_0 s)} - \sum_{i=1}^N \frac{2^{N-i}}{\left(1+\frac{\bar{\gamma}_0}{2}s\right)^i}. \tag{9.14}$$

Substituting (9.14) into (9.11) and solving some required integrals, the closed-form expression of $I(x, \varphi)$ for MRC scheme can be obtained as

$$I(x, \varphi) = 2^N I_1\left(\varphi; \frac{x^2}{2}\bar{\gamma}_0\right) - \sum_{i=1}^N 2^{N-i} I_i\left(\varphi; \frac{x^2}{4}\bar{\gamma}_0\right), \tag{9.15}$$

where $I(.;.)$ is given in [17, (5A.24)] as

$$I(\varphi; c) = \frac{1}{\pi} \int_0^\varphi \left(\frac{\sin^2\theta}{\sin^2\theta + c}\right)^i d\theta$$
$$= \frac{\varphi}{\pi} - \frac{\beta}{\pi}\left\{\left(\frac{\pi}{2} + \arctan(\alpha_0)\right) \times \sum_{k=0}^{i-1}\binom{2k}{k}\frac{1}{(4(1+c))^k} + \right.$$
$$\sin(\arctan(\alpha_0))\sum_{k=1}^{i-1}\sum_{j=1}^k \frac{T_{jk}}{(1+c)^k}[\cos(\arctan(\alpha_0))]^{2(k-j)+1}\right\}, \tag{9.16}$$

wherein $\beta_0 = \sqrt{\frac{c}{1+c}}\,\text{sgn}(\varphi)$, $\alpha_0 = -\beta_0\cot(\varphi)$ and $T_{jk} = \frac{\binom{2k}{k}}{\binom{2(k-j)}{(k-j)}4^j[2(k-j)+1]}$. The value of $I_1(.;.)$ can be obtained from (9.16) by considering, $i = 1$. Further, substituting (9.15) into (9.10), we obtain the closed-form expression of $I(a, \pi/2)$, $I(b, \pi/2)$, $I(a, \frac{\pi}{2} - \arctan(b/a))$ and $I(b, \arctan(b/a))$. Furthermore, by substituting these values into (9.10), a closed-form lower-bound expression of ASER for RQAM using MRC scheme at receiver is given as

$$P_s^{RQAM,MRC}(e) = 2^N \left[2pI_1\left(\frac{\pi}{2}; \frac{a^2}{2}\bar{\gamma}_0\right) + 2qI_1\left(\frac{\pi}{2}; \frac{b^2}{2}\bar{\gamma}_0\right) - 2pq\left(I_1\left(\frac{\pi}{2} - \arctan\left(\frac{b}{a}\right); \frac{a^2}{2}\bar{\gamma}_0\right) + \right. \right.$$
$$\left. I_1\left(\arctan\left(\frac{b}{a}\right); \frac{b^2}{2}\bar{\gamma}_0\right)\right) \right] - \sum_{i=1}^{N} 2^{N-i} \left[2pI_i\left(\frac{\pi}{2}; \frac{a^2}{4}\bar{\gamma}_0\right) + 2qI_i\left(\frac{\pi}{2}; \frac{b^2}{4}\bar{\gamma}_0\right) - 2pq\left(I_i\left(\frac{\pi}{2} - \right.\right.\right.$$
$$\left.\left.\left. \arctan\left(\frac{b}{a}\right); \frac{a^2}{4}\bar{\gamma}_0\right) + I_i\left(\arctan\left(\frac{a}{b}\right); \frac{b^2}{4}\bar{\gamma}_0\right)\right)\right]. \tag{9.17}$$

The expression (9.17) represents lower-bound of ASER for general order RQAM scheme. For the special case of M-ary SQAM scheme, lower-bound expression of ASER can be obtained by substituting $M_I = M_Q = \sqrt{M}$ and $\beta = 1$ in (9.17). Similarly, for BPSK scheme, $M_I = 2$, $M_Q = 1$, $p = 0.5$, $q = 0$, $a = \sqrt{2}$ and $\beta = 0$ are substituted in (9.17) to derive the ASER expression.

9.3.2. ASER of General Order XQAM for MRC Scheme

The conditional SER for M-ary XQAM modulation scheme is presented in [31, (9.15)] as

$$P_s(e|\gamma) = g_1 Q_z\left(A_0\sqrt{\gamma}, \frac{\pi}{2}\right) + \frac{4}{M} Q_z\left(A_1\sqrt{\gamma}, \frac{\pi}{2}\right) - g_2 Q_z\left(A_0\sqrt{\gamma}, \frac{\pi}{4}\right) -$$
$$\frac{8}{M} \sum_{m=1}^{v-1} Q_z(A_0\sqrt{\gamma}, \alpha_m) - \frac{4}{M} \sum_{m=1}^{v-1} Q_z(A_m\sqrt{\gamma}, \beta_m^+) + \frac{4}{M} \sum_{m=2}^{v} Q_z(A_m\sqrt{\gamma}, \beta_m^-), \tag{9.18}$$

where $g_1 = 4 - \frac{6}{\sqrt{2M}}$, $g_2 = 4 - \frac{12}{\sqrt{2M}} + \frac{12}{M}$, $v = \frac{1}{8}\sqrt{2M}$, $A_0 = \sqrt{\frac{96}{31M-32}}$,

$A_m = \sqrt{2}mA_0$, $m = 1, ..., v$, $\alpha_m = \arctan\left[\frac{1}{2m+1}\right]$, $m = 1, ..., v-1$,

$\beta_m^- = \arctan\left[\frac{m}{m-1}\right]$, $m = 2, ..., v$, $\beta_m^+ = \arctan\left[\frac{m}{m+1}\right]$, $m = 1, ..., v-1$.

Substituting (9.18) into (9.8), the closed-form expression of ASER for XQAM using MRC scheme can be obtained as

$$P_s^{XQAM,MRC}(e) = g_1 I\left(A_0, \frac{\pi}{2}\right) + \frac{4}{M} I\left(A_1, \frac{\pi}{2}\right) - g_2 I\left(A_0, \frac{\pi}{4}\right) - \frac{8}{M} \sum_{m=1}^{v-1} I(A_0, \alpha_m) -$$
$$\frac{4}{M} \sum_{m=1}^{v-1} I(A_m, \beta_m^+) + \frac{4}{M} \sum_{m=2}^{v} I(A_m, \beta_m^-). \tag{9.19}$$

By using (9.11), we can evaluate the value of $I\left(A_0, \frac{\pi}{2}\right)$, $I\left(A_1, \frac{\pi}{2}\right)$, $I\left(A_0, \frac{\pi}{4}\right)$, $I(A_0, \alpha_m)$, $I(A_m, \beta_m^+)$ and $I(A_m, \beta_m^-)$, and by substituting these values into (9.19), a closed-form lower-bound expression of ASER for XQAM using MRC scheme can be obtained as

$$P_s^{XQAM,MRC}(e) = 2^N \left[g_1 I_1 \left(\frac{\pi}{2}; \frac{A_0^2}{2}\bar{\gamma}_0 \right) + \frac{4}{M} I_1 \left(\frac{\pi}{2}; \frac{A_1^2}{2}\bar{\gamma}_0 \right) - g_2 I_1 \left(\frac{\pi}{4}; \frac{A_0^2}{2}\bar{\gamma}_0 \right) - \right.$$
$$\frac{8}{M}\sum_{m=1}^{v-1} I_1 \left(\alpha_m; \frac{A_0^2}{2}\bar{\gamma}_0 \right) - \frac{4}{M}\sum_{m=1}^{v-1} I_1 \left(\beta_m^+; \frac{A_m^2}{2}\bar{\gamma}_0 \right) + \frac{4}{M}\sum_{m=2}^{v} I_1 \left(\beta_m^-; \frac{A_m^2}{2}\bar{\gamma}_0 \right) \right] -$$
$$\sum_{i=1}^{N} 2^{N-i} \left[g_1 I_i \left(\frac{\pi}{2}; \frac{A_0^2}{4}\bar{\gamma}_0 \right) + \frac{4}{M} I_i \left(\frac{\pi}{2}; \frac{A_1^2}{4}\bar{\gamma}_0 \right) - g_2 I_i \left(\frac{\pi}{4}; \frac{A_0^2}{4}\bar{\gamma}_0 \right) - \right.$$
$$\left. \frac{8}{M}\sum_{m=1}^{v-1} I_i \left(\alpha_m; \frac{A_0^2}{4}\bar{\gamma}_0 \right) - \frac{4}{M}\sum_{m=1}^{v-1} I_i \left(\beta_m^+; \frac{A_m^2}{4}\bar{\gamma}_0 \right) + \frac{4}{M}\sum_{m=2}^{v} I_i \left(\beta_m^-; \frac{A_m^2}{4}\bar{\gamma}_0 \right) \right]. \quad (9.20)$$

9.3.3. ASER of General Order XQAM for BRS Scheme

To evaluate the closed-form expression of lower-bound for ASER using XQAM, we need an expression of MGF for BRS scheme which is presented in [28, (9.24)] as

$$M_{\gamma_t}(s) = \sum_{i=1}^{N} \binom{N}{i} \frac{i\,(-1)^{i-1}}{(1+\bar{\gamma}_c s)(1+\bar{\gamma}_{SD}s)}. \quad (9.21)$$

By substituting partial fraction expansion of (9.21) into (9.11), the closed-form expression of $I(z, \varphi)$ for BRS scheme can be obtained as

$$I(z, \varphi) = \sum_{i=1}^{N} \binom{N}{i} \frac{(-1)^{i-1}}{1-2i} \left[I_1 \left(\varphi; \frac{z^2}{4i}\bar{\gamma}_0 \right) - 2i\,I_1(\varphi; \frac{z^2}{2}\bar{\gamma}_0) \right], \quad (9.22)$$

where $I_1(.;.)$ is evaluated by putting $i = 1$ in (9.16).

With the aid of (9.22) and using the similar approach as for MRC in (9.19), we obtain the closed-form expression of ASER for XQAM using BRS scheme as

$$P_s^{XQAM,BRS}(e) = \sum_{i=1}^{N} \binom{N}{i} \frac{(-1)^{i-1}}{(1-2i)} \left[g_1 \left(I_1 \left(\frac{\pi}{2}; \frac{A_0^2}{4i}\bar{\gamma}_0 \right) - 2\,i\,I_1 \left(\frac{\pi}{2}; \frac{A_0^2}{2}\bar{\gamma}_0 \right) \right) + \right.$$
$$\frac{4}{M} \left(I_1 \left(\frac{\pi}{2}; \frac{A_1^2}{4i}\bar{\gamma}_0 \right) - 2\,i\,I_1 \left(\frac{\pi}{2}; \frac{A_1^2}{2}\bar{\gamma}_0 \right) \right) - g_2 \left(I_1 \left(\frac{\pi}{4}; \frac{A_0^2}{4i}\bar{\gamma}_0 \right) - 2\,i\,I_1 \left(\frac{\pi}{4}; \frac{A_0^2}{2}\bar{\gamma}_0 \right) \right) -$$
$$\frac{8}{M}\sum_{m=1}^{v-1} \left(I_1 \left(\alpha_m; \frac{A_0^2}{4i}\bar{\gamma}_0 \right) - 2\,i\,I_1 \left(\alpha_m; \frac{A_0^2}{2}\bar{\gamma}_0 \right) \right) - \frac{4}{M}\sum_{m=1}^{v-1} \left(I_1 \left(\beta_m^+; \frac{A_m^2}{4i}\bar{\gamma}_0 \right) - \right.$$
$$\left. 2\,i\,I_1 \left(\beta_m^+; \frac{A_m^2}{2}\bar{\gamma}_0 \right) \right) + \frac{4}{M}\sum_{m=2}^{v} \left(I_1 \left(\beta_m^-; \frac{A_m^2}{4i}\bar{\gamma}_0 \right) - 2\,i\,I_1 \left(\beta_m^-; \frac{A_m^2}{2}\bar{\gamma}_0 \right) \right) \right]. \quad (9.23)$$

9.4. Numerical and Simulation Results

In this section, theoretical (lower-bound) results of ASER for RQAM and XQAM with MRC scheme obtained from (9.17) and (9.20), respectively, and theoretical result of ASER for XQAM with BRS scheme obtained from (9.23) are compared with Monte Carlo simulation (exact) results to verify the correctness of derived closed-form expressions.

Fig. 9.2. shows comparison between theoretical and simulated ASER curves versus average SNR, E_s/N_0 for 8×4-RQAM and 4×2-RQAM constellations for MRC scheme with $\beta = 1$ and $N = 1, 2, 3$. It is observed that both the theoretical and simulated curves obtained from (9.17) and Monte Carlo almost overlap, especially at medium and high SNRs. It can also be observed that the ASER performance is degraded by increasing the constellation points from 4×2 to 8×4-RQAM, due to the presence of more number of symbols in a quadrant which causes poor ASER performance for 8×4-RQAM. Further, it is also observed that the ASER performance improved significantly at medium and high SNRs by increasing the number of relay, N. This is due to increasing the diversity order of the system. For example, to maintain an ASER of 10^{-3} in 8×4-RQAM constellation, the SNR gain of 4.66 dB (approx.) is observed when number of relays increases from $N = 1$ to $N = 2$, and 2.6 dB (approx.) is observed when number of relays increase from $N = 2$ to $N = 3$. Similarly, for 4×2-RQAM constellation, to maintain an ASER of 10^{-3}, the SNR gain of 4.55 dB (approx.) is observed when number of relays increases from $N = 1$ to $N = 2$, and 2.5 dB (approx.) is observed when number of relays increase from $N = 2$ to $N = 3$. From this observation, it can be seen that for a given ASER value the increment in SNR gain gradually decreases with increase in N. Furthermore, it is observed that the improvement in ASER performance is not noticeable at low SNRs ($E_s/N_0 < 10$ dB). Moreover, percentage of relative difference between theoretical and simulated values for 4×2-RQAM constellation can also be calculated by $\frac{\text{Simulated}-\text{Theoretical}}{\text{Simulated}} \times 100$ as a function of E_s/N_0 and N which is shown in Table 9.1. From Fig. 9.2 and Table 9.1 it is observed that the gap between theoretical and simulated curves increases with increase in N for a given value of SNR, however, an improvement occurs at high SNRs. This indicates that the derived lower-bound expressions can give good approximation for small values of N.

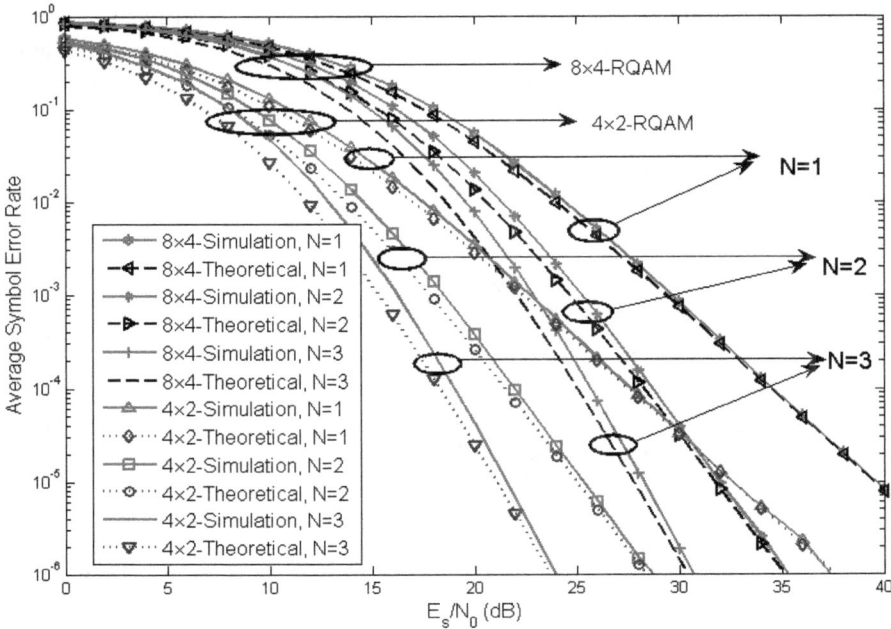

Fig. 9.2. ASER versus E_s/N_0 for 8×4-RQAM and 4×2-RQAM constellations for MRC scheme with $\beta = 1$.

Table 9.1. Relative Difference between simulated and theoretical values of ASER for 4×2-RQAM for MRC scheme with $\beta = 1$.

E_s/N_0 (dB)	N	ASER		
		Simulated	**Theoretical**	**Relative Difference (%)**
0	1	0.57876	0.54111	6.5052
	2	0.54238	0.47800	11.8699
	3	0.50999	0.4252	16.6258
14	1	0.03794	0.03028	20.1897
	2	0.01336	0.008598	35.6717
	3	0.00467	0.002503	46.4025
25	1	3.3760×10^{-4}	3.0709×10^{-4}	9.03732
	2	11.99×10^{-6}	9.6312×10^{-6}	19.6729
	3	4.4×10^{-7}	3.1207×10^{-7}	29.0747

Fig. 9.3 illustrates comparison between theoretical and simulated ASER curves versus average SNR, E_s/N_0 for 128-XQAM and 32-XQAM constellations for MRC scheme with $N = 1, 2$ and 3. Percentage of relative difference between theoretical and simulated values for 32-XQAM constellation for MRC scheme can also be calculated by $\frac{\text{Simulated} - \text{Theoretical}}{\text{Simulated}} \times 100$ as a function of E_s/N_0 and N which is presented in Table 9.2. It can be observed that the ASER curves obtained from (9.20) and the impact of N on ASER performance follow similar trend as the ASER in Fig. 9.2. Further, to maintain an ASER of 10^{-3}, the SNR gain can be achieved by increasing N from 1 to 2 and 2 to 3 for 128-XQAM constellation are 4.75 dB and 2.65 dB, respectively. Similarly, to maintain an ASER of 10^{-3} for 32-XQAM constellation, the SNR gain of 4.5 dB (approx.) and 2.6 dB (approx.) are observed when N increase from 1 to 2 and 2 to 3, respectively.

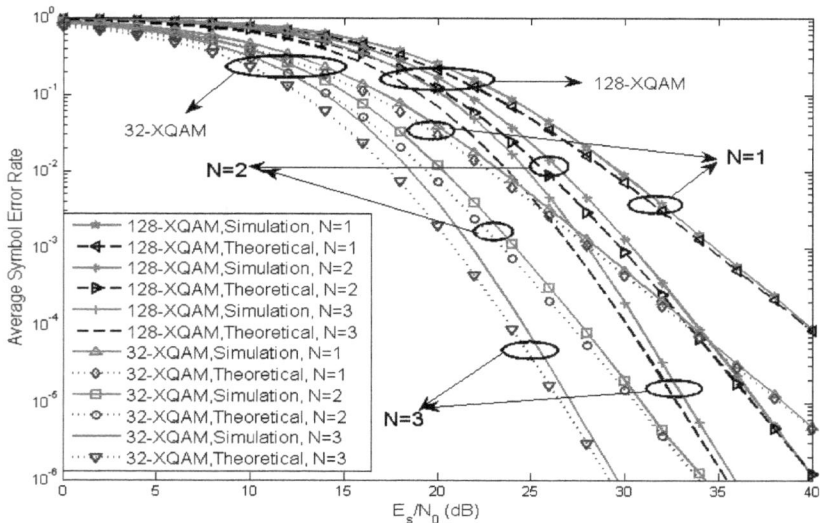

Fig. 9.3. ASER versus E_s/N_0 for 128-XQAM and 32-XQAM constellation for MRC scheme.

Table 9.2. Relative Difference between simulated and theoretical values of ASER for 32-XQAM for MRC scheme.

E_s/N_0 (dB)	N	ASER		
		Simulated	Theoretical	Relative Difference (%)
0	1	0.8620	0.82848	3.8886
	2	0.8464	0.79548	6.0161
	3	0.8340	0.76532	8.235
16	1	0.1305	0.11005	15.6705
	2	0.0711	0.04983	29.6186
	3	0.0385	0.0229	40.5194
34	1	7.6023×10^{-5}	7.13643×10^{-5}	6.1280
	2	10.55×10^{-7}	9.65691×10^{-7}	8.4654
	3	1.555×10^{-8}	1.34411×10^{-8}	13.5621

Fig. 9.4 shows comparison between theoretical and simulated ASER curves versus average SNR, E_s/N_0 for 128-XQAM and 32-XQAM constellations for BRS scheme with $N = 1, 2$ and 3. Further, percentage of relative difference between theoretical and simulated values for 32-XQAM constellation for BRS scheme can also be calculated by $\frac{\text{Simulated}-\text{Theoretical}}{\text{Simulated}} \times 100$ as a function of E_s/N_0 and N as shown in Table 9.3. Furthermore, it can also be observed that the ASER curves obtained from (9.23) and the impact of N on ASER performance follow similar trend as the ASER in Fig. 9.2 and Fig. 9.3. The SNR gain for 128-XQAM constellation that can be obtained by increasing N from 1 to 2 and 2 to 3 are 3.9 dB and 1.75 dB, respectively, to maintain an ASER of 10^{-3}. Similarly for 32-XQAM constellation, to maintain an ASER of 10^{-3}, the SNR gain of 3.6 dB (approx.) and 1.7 dB (approx.) are observed by increasing N from 1 to 2 and 2 to 3, respectively.

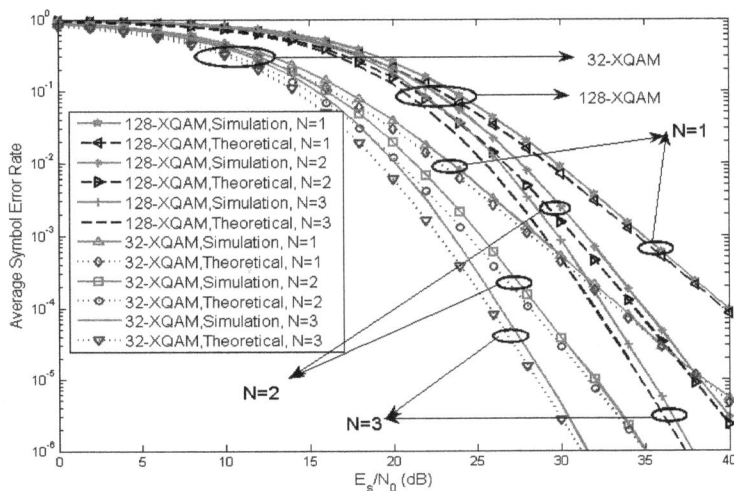

Fig. 9.4. ASER versus E_s/N_0 for 128-XQAM and 32-XQAM constellation for BRS scheme.

Table 9.3. Relative Difference between simulated and theoretical values of ASER for 32-XQAM for BRS scheme.

E_s/N_0 (dB)	N	ASER		
		Simulated	**Theoretical**	**Relative Difference (%)**
0	1	0.8617	0.82848	3.8550
	2	0.8538	0.811151	4.9951
	3	0.8484	0.8008	5.6105
16	1	0.1313	0.11005	16.1843
	2	0.0931	0.06836	26.5735
	3	0.0755	0.04886	35.2847
34	1	7.5669×10^{-5}	7.13643×10^{-5}	5.6888
	2	2.1314×10^{-7}	1.90488×10^{-7}	10.6277
	3	9.1429×10^{-8}	7.7323×10^{-8}	15.4283

Fig. 9.5 demonstrates ASER performance versus average SNR, E_s/N_0 for 8×4-RQAM and 16×2-RQAM constellations with several values of β and N. To avoid the overlapping of curves, we have considered for $\beta=1$ and $\beta=6$ with $N = 1$ and $N = 3$. It can be observed that the ASER performance is improved by increasing the value of N due to improved diversity gain. Further, it can also be observed that for a fixed value of N, the ASER performance depends on the parameter β, and performance of 8×4-RQAM constellation is always better than 16×2-RQAM constellation for all SNRs when $\beta = 1$. However, by increasing the value of β, performance of 8×4 constellation degrades fast. This is because of relation between M_I and M_Q, and the definition of β which is the ratio of quadrature phase decision distance to in-phase decision distance. Further, at fixed value of β and N, ASER performance depends on the magnitude of the difference between M_I and M_Q. The constellation with large difference gives poor performance than the constellation with small difference. This is because of the definition of β i.e., more number of symbols in a quadrant tending towards poor ASER performance.

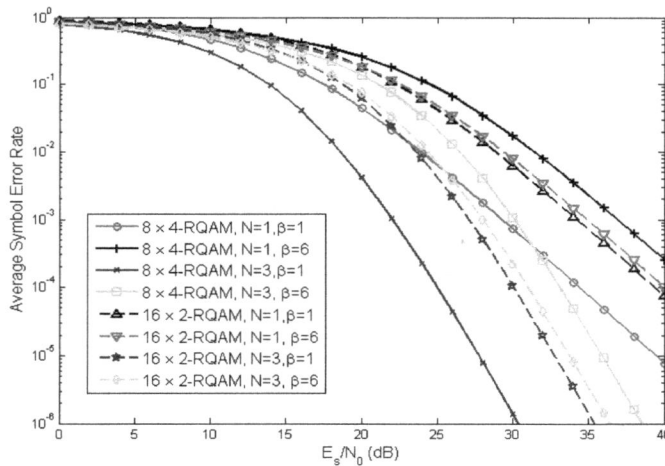

Fig. 9.5. ASER for 8×4-RQAM and 16×2-RQAM constellations for MRC scheme.

Fig. 9.6 illustrates comparison between ASER performance for M-ary RQAM and M-ary XQAM for MRC scheme with $\beta = 1$ and $N = 2$, using different constellations points such as $M = 32, 128$ and 512 (as shown in the legend of figure). We can see that XQAM outperforms RQAM in terms of SNR for a given value of ASER. For example, to maintain an ASER of 10^{-3}, the SNR gain of 1.13 dB, 1.11 dB and 1.10 dB are obtained for $M = 32$, $M = 128$ and $M = 512$, respectively. It is also observed that the SNR gain is decreased gradually by increasing value of M.

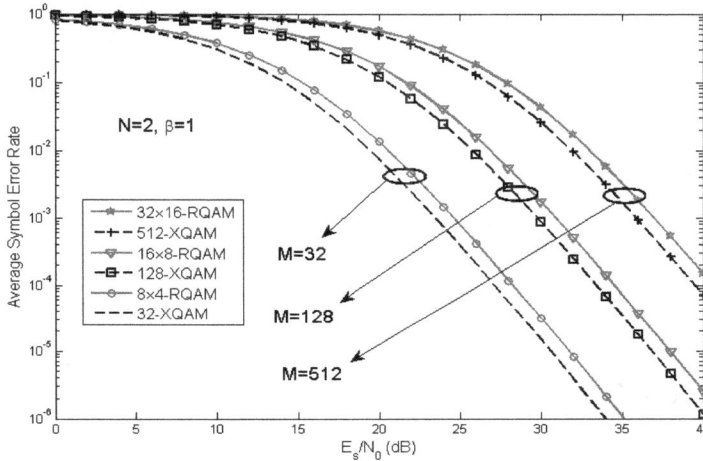

Fig. 9.6. Comparison of ASER results of M-ary RQAM and M-ary XQAM for MRC with $\beta = 1$ and $N = 2$.

Fig. 9.7 shows comparison of ASER performance for M-ary XQAM between MRC and BRS schemes with $N = 2$ and varying constellations points. It is observed that the MRC exhibits SNR gain over BRS scheme for a given value of ASER. For example, to maintain an ASER of 10^{-3}, the SNR gain of 0.95 dB, 0.93 dB and 0.92 dB are obtained for $M = 32$, $M = 128$ and $M = 512$, respectively. From this observation, it can be concluded that MRC outperforms BRS scheme for all considered constellations points, M.

9.5. Conclusion

In this chapter, we present performance analysis of general order RQAM and XQAM schemes over Rayleigh fading channel in cooperative variable-gain AF dual-hop relaying networks over i.i.d. Rayleigh fading channels. Closed-form expressions for lower-bound of ASER with MRC and BRS scheme are derived by MGF approach. By comparing numerical and simulated results, correctness of the derived expressions is verified. The impact of number of relays, constellation size and the decision distance ratio (β) on ASER performance are highlighted. The advantage of MRC over BRS scheme in terms of SNR gain is discussed. Further, advantage of XQAM over RQAM constellation for transmission of odd number of bits per symbol is also highlighted in terms of SNR gain.

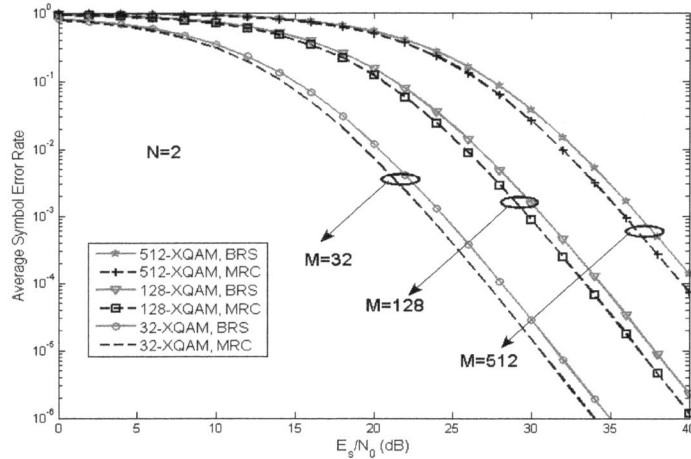

Fig. 9.7. Comparison of ASER results for *M*-ary XQAM between MRC and BRS scheme.

Acknowledgements

The authors thank MHRD and IIT Indore for all the support.

References

[1]. J. N. Laneman, D. N. C. Tse, G. W. Wornell, Cooperative diversity in wireless networks: efficient protocols and outage behavior, *IEEE Trans. Inf. Theory*, Vol. 50, Issue 12, 2004, pp. 3062-3080.

[2]. N. Kumar, V. Bhatia, Performance analysis of amplify-and-forward cooperative networks with best-relay selection over Weibull fading channels, *Wireless Pers. Commun.*, Vol. 85, Issue 3, 2015, pp. 641-653.

[3]. N. Kumar, V. Bhatia, Outage probability and average channel capacity of amplify-and-forward in conventional cooperative communication networks over Rayleigh fading channels, *Wireless Pers. Commun.*, Vol. 88, Issue 4, 2016, pp. 943-951.

[4]. Y. Yang, H. Hu, J. Xu, G. Mao, Relay technologies for WiMAX and LTE advanced mobile systems, *IEEE Commun. Mag.*, Vol. 47, Issue 10, 2009, pp. 100-105.

[5]. T. Wang, A. Cano, G. B. Giannakis, J. N. Laneman, High-performance cooperative demodulation with decode-and-forward relays, *IEEE Trans. Commun.*, Vol. 55, Issue 7, 2007, pp. 1427-1438.

[6]. G. Pan, Q. Feng, Performance analysis of DF relaying multi-hop systems over log-normal fading channels, *Int. J. Electron. Commun. (AEU)*, Vol. 67, Issue 6, 2013, pp. 457-462.

[7]. E. Soleimani-Nasab, M. Ardebilipour, A. Kalantari, B. Mahboobi, Performance analysis of multi-antenna relay networks with imperfect channel estimation, *Int. J. Electron. Commun. (AEU)*, Vol. 67, Issue 1, 2013, pp. 45-57.

[8]. P. Kalansuriya, C. Tellambura, Performance analysis of decode-and-forward relay network under adaptive M-QAM, in *Proceedings of the IEEE Int. Conf. Commun. (ICC'09)*, Dresden, Germany, 2009, pp. 1-5.

[9]. C. K. Datsikas, K. P. Peppas, F. I. Lazarakis, G. S. Tombras, Error rate performance analysis of dual-hop relaying transmissions over generalized-K fading channels, *Int. J. Electron. Commun. (AEU)*, Vol. 64, Issue 11, 2010, pp. 1094-1099.

[10]. I. Altunbas, A. Yılmaz, S. S. Kucur, O. Kucur, Performance analysis of dual-hop fixed-gain AF relaying systems with OSTBC over Nakagami-m fading channels, *Int. J. Electron. Commun. (AEU)*, Vol. 66, Issue 10, 2012, pp. 841-846.

[11]. N. Bissias, G. P. Efthymoglou, V. A. Aalo, Performance analysis of dual-hop relay systems with single relay selection in composite fading channels, *Int. J. Electron. Commun. (AEU)*, Vol. 66, Issue 1, 2012, pp. 39-44.

[12]. C. Zhong, M. Matthaiou, G. K. Karagiannidis, A. Huang, Z. Zhang, Capacity bounds for AF dual-hop relaying in fading channels, *IEEE Trans. Veh. Technol.*, Vol. 61, Issue 4, 2012, pp. 1730-1740.

[13]. D. G. Brennan, Linear diversity combining techniques, *Proceedings of the IRE*, Vol. 47, Issue 6, 1959, pp. 1075-1102.

[14]. H. Katiyar, R. Bhattacharjee, Performance of MRC combining multi-antenna cooperative relay network, *Int. J. Electron. Commun. (AEU)*, Vol. 64, Issue 10, 2010, pp. 988-991.

[15]. S. S. Ikki, M. H. Ahmed, Exact error probability and channel capacity of the best-relay cooperative-diversity networks, *IEEE Signal Process. Lett.*, Vol. 16, Issue 12, 2009, pp. 1051-1054.

[16]. N. Kumar, V. Bhatia, Performance analysis of OFDM based AF cooperative systems in selection combining receiver over Nakagami-m fading channels with nonlinear power amplifier, *Int. J. Commun. Syst.*, Vol. 30, Issue 7, 2017, pp. 1-17.

[17]. M. K. Simon, M.-S. Alouini, Digital Communication over Fading Channels, 2nd Ed., *Wiley*, New York, 2005.

[18]. A. Maaref, S. Aïssa, Exact error probability analysis of rectangular QAM for single-and multichannel reception in Nakagami-m fading channels, *IEEE Trans. Commun.*, Vol. 57, Issue 1, 2009, pp. 214-221.

[19]. J. G. Smith, Odd-bit quadrature amplitude-shift keying, *IEEE Trans. Commun.*, Vol. 23, Issue 3, 1975, pp. 385-389.

[20]. P. K. Vitthaladevuni, M.-S. Alouini, J. C. Kieffer, Exact BER computation for cross QAM constellations, *IEEE Trans. Wireless Commun.*, Vol. 4, Issue 6, 2005, pp. 3039-3050.

[21]. H. Yu, G. Wei, F. Ji, X. Zhang, On the error probability of cross-QAM with MRC reception over generalized fading channels, *IEEE Trans. Veh. Technol.*, Vol. 60, Issue 6, 2011, pp. 2631-2643.

[22]. S. Abrar, I. M. Qureshi, Blind equalization of cross-QAM signals, *IEEE Signal Process. Lett.*, Vol. 13, Issue 12, 2006, pp. 745-748.

[23]. S. Panigrahi, T. Le-Ngoc, Fine-granularity loading schemes using adaptive Reed-Solomon coding for discrete multitone modulation systems, in *Proceedings of IEEE Int. Conference Commun. (ICC'05)*, Seoul, Korea, 16-20 May 2005, pp. 1352-1356.

[24]. S. Panigrahi, T. Le-Ngoc, Non-iterative bit-loading algorithm for ADSL-type DMT applications, *Proceedings Inst. Electr. Eng. Commun.*, Vol. 150, 2003, pp. 414-418.

[25]. X.-C. Zhang, H. Yu, G. Wei, Exact symbol error probability of cross-QAM in AWGN and fading channels, *EURASIP J. Wireless Commun. Netw.*, Vol. 2010, 2010, pp. 1-9.

[26]. H. Yu, G. Wei, Symbol error probability of cross QAM in Rayleigh fading channels, *IEEE Commun. Lett.*, Vol. 14, Issue 5, 2010, pp. 375-377.

[27]. H. Yu, Y. Zhao, J. Zhang, Y. Wang, SEP performance of cross QAM signaling with MRC over fading channels and its arbitrarily tight approximation, *Wireless Pers. Commun.*, Vol. 69, Issue 4, 2013, pp. 1567-1581.

[28]. M. Torabi, W. Ajib, D. Haccoun, Performance analysis of amplify-and-forward cooperative networks with relay selection over Rayleigh fading channels, in *Proceedings of IEEE Veh. Tech. Conference-Spring (VTC'09)*, Barcelona, Spain, 26-29 April 2009, pp. 1-5.

[29]. D. Dixit, P. R. Sahu, Symbol error rate of rectangular QAM with best-relay selection in cooperative systems over Rayleigh fading channels, *IEEE Commun. Lett.,* Vol. 16, Issue 4, 2012, pp. 466-469.

[30]. P. A. Anghel, M. Kaveh, Exact symbol error probability of a cooperative network in a Rayleigh-fading environment, *IEEE Trans. Wireless Commun.,* Vol. 3, Issue 5, 2004, pp. 1416-1421.

[31]. D. Dixit, P. R. Sahu, Performance of QAM signaling over TWDP fading channels, *IEEE Trans. Wireless Commun.,* Vol. 12, Issue 4, 2013, pp. 1794-1799.

Chapter 10
High-Gain Low-Cost Microstrip Antennas and Arrays Based on FR4 Epoxy

Babak Honarbakhsh

10.1. Introduction

Beside valuable properties such as being low-profile, light-weight, and simple-to-realize, microstrip antennas (MSAs) suffer from low gain. This shortcoming is mainly due to their small aperture and surface wave excitation in their dielectric. Thus far, many attempts have been made for gain enhancement of MSAs by either increasing the gain of a single MSA or by arraying low/moderate gain elements [1-16]. Current solutions for enhanced gain MSAs are mainly based on either multi-layer or low-loss dielectrics with small electric permittivity. Designing large arrays of MSAs is also a challenging problem, noting that increase in the array size, increases the loss in the feed network. This loss is due to finite conductivity, unwanted radiation of transmission lines (TLs), and surface wave excitation in dielectric media [17]. It should be pointed out that increase in ohmic losses leads to increase in antenna noise temperature, T_A. As a result, arraying of even high-gain elements, if not performed in a wise manner, does not lead to an efficient design. Yet, some successful arraying techniques are reported for composition of high-efficiency large MSA arrays [18-20]. Specially, in [20], undesirable effects of the feed network are defeated by replacing the microstrip TLs by a CNC milled metallic waveguide with the same dielectric structure suggested in [7].

On the other hand, due to increasing development of wireless communication systems, the realization cost has become an important design parameter. Up to now, many efforts are taken to reach to cost-effective MSAs. The difficulty stems from the inverse relation between the price and loss tangent of microstrip substrates. Most of the reported low-cost designs are based on FR4 epoxy [7], [20-40]. Therefore, designing a high-gain and low-cost MSA is a challenging problem. The said problem becomes more difficult for array antennas. Specifically, the FR4 epoxy exhibits $\varepsilon_r \sim 4.4$ and $tg\delta \sim 0.02$ at microwave frequencies [41]. Thus, it is not only lossy, but also it has a relatively large electric

Babak Honarbakhsh
Electrical Engineering Department, Shahid Beheshti University, Iran

permittivity. The first property degrades the system performance from two aspects. First, it dissipates the EM energy by converting it to the thermal energy. This leads to waste of transmitter power in transmitter and fading in receiver which in turn, decreases the antenna gain and efficiency. Second, it increases the noise temperature of the system and decreases its dynamic range. May be the only attractive aspect of its lossy nature, is bandwidth enhancement [42]. The second property is undesirable, especially for high gain designs, again from two aspects. First, the large dielectric constant leads to gain drop due to size reduction [43]. Second, increase in ε_r increases the surface wave excitation in the dielectric which decreases the antenna gain [44]. The gain lowering effect of surface waves cannot be compensated by increasing the array size because the large amount of the EM energy, when propagating along the feed network, leaks away from TLs. Thus, a high-gain MSA cannot be reached using, only, an FR4 epoxy as the substrate.

A trivial solution for the problem of low-cost high-gain MSA design is to remove the substrate and to suspend the metallization by a spacer above the ground plane [45]. This strategy requires cut-plotting the metallization from a thin metal sheet and thus, is impractical for complex and fine geometries, for instance large arrays. Compared to usual realization methods such as photolithography and chemical etching, the said solution is less accurate, more expensive and less available.

Hence, if an antenna is to be simple-to-realize, air cannot be the only employed dielectric. An innovative usage of air and FR4 epoxy may lead to satisfactory solution. At present, two such solutions are available [7, 46-47]. In [7], a layer of foam and a layer of FR4 are stacked to comprise the substrate. The foam is placed on the ground plane and the metallization is realized on the exterior side of the FR4. Thus, the need to cut-plotting is removed. This technique is successfully used in design of a 2×16 MSA array [20]. The dielectric structure used in [46-47] is the same as [7] but in contrast, the metallization is realized on the interior side of the FR4.

In this chapter, the two aforementioned techniques are compared and the effect of the FR4 thickness on antenna parameters including gain, efficiency, pattern symmetry and size are addressed. In fact, placing the metallization on the interior side of the FR4 layer, improves antenna gain and efficiency which proves superiority of [46-47] over [7] for low-cost high-gain applications. The reported method in [7], leads to more symmetric pattern not only compared to [46-47], but also to the ideal MSA with air substrate. Besides, achieving to low-cost high-gain MSA arrays is possible without the need to metallic waveguide feed network, as suggested in [20]. Hereafter, the proposed methods in [46-47] and [7] will be denoted by "method 1" and "method 2", respectively. The results reported here are generated using method of moments (MoM) engines embedded in Agilent® Advanced Design System (ADS) and FEKO® suite software.

10.2. The Studied Methods

Dielectric structures of methods 1 and 2 are described in Fig. 10.1. In both methods, the first layer is air to decrease the contribution of surface waves. Thus, arraying techniques can expectantly be applied for gain enhancement. It can be predicted that the amount of surface wave excitation in method 1 be less than method 2, since the FR4 layer plays the role of superstrate in method 1 but in method 2, it is a part of the substrate. Mechanical robustness demands for large h. On the other hand, increasing h decreases the antenna gain and increases its T_A. Consequently, there is an optimum value for h which is application dependent. The antennas designed by the aforementioned methods compared to the air-only substrate can be predicted to have less physical dimensions and gain with more T_A. Conversely, compared to the FR4-only substrate, the said methods are predicted to lead to larger size, more gain and less T_A. Since the FR4 epoxy is used as the second layer in the dielectric medium, common printed circuit board (PCB) technologies can be applied. Finally, realization of method 1, which requires injecting the EM energy into interior side of the structure, is not as simple as method 2. Although, feeding in both of the said methods are considerably simpler than what is proposed in [20].

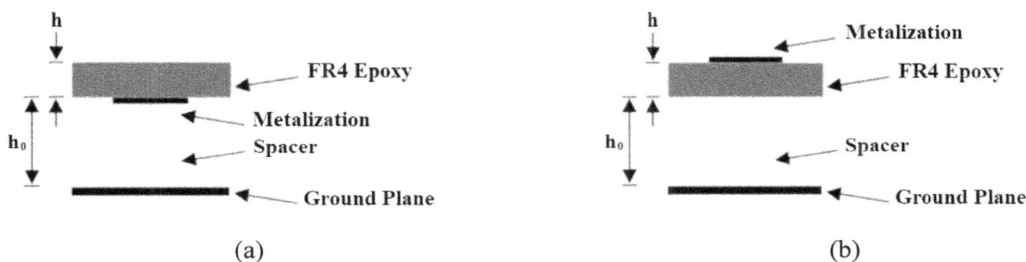

(a)   (b)

Fig. 10.1. Schematic description of the studied structures: (a) method 1; (b) method 2.

10.3. Parameter Study and Performance Comparison

In this part, theoretical predictions of the previous section are verified numerically based on MoM using Agilent® Momentum® and a detailed parameter study is carried out over the FR4 thickness. The test antenna is a probe-fed square patch with side length L. In all simulations, the thickness of the spacer and the working frequency are set to, respectively, $h_0 = 1$ mm and $f_0 = 5.8$ GHz. The antenna side length vs. h is reported in Fig. 10.2 (a). As can be seen, the resonant length for method 1 is less sensitive to h, compared to method 2. This can be justified by noting that the interaction of FR4 layer with EM waves in method 2 is more pronounced compared to method 1. From this figure, an effective relative permittivity, $\varepsilon_{r,\ eff}$, for each structure can be estimated by equating the resonant length, L, to $\lambda_g/2$ and assuming $\lambda_g = \lambda_0/\sqrt{\varepsilon_{r,\ eff}}$. Dependence of $\varepsilon_{r,\ eff}$, gain and efficiency on h are reported in Figs. 10.2 (b) through (d). From Fig. 10.2 (b), it can be observed that as h approaches zero, the gain of a single patch approaches the maximum possible of 10 dB. As expected, increasing of h decreases size, gain and efficiency and increases $\varepsilon_{r,\ eff}$. Note that variations of these parameters for the second method are more dramatic with respect to the first one. Remarkably, for the very thick case of $h = 3$ mm, decrease in gain for method 1 with respect to its maximum is less than 1 dB. As a result, method 1 is a promising solution for out-door wireless links since, FR4 epoxy is a sturdy material and can properly shield the metallization against damaging environmental factors. Thus, antennas designed by method 1 are self-covered and do not need radome.

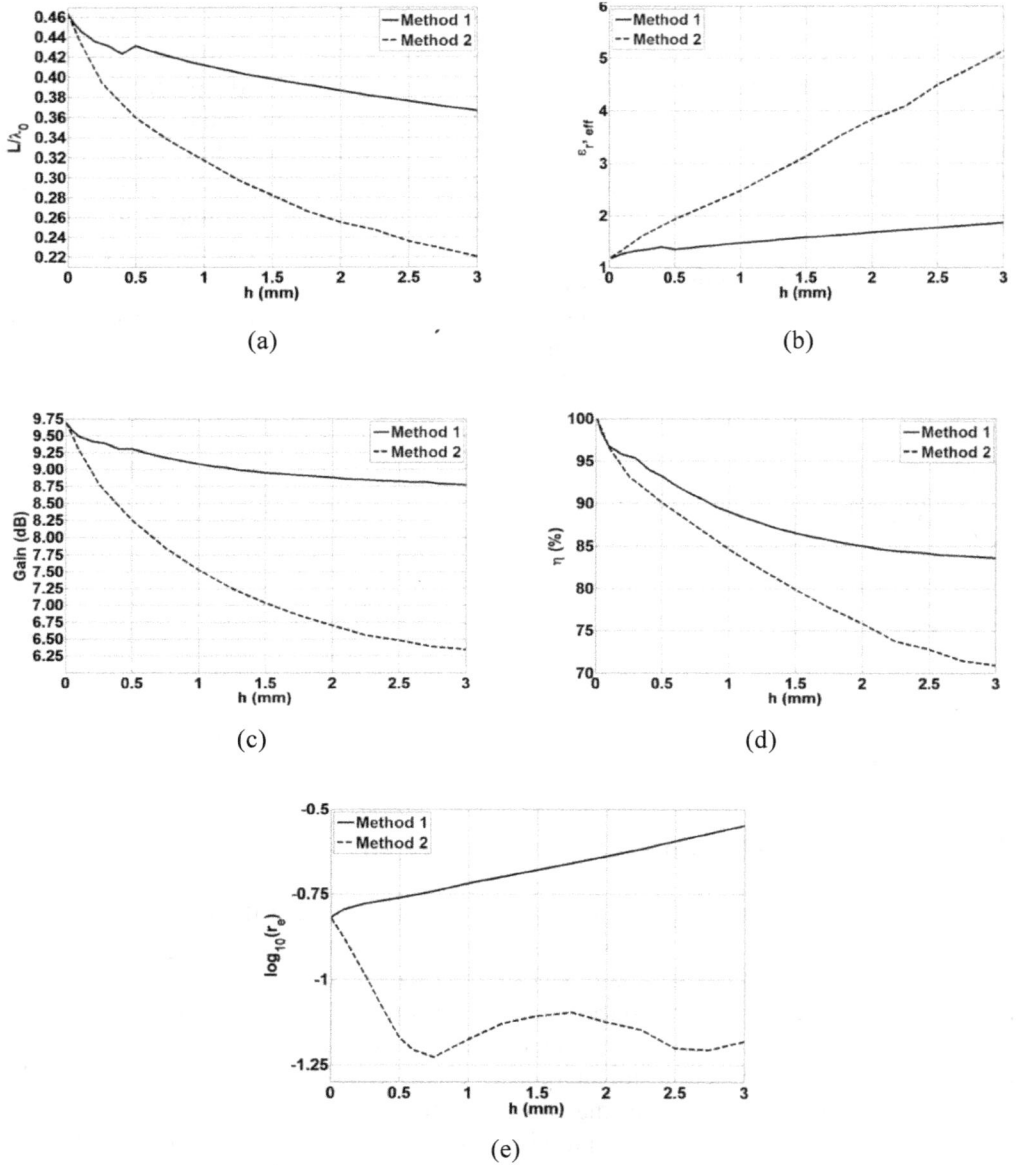

Fig. 10.2. Antenna parameters vs. FR4 thickness: (a) Side length; (b) Estimated relative effective permittivity; (c) Gain; (d) Efficiency, and (e) Pattern symmetry.

The effect of h on pattern symmetry is studied by computing the mean value of relative error between E- and H-plane patterns between ideal and two-layer structures and is reported in Fig. 10.2 (e). As can be seen, for method 1, the pattern becomes more asymmetric as h increases. In contrast, as the FR4 layer becomes thicker, the pattern of method 2 becomes more symmetric. Thus, method 2 is suited when a symmetric pattern is required. In Figs. 10.3 and 10.4, radiation pattern and scattering parameter of four square patch antennas with different dielectric media including air, FR4, method 1 and method 2, are depicted, wherein h_0 is set to 1 mm. For the third and fourth cases, h is set to 0.5 mm. It is interesting that the radiation pattern of method 2 is more symmetric than the ideal case. Noting to Fig. 10.4, the 10 dB bandwidth of the first and the third cases are close to each other. The wider bandwidth of the second case is due to its lossy substrate. Input impedance of methods 1 and 2 are reported in Fig. 10.5.

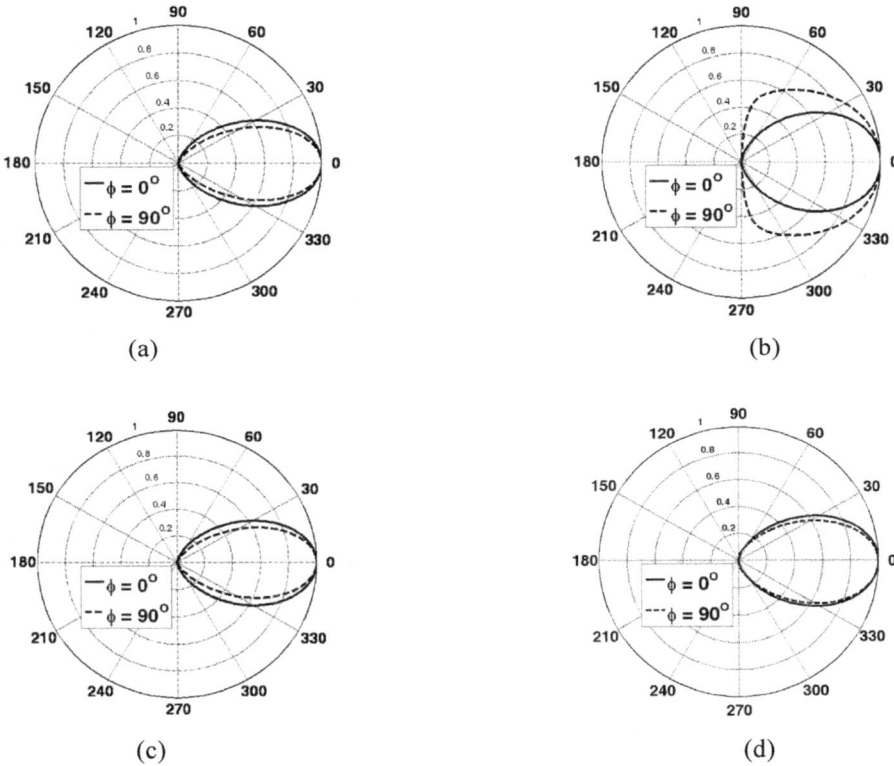

(a)

(b)

(c)

(d)

Fig. 10.3. Radiation patterns for square patch antennas with different dielectric structures: (a) Air-only; (b) FR4-only; (c) Method 1, and (d) Method 2.

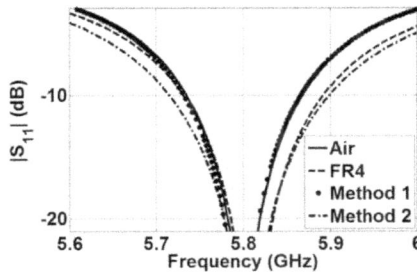

Fig. 10.4. Scattering parameter of square patch antennas with different dielectric structures: (a) Air-only; (b) FR4-only; (c) Method 1, and (d) Method 2.

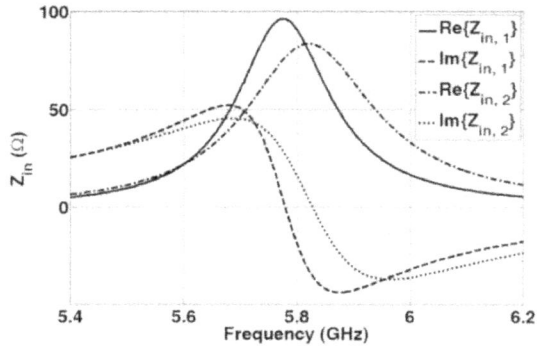

Fig. 10.5. Input impedance of square patch antennas for methods 1 and 2, assuming $h = 0.5$ mm.

To verify arraying capability of methods 1 and 2, seven square arrays with corporate feed network are designed with the said four dielectric structures with 2×2 to 8×8 elements and their gain and radiation efficiency are reported in Fig. 10.6. Noting to Fig. 10.6(a), as the size of the first case is progressively doubled, its gain also doubles. In contrast, the second case not only exhibits considerable less gain than the others, but also the overall variation of its gain is decreasing, which is due to the large amount of surface wave excitation in its substrate. The third case, i.e., method 1, although leads to less gain compared to the ideal case, it provides much more gain than the FR4 case. Furthermore, the slope of its gain curve is significantly greater than that of the second case and has not become decreasing. Method 2 provides less gain than the first one but considerably more gain with respect to the FR4 case. At last, radiation efficiency for theses arrays is reported in Fig. 10.6 (b).

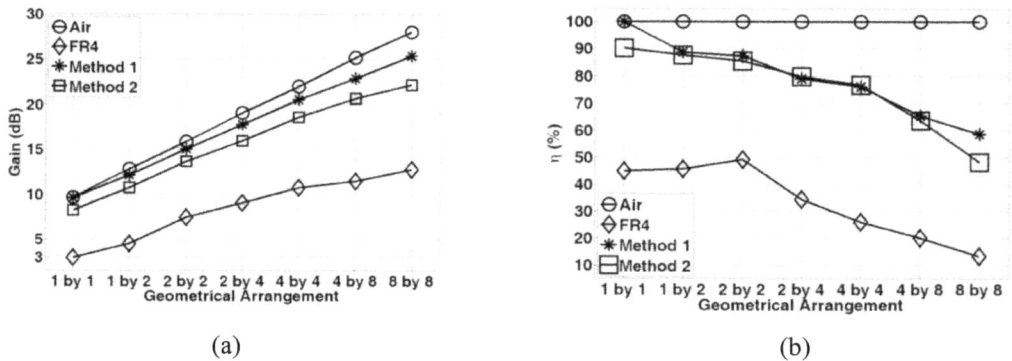

(a)   (b)

Fig. 10.6. Characteristics of corporate feed arrays with different dielectric structure: (a) Gain, and (b) Efficiency.

Ability of method 1 in providing MSA arrays and its superiority to method 2 can be justified by comparing the average per-element gain of the 32 and 64 element designs and the corresponding value for the arrays proposed in [20]. For the 32 element array, these values are, respectively, 7.8 dB, 5.6 dB and 8.0 dB and for the 64 element case, the said values become, respectively, 7.2 dB, 4.1 dB and 7.6 dB. Thus, method 1 can be considered as a pure microstrip alternate to the hybrid waveguide-microstrip structures proposed in [20]. As a result, method 1 is intrinsically superior to method 2, in the sense of gain, for large arrays. Specifically, gain per square centimeter for a 32 element array based on methods 1 and 2 is, respectively, -5.26 dB/cm^2 and -5.25 dB/cm^2 (too close). However, for 64 element arrays, the corresponding values become -5.84 dB/cm^2 and -6.78 dB/cm^2. This justifies superiority of method 1 over method 2, in the sense of gain. As a conclusion, method 1 compared to method 2 leads to larger size, a bit less bandwidth and more gain. Due to exploitation of the sturdy FR4 epoxy as a superstrate, it is more robust and can be used for out-door applications. Method 1 has a superior performance when applied to MSA arrays and is a promising technique to achieve high-gain and low-cost pure microstrip arrays. Realization of method 2 is simpler than method 1, since soldering of the interior conductor of SMA connector to the exterior surface of the FR4 layer is straightforward. Thus, priority of each of the methods 1 and 2 over the other is application dependent.

10.4. Design Examples

To demonstrate capability of methods 1 and 2 in providing high-gain MSA arrays, two linearly polarized arrays with corporate feed networks are designed and analyzed based on these methods. The first one is an 8 × 8 array with pencil-beam at 5.8 GHz that can be used in WiMAX® wireless bridges. The second one is a 2 × 16 array with fan-beam at 12.45 GHz that can be used in digital video broadcast (DVB) applications. Full-wave analysis is performed using MoM engines embedded in Agilent® ADS and FEKO® suite software. The layout, S-parameter and gain patterns of the pencil- and fan-beam arrays are depicted in Figs. 10.7 and 10.8, respectively. Thickness of air and FR4 layers for pencil-beam arrays are, respectively, 1 and 0.5 mm. To compare the studied methods with the one proposed in [20], thickness of the FR4 layer in the fan-beam designs is set to 0.1 mm, accordingly. Yet, since the feed network in methods 1 and 2 is microstrip, height of the air layer need not be as thick as suggested in [20] and is set to 0.5 mm, here. Various parameters of designed arrays are reported in Tables 10.1 and 10.2. In accordance with the previous results, method 2 leads to less size compared to method 1. As well, radiation patterns show superiority of method 1 compared to method 2 in the sense of maximum gain. Finally, there are two important points to be mentioned. First, in Table 10.2 there is a great difference in bandwidth between [20] and simulated results. This is because in [20] a metallic waveguide structure is used as the feed network and it is well-known that bandwidth of such structures are much more compared to their microstrip counterparts, exploited in methods 1 and 2. Second, while it is unfair to compare simulated results with another experimental one as shown in Table 10.2, it is unlikely that the error of both software in gain computation be more than 1 dB. Thus, it is not hard to accept methods 1 and 2 as simpler and less-costly alternates to the one proposed in [20].

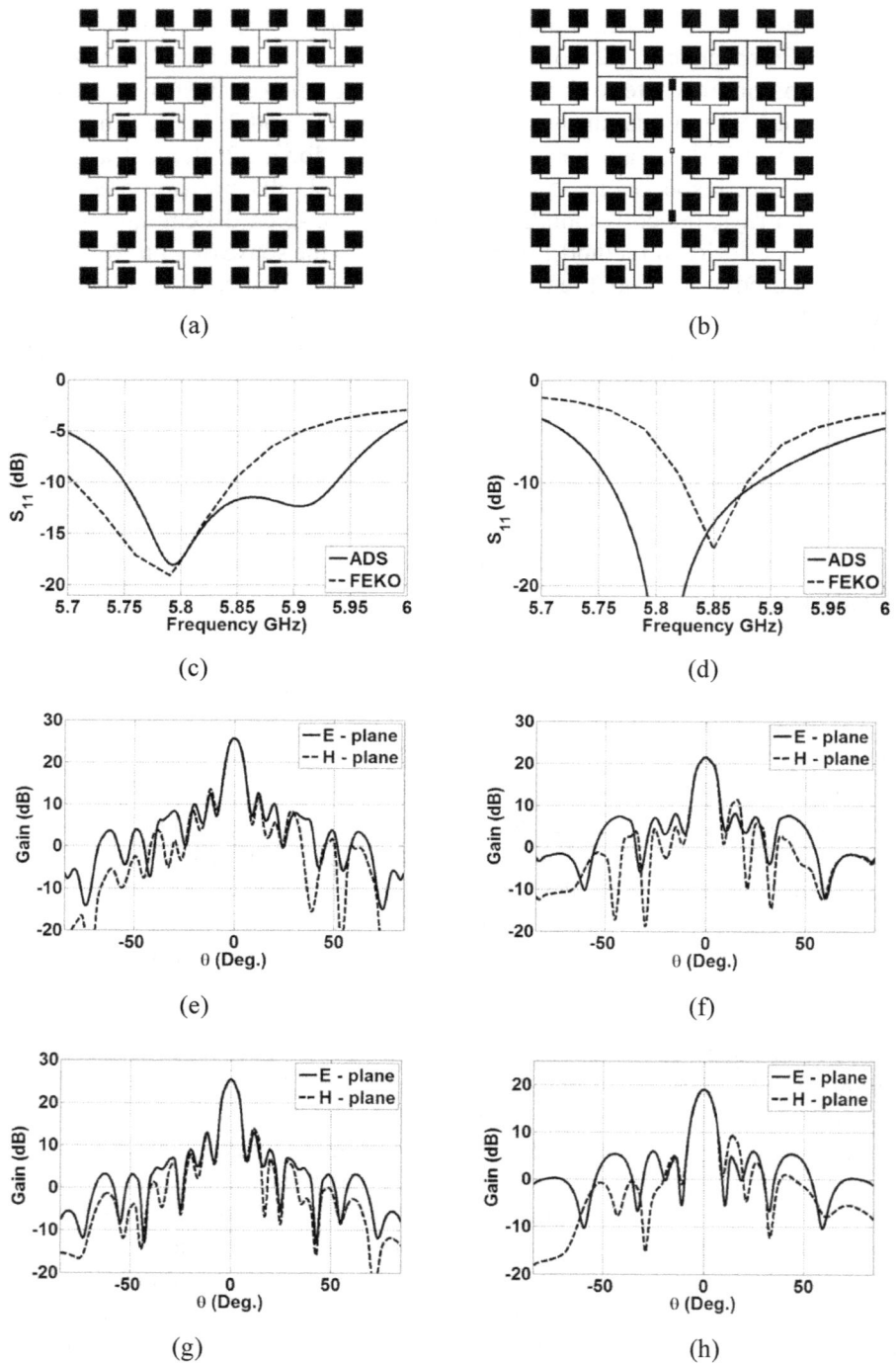

Fig. 10.7. Pencil-beam array: (a) Layout (method 1); (b) Layout (method 2); (c) |S11| (method 1); (d) |S11| (method 2); (e) Gain (method 1 - ADS); (f) Gain (method 2 - ADS); (g) Gain (method 1 - FEKO); (h) Gain (method 2 - FEKO).

(a)

(b)

(c)

(d)

(e)

(f)

(g)

(h)

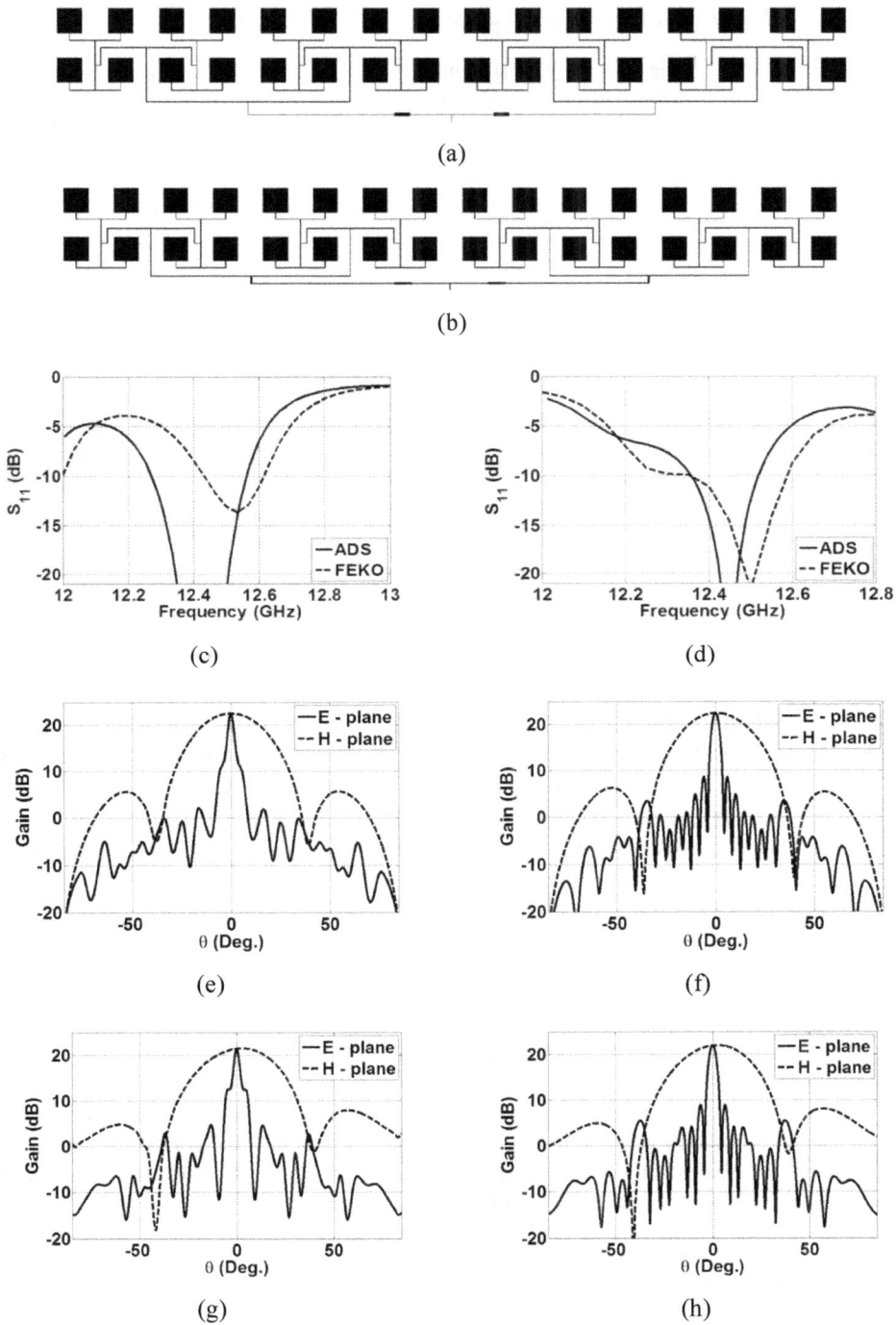

Fig. 10.8. Fan-beam array: (a) Layout (method 1); (b) Layout (method 2); (c) $|S_{11}|$ (method 1);
(d) $|S_{11}|$ (method 2); (e) Gain (method 1 - ADS); (f) Gain (method 2 - ADS);
(g) Gain (method 1 - FEKO); (h) Gain (method 2 - FEKO).

Table 10.1. Parameters of the sample pencil-beam MSA array.

		Gain (dB)	HPBW (Deg.) E-plane	HPBW (Deg.) H-plane	S_{11} (dB)	BW (MHz)	Efficiency (%)	Dimensions (cm^2)
ADS	Method 1	25.6	6.9	6.9	-18	186	62.7	35×35
	Method 2	21.5	9.6	9.6	-40	126	47.5	28×28
FEKO	Method 1	25.4	7.2	7.2	-18	160	55.5	35×35
	Method 2	19.1	9.0	8.9	-16	54	20.1	28×28

Table 10.2. Parameters of the sample fan-beam MSA array.

		Gain (dB)	HPBW (Deg.) E-plane	HPBW (Deg.) H-plane	S_{11} (dB)	BW (MHz)	Efficiency (%)	Dimensions (cm^2)
ADS	Method 1	22.6	2.7	31	-40	287	63	32×4.8
	Method 2	21.6	2.7	34	-37	150	61	30×4.1
FEKO	Method 1	22.5	3.7	31	-14	167	63	32×4.8
	Method 2	22.1	3.9	33	-22	150	64	30×4.1
Measured	[20]	20.0	5.5	39	-20	> 500	63	37×4.0

10.5. Conclusions

Two low-cost high-gain techniques were studied, which both utilizes a suspended metallization etched on an FR4 layer. These methods are discriminated based on the position of the metallization that is placed, respectively, on the interior and exterior sides of the FR4 epoxy. A detailed parameter study on the FR4 thickness shows superiority of the first method over the second one in the sense of gain and efficiency, especially for array designs. Additionally, the second method leads to a more symmetric pattern and less size with respect to the second one.

References

[1]. S. Assailly, C. Terret, J. P. Daniel, Some results on broad-band microstrip antenna with low cross polar and high gain, *IEEE Trans. Antennas Propagat.*, Vol. 39, Issue 3, 1991, pp. 413-415.

[2]. H. Legay, L. Shafai, New stacked microstrip antenna with large bandwidth and high gain, *IEE Proceedings Microw. Antennas Propag.*, Vol. 141, Issue 3, 1994, pp. 199-204.

[3]. S. Egashira, E. Nishiyama, Stacked microstrip antenna with wide bandwidth and high gain, *IEEE Trans. Antennas Propagat.*, Vol. 44, Issue 11, 1991, pp. 1533-1534.

[4]. C. Y. Huang, J. Y. Wu, K. L. Wong, High-gain compact circularly polarised microstrip antenna, *Electron. Lett.*, Vol. 34, Issue 8, 1998, pp. 712-713.

[5]. T. H. Hsieh, C. S. Lee, Double-layer high-gain microstrip array antenna, *IEEE Trans. Antennas Propagat.*, Vol. 48, Issue 7, 2000, pp. 1033-1035.

[6]. H. S. Shin, N. Kim, Wideband and high-gain one-patch microstrip antenna coupled with H-shaped aperture, *Electron. Lett.*, Vol. 38, Issue 19, 2002, pp. 1072-1073.

[7]. R. Gardelli, G. L. Cono, M. Albani, A low-cost suspended patch antenna for WLAN access points and point-to-point links, *IEEE Trans. Antennas Wireless Propagat. Lett.*, Vol. 3, 2004, pp. 90-93.

[8]. C. L. Mak, H. Wong, K. M. Luk, High-gain and wide-band single-layer patch antenna for wireless communications, *IEEE Trans. Veh. Technol.*, Vol. 54, Issue 1, 2005, pp. 33-40.

[9]. G. R. DeJean, M. M. Tentzeris, A new high-gain microstrip Yagi array antenna with a high front-to-back ratio for WLAN and millimeter-wave applications, *IEEE Trans. Antennas Propagat.*, Vol. 55, Issue 2, 2007, pp. 298-304.

[10]. H. Wang, D. G. Fang, A single layer wideband U-slot microstrip patch antenna array, *IEEE Trans. Antennas Wireless Propagat. Lett.*, Vol. 7, 2008, pp. 9-12.

[11]. S. Chattopadhyay, J. Y. Siddiqui, D. Guha, Rectangular microstrip patch on a composite dielectric substrate for high-gain wide-beam radiation patterns, *IEEE Trans. Antennas Propagat.*, Vol. 57, Issue 10, 2009, pp. 3324-3327.

[12]. R.V.S. Ram Krishna, R. Kumar, Slotted ground microstrip antenna with FSS reflector for high-gain horizontal polarization, *Electron. Lett.*, Vol. 51, Issue 8, 2015, pp. 599-600.

[13]. X. J. Gao, T. Cai, L. Zhu, Enhancement of gain and directivity for microstrip antenna using negative permeability metamaterial, *AEU – Int. J. Electron. C.*, Vol. 70, Issue 7, 2016, pp. 880-885.

[14]. Q. U. Khan, M. B. Ihsan, Higher order mode excitation for high gain microstrip patch antenna, *AEU – Int. J. Electron. C.*, Vol. 68, Issue 11, 2014, pp. 1073-1077.

[15]. K. Mandal, P. P. Sarkar, A compact high gain microstrip antenna for wireless applications, *AEU – Int. J. Electron. C*, Vol. 67, Issue 12, 2013, pp. 1010-1014.

[16]. S. Park, C. Kim, Y. Jung, H. Lee, D. Cho, M. Lee, Gain enhancement of a microstrip patch antenna using a circularly periodic EBG structure and air layer, *AEU – Int. J. Electron. C*, Vol. 64, Issue 7, 2010, pp. 607-613.

[17]. E. Levine, G. Malamud, S. Shtrikman, D. Treves, Study of microstrip array antennas with the feed network, *IEEE Trans. Antennas Propagat.*, Vol. 37, Issue 4, 1989, pp. 426-434.

[18]. S. Demir, Efficiency calculation of feed structures and optimum number of antenna elements in a subarray for highest G/T, *IEEE Trans. Antennas Propagat.*, Vol. 52, Issue 4, 2004, pp. 1024-1029.

[19]. M. Yousefbeigi, A. Enayati, M. Shahabadi, D. Busuioc, Parallel– series feed network with improved G/T performance for high-gain microstrip antenna arrays, *Electron. Lett.*, Vol. 44, Issue 3, 2008, pp. 180-182.

[20]. M. Shahabadi, D. Busuioc, A. Borji, S. Safavi-Naeini, Low-cost, high-efficiency quasi-planar array of waveguide-fed circularly polarized microstrip antennas, *IEEE Trans. Antennas Propagat.*, Vol. 53, Issue 6, 2004, pp. 2036-2043.

[21]. S. Y. Chen, P. Hsu, CPW-fed horn-shaped slot array antenna for 5-GHz WLAN access point, in *Proceedings of the European Conference on Wireless Technology (ECWT'85)*, Amsterdam, Netherlands, 1985, pp. 281-284.

[22]. M. Niroojazi, M. N. Azarmanesh, Practical design of single feed truncated corner microstrip antenna, in *Proceedings of the Conference on Communication Networks and Services Research (CNSR'04)*, 21 May 2004.

[23]. L. N. Sharma, S. Das, A. K. Gogoi, ISM band triangular patch antenna on FR4 substrate with 'U' pattern slots, in *Proceedings of the IEEE Annual India Conference (INDICON'05)*, Chennai, India, 11-13 December 2005, pp. 184-187.

[24]. M. Nashaat, H. Elsadek, Miniturized E-shaped dual band PIFA on FR4 substrates, in *Proceedings of the Conference on Radio Science (NRSC'06)*, 14-16 March 2006.

[25]. D. Bhardwaj, S. Sancheti, D. Bhatnagar, Design of square patch antenna with a notch on FR4 substrate, in *Proceedings of the Asia-Padific Microwave Conference (APMC'07)*, December 2007, pp. 11-14.

[26]. C. F. Yang, W. C. Tzou, J. H. Tsai, H. M. Chen, Y. F. Lin, Enhance antenna bandwidth by using high permittivity ceramic and FR4 stacked structure, in *Proceedings of the IEEE International Conference on Antennas and Propagation (AP-S'07)*, Honolulu, USA, 9-15 June 2007, pp. 2012-1015.

[27]. D. C. Nascimento, R. Schildberg, J. C. da S. Lacava, New considerations in the design of low-cost probe-fed truncated corner microstrip antennas for GPS applications, in *Proceedings of the IEEE International Conference on Antennas and Propagation (AP-S'07)*, Honolulu, USA, 9-15 June 2007, pp. 749-752.

[28]. G. C. Hock, C. K. Chakrabarty, Emilliano, M. H. Badjian, Dielectric verification of FR4 substrate using microstrip bandstop resonator and CAE tool, in *Proceedings of the Malaysia International Conference on Communications (MICC'09)*, Kuala Lumpur, Malaysia, 15-17 December 2009, pp. 894-898.

[29]. T. Abdul Rahman, I. M. Ibrahim, P. S. Wei, J. A. Ab Ghani, C. Wahab, A study on effectiveness of FR4 as an electric material for radial line slot array antenna for wireless backhaul application, in *Proceedings of the Asia Pacific Conference on Communication (APCC'11)*, Sabah, Malaysia, 2-5 October 2011, pp. 385-388.

[30]. G. C. Hock, N. T. N. Tho, C. K. Charabarty, T. S. Kiong, Design of patch antennas array at low frequency application by using unknown FR4 material, in *Proceedings of the IEEE Conference on Control System, Computing and Engineering (ICCSCE'11)*, Penang, Malaysia, 25-27 November 2011, pp. 302-305.

[31]. Z. Ma, G. A. E. Vandenbosch, Optimal design of wideband microstrip arrays with high aperture efficiency on FR4 substrate, in *Proceedings of the European Conference on Antennas and Propagation (EUCAP'12)*, Prague, Czech Republic, 26-30 March 2012, pp. 2924-2927.

[32]. D. C. Nascimento, J .C. da S. Lacava, Cavity-backed hybrid microstrip antenna on FR4 substrate, in *Proceedings of the IEEE International Conference on Antennas and Propagation (AP-S'13)*, Orlando, USA, 7-13 July 2013, pp. 166-167.

[33]. M. M. Islam, M. T. Islam, M. R. I. Faruque, Design of a triple frequency band patch antenna on FR4 substrate material, in *Proceedings of the IEEE International Conference on RF and Microwave (RFM'13)*, Penang, Malaysia, 9-11 December 2013, pp. 349-352.

[34]. Z. Chen, Y. P. Zhang, FR4 PCB grid array antenna for millimeter-wave 5G mobile communications, in *Proceedings of the Microwave Workshop Series on RF and Wireless Technologies for Biomedical and Healthcare Applications (IMWS-BIO'13)*, 9-11 December 2013, Singapore, pp. 1-3.

[35]. I. M. Ibrahim, T. A. Rahman, M. I. Sabran, M. F. Jamlos, Comparison of polypropylene and FR4 dielectric cavity for RLSA antenna design at 5.8 GHz, in *Proceedings of the IEEE Region 10 Symposium*, 14-16 April 2014, Kuala Lumpur, Malaysia, pp. 232-234.

[36]. N. Iizasa, R. Pokharel, H. Kanaya, K. Yoshitomi, High gain 4 × 4 slot dipole antenna array in the 5 GHz band, in *Proceedings of the Asia-Pacific Conference on Antennas and Propagation (APCAP'14)*, 26-29 July 2014, Harbin, China, pp. 197-200.

[37]. R. Lelaratne, R. J. Langley, Dual-band patch antenna for mobile satellite systems, *IEE Microwave Antennas Propag.*, Vol. 147, Issue 6, 2000, pp. 427-430.

[38]. D. C. Nascimento, R. Schildberg, J. C. da S. Lacava, Design of low-cost microstrip antennas for Glonass applications, *PIERS Online*, Vol. 4, No. 7, 2008, pp. 767-770.

[39]. W. Hong, K. H. Baek, A. Goudelev, Multilayer antenna package for IEEE 802.11ad employing ultralow-cost FR4, *IEEE Trans. Antennas Propagat.*, Vol. 60, Issue 12, 2012, pp. 5932-5938.

[40]. D. C. Nascimento, J. C. da S. Lacava, Design of arrays of linearly polarized patch antennas on an FR4 substrate, *IEEE Antennas Propagat. Mag.*, Vol. 57, Issue 4, 2015, pp. 12-22.

[41]. N. Nasimuddin, Microstrip Antennas, *InTech*, 2011.

[42]. R. E. Collin, Antennas and Radiowave Propagation, *McGraw-Hill*, 1985.

[43]. C. A. Balanis, Antenna Theory: Analysis and Design, *Wiley*, 2005.

[44]. D. M. Pozar, D. H. Schaubert, Microstrip Antennas, *Wiley-IEEE Press*, 1995.

[45]. J. R. James, P. H. Hall, Handbook of Microstrip Antennas, *IET Press*, 1989.

[46]. J. B. Ness, Microstrip antenna, US4697189A, *US Patent*, 1987.

[47]. F. Lalezari, Microstrip antenna structure having an air gap and method of constructing same, US5444453 A, *US Patent*, 1995.

Chapter 11

Interference Management and System Optimisation for Femtocells Technology in LTE and Future 4G/5G Networks

Rand Raheem, Aboubaker Lasebae, Mahdi Aiash, Jonathan Loo, Robert H. Colson and Ali Raheem

11.1. Introduction

3GPP had standardised the Long Term Evolution (LTE) to improve the spectral efficiency and the speed of data rate of a cellular network. It also, introduced an intelligent Base Station (BS), which is the evolved Node B (eNB), and that was an important complement to simplify the system architecture and minimise the control plane and UE plane latency. In addition, LTE offered significant improvements over previous technologies such as UMTS and High-Speed Packet Access (HSPA) by introducing a novel Physical Layer and reforming the Core Network. The main reasons for these changes in the Radio Access Network (RAN) system design are the need to provide higher spectral efficiency, lower delay, and more multi-user flexibility than the currently deployed networks. Besides that, 3GPP has extended the original proposed LTE to LTE-Advance (LTE-A) by employing advanced Multi-antenna Multiple-Input, Multiple-Output (MIMO) techniques, Carrier Aggregation (CA), Relay stations, Heterogeneous Network (HetNet) and many other schemes [1]. However, since the LTE-A standard creates a bridge between 4G and 5G worlds, in many ways; the notion of HetNet is serving as glue between LTE-A and 5G worlds. That is why many wireless industry observers call 5G wireless an enhanced form of LTE-A. In LTE-A, HetNet is a gradual evolution of the cellular architecture, is a vastly more complex network as small cells add hundreds or even thousands of entry points into the cellular system.

The Self-Organizing Network (SON) concept is also one of the key enabling technologies that are considered in LTE-A applications in order to organise the communication between these small cells. That makes sense because the main concept behind developing

Rand Raheem
Faculty of Science & Technology, Middlesex University, London, United Kingdom

5G systems is to expand the idea of small cell network to a completely new level and create a super dense network that will put tiny cells in every room and vehicle. This leads to a conclusion that LTE-A is the foundation of 5G RAN. The 5G Network raises the bar to higher frequency usage from 6 GHz to 100 GHz and Massive MIMO technology. The Massive MIMO technology acts as a large array of radiating elements that extends the antenna matrix to a new level from 16×16 to 256×256 MIMO and this is considered as a leap in the wireless network speed and coverage [2]. Therefore, it is expected that by 2020 the 5G systems will be available for users in order to enhance the performance of current technologies [3].

11.1.1. Motivation Towards Small Cells

With the increasing number of users and applications, changes are required in any technology to cope with the users' needs. In the past, mobile communication networks were dominated by phone calls and data transfer requirements with limited number of users, while a few years ago, emails, web pages, data files and applications dominated these networks as well as the increased number of users who require to be connected to the Internet everywhere and at any time. This has led to a huge growth in the amount of data and pressure on the network Operators that are trying to transfer data across mobile communication networks. In response to this problem, network Operators have greatly increased the capacity of the mobile communication networks. One way of achieving this is by improving the used technology so that information between mobile phones and the BSs can be transmitted and received faster than ever before. Therefore, LTE technology was associated to releases 8 and 9, and now LTE- Advanced supports wider Bandwidths (BWs) with low latencies to maximise the data rate and provide better Quality of Service (QoS) to end User Equipment (UE). In addition, advance modulations techniques are used, such as 64 QAM, to support high data rates with multiple services that will open the door towards 5G Networks and onwards. However, these high data rates may cause high traffic load and burden on the serving eNB that may take the whole network down as Fig. 11.1 shows.

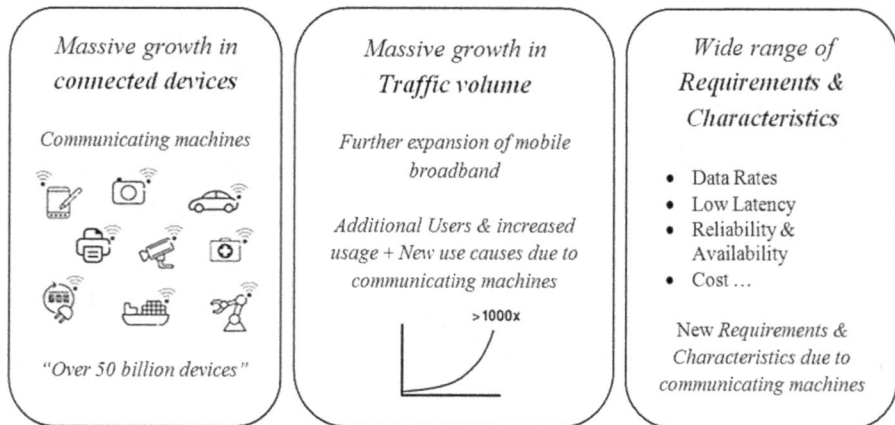

Fig. 11.1. Key challenges to the evolution of LTE.

Therefore, it was required to come up with small BSs to improve the network capacity and to offload the load from the eNB to these small BSs. Besides that, LTE high penetration loss has created an obstacle for indoor UEs to be connected to the outdoor BSs. Consequently, Femtocells have been deployed as a solution to mitigate the LTE high penetration loss in an indoor environment like houses, offices, small organisations to meet the users' needs. In contrast, the European 5G project (METIS) predicts that by 2020, a large number of mobile UEs will be vehicular. The number of those vehicular UEs will be 50 active vehicular UE devices per bus while up to 300 active vehicular UE devices per train [4]. This has created a significant motivation towards improving the performance of cell-edge vehicular UEs inside public transportations like buses and trains, reducing outage probabilities, penetration loss, path-loss and mitigating the interference in LTE and 5G Networks. However, due to the high mobility of those vehicular UEs, deploying fixed Femtocells may not be a reasonable solution for those UEs as the excessive penetration loss and the unnecessary number of Handovers (HOs) plays a significant role in this case. Thus, an efficient technology for LTE cell-edge vehicular users is required as the performance of those UEs in LTE network has been always a motivation for many researches to improve their connection and data rates as shown in [5-7]. The previous motivations were the enlightenments for many technologies to be developed like; Relay nodes, Femtocells, Pico-cells and Microcells to improve the end UE connection for indoor environments. While a few studies have considered the vehicular environment as a challenge issue to improve the end vehicular UEs' connection and performance as Fig. 11.2 shows [8].

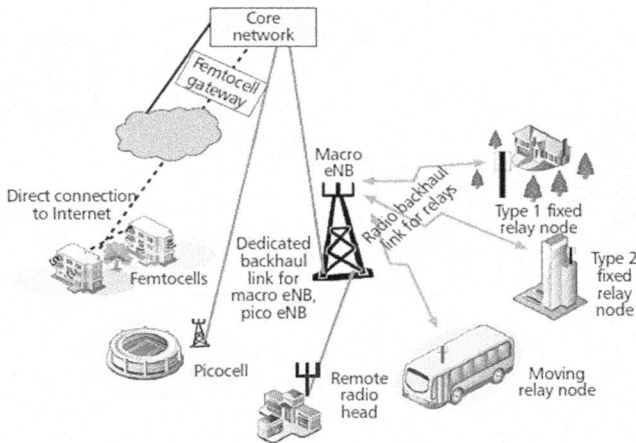

Fig. 11.2. LTE small base stations.

Moreover, the growth in traffic requires an improvement in the spectrum and its efficiency in order to cope with the increased number of small mobile cells. The latter will help in increasing the network density and decreasing the load burden on the Macrocell (eNB) as stated before. This provides a strong motivation for developing Femtocells and provides short-range coverage. Needless to say that, the most typical deployment of Femtocells is the co-channel deployment where the same carriers are shared between the Femtocell and the Macrocell. Consideration must be taken into account here to overcome the interference issue in co-channel deployment, which cannot be mitigated efficiently by traditional network planning techniques [9]. Several studies have proved that vehicular UEs are more affected by interference, dropped calls and unnecessary number of HOs [10]. Therefore, managing interference in a shared spectrum makes mobile broadband sustainable and ensures that data throughput is improved with low signals leakage so that each UE gets the most BW without the need for manual intervention [11].

11.1.2. Challenges of Vehicular Environment

With the evolution of communication services and wireless diversity, mobile networks require to support the increasing demands in order to provide UEs with better QoS. The LTE ecosystem for the next decade is expected to meet the market demands with a wide range of services with multiple devices, higher capacity and coverage areas. Thus, the evolution of LTE networks created many obstacles that are required to be considered continuously such as LTE high penetration losses and traffic loads. The high traffic loads occur because of the many applications with high data throughput that are required to be handled in a much more efficient way as Fig. 11.3 shows [12]. This may not be a big issue in the case of fixed and slow mobility UEs while it may be a concern in the case of vehicular UEs since the latter is more affected by the high penetration loss, path-loss, interference and mobility. However, addressing these issues will be the evolution towards developing 5G Networks.

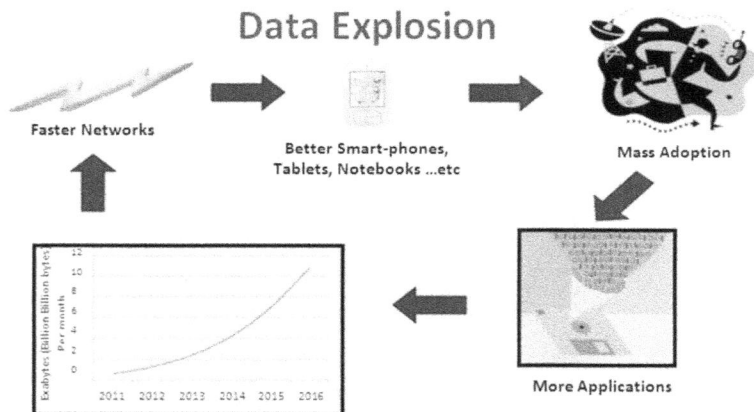

Fig. 11.3. LTE Data Explosion challenges.

Having fixed Femtocells have helped the network in expanding the coverage areas of indoors environment and overcoming the issue of high penetration loss. This also has helped in improving the performance of those indoor users in LTE and future networks. While the challenge issue here is still the quality of connection and performance of vehicular UEs in outdoor environment such as public transportations like buses. This means that the ability of having outdoor Femtocells to serve vehicular UEs in buses has become a concern in every research. Therefore, this work reviews the impact of having fixed Femtocells to serve outdoor vehicular UEs. It also examines the need of having an efficient technology that serves the vehicular UEs from inside the vehicle and aims to improve the throughput and Signal to Interference plus Noise Ratio (SINR) while ultimately reducing outage probabilities. This technology needs to satisfy the users' needs and reduce the traffic load on the eNB. Added to this, with the increased number of UEs and signaling, the serving cell should assign the UEs to the 'always online' state which means a longer connection time with better link quality and hence dynamic scheduling and dynamic resource allocation [13].

Moreover, there are various forms of interference that are a matter of major concern as they make the LTE and 5G systems vulnerable, which in turn cause communication failures and could take down the whole network [14]. One of these interference forms is the co-channel interference originating from the presented Femtocell (Mobile-Femto) with the nearby eNB and the second interference form is the co-tier interference between the Mobile-Femto and nearby Mobile-Femto. To mitigate this issue, the impact of transmission power and network deployment has been tackled to evaluate the need of having an efficient frequency reuse scheme. This dynamic technique dose not waste frequency since the same frequency can be reused several times over distance to mitigate the generated interference.

Addressing these challenges was therefore the main drive of this work as Femtocell technology and methods are required to improve the existing networks and smart phones' performance to meet the growing demands for data and subscribers. Therefore, the aim of this chapter is to study the LTE standard and provide comprehensive solutions to mitigate outlined challenges that can be overcome in 5G and future networks. In an LTE system and due to technology evolution, it is required to introduce self-optimising and dynamic solutions to endow the current standard with more intelligence. This will make the network more adaptive and able to handle peak data demands and cope with the increasing capacity requirements especially in the case of the vehicular UEs who are considered as a significant percentage in LTE networks. Additionally, by 2020 it is predicted that this percentage will be increased especially with the evolution of 5G Networks [4].

11.2. Related Work

Mobile computing is fast becoming a vital part of everyday life in which UEs demand being reachable anywhere and anytime as they spend much time traveling from one place to another, often by trains or buses. The ultimate purpose of passengers is the ability to be connected to the Internet while they are moving from one place to another with their mobile devices. Providing indoor coverage on trains and buses directly with the outdoor BSs may not be a good solution due to the high penetration losses in the LTE and LTE-A networks, especially for high speed trains caused by the Faraday cage characteristics of the railcars [15]. This fact leads to a poor signal quality inside the train and offering broadband services is not always possible. Although, the broadband access on buses and trains could be achieved by installing BSs close to the railways, however, this solution is not quite convenient to trains and buses' Operators due to the high investment needed to deploy such BSs. In additions, this will increase the number of unnecessary HOs. This issue has focused the research community effort on offering solutions that take advantages of the existing wireless infrastructure in order to propose efficient methods to manage the mobility in seamless way for the end-UEs. Therefore, this literature reviews the related work in terms of the vehicular UEs performance, the generated interference and the suggested solutions to improve their performance in LTE networks.

11.2.1. LTE Vehicular UEs Penetration Loss

The indoor coverage can badly be degraded by the penetration losses through the walls of buildings. If the BS is outdoors but the mobile is indoors, then the penetrations losses typically reduce the received signal power by 10 to 20 decibels (dB) - a factor of 10 to 100 - which can greatly reduce the indoor coverage. This is one of the motivations behind the progressive introduction of Femtocells [16]. It is worth noting that the previous fact applies to vehicular UEs as well, as those UEs experience high penetration loss since there is a barrier, the vehicle's chassis between those UEs and the outdoor BSs that affects the quality of the transmitted and received signals. Thus, in order to solve the previous issue of the high LTE Vehicular Penetration Loss (VPL), the authors in [17] present a series of VPL measurements performed in 800 MHz-frequency band. These measurements were conducted for three different vehicle types, mini-van, full size car and sports car with different types of environments e.g. urban or suburban. The statistical properties of the VPL have been examined in order to determine the benchmark parameters to be used in the design of wireless communication systems. While the achieved results have shown that the vehicle's chassis, speed and distance from the serving BS play very effective roles in terms of the quality of the transmitted signal. Hence, this study has made it clear why the vehicular UEs in buses and trains suffer from the worst link connection with the outdoor BS. According to [18], the high-speed trains can be a fruitful environment for mobile services as users are concentrated in relatively small areas. In the train environments, the trains' paths are always known and the railway environment itself has large tunnels, wide cuttings and curves. However, several issues arise in such an environment like fading, Doppler, transients, and penetration loss into carriages, as well as special situations such as cuttings and tunnels. This creates a problem with the operation of the physical layer as this may affect the link between the UE and the outdoor serving BS that causes performance degradation at high speeds.

Obviously, this has shown that vehicular UEs are most affected by the high penetration loss due to the signal strength fluctuation and radio link failures between the vehicular UEs and the outdoor BSs as the mobility aspect plays a negative role in this case. Therefore, the coming sub-section discusses the impact of the VPL on the vehicular UEs performance, besides analysing the previous implemented solutions and technologies to improve the vehicular UEs performance.

11.2.2. Vehicular UEs Performance

The signal quality inside vehicles is very poor due to the high penetration loss, path-loss and fading. However, poor signal means poor SINR as in wireless communication the

SINR is being used to measure the signal quality of wireless connections. Therefore, in order to improve the SINR inside a vehicle, indoor coverage needs to be deployed as study [19] shows. In this study, the authors discuss the ability of improving the QoS of vehicular UEs and solving the issues behind the low SINR by deploying mobile Femtocells in the Macrocell. The proposed study has stated that the short distance between the UE and the Femtocell provides better signal quality. The inside Femtocells have been connected to the Core Network or the Macrocell through the satellite networks. One stronger transceiver is installed outside the roof of the vehicle. This transceiver is connected to the Femtocells using wired connection and to the Macrocell and the satellite access networks using wireless link. Another study however has discussed the cell-edge vehicular UEs who suffer from low SINR and performance degradation in general [20]. The authors have considered the feasibility of Decode and Forward (DF) Relay nodes from the 3GPP LTE-A perspective as an attractive solution to solve the SINR reduction. The proposed solution is based on finding the relation between the relay node transmission power, ratio between number of BSs and Relay nodes and the performance of the system. The achieved results have shown a good performance in terms of the signal strength after the deployment of the Relay nodes. However, the proposed solution was not practical from the network Operator point of view as the interference issue might occur in the case of increased traffic. This is because with the high traffic load, increasing the transmission power to cover wider range can increase the interference, while reducing the transmission power to eliminate the interference can increase the number of HOs especially in the case of vehicular UEs. This has its negative impact on the spectrum efficiency of served UEs and the utilisation of network resources.

The spectrum/spectral efficiency or the BW efficiency is more affected by the UE's mobility and speed. The authors in [21] have discussed the ability of improving the spectrum efficiency via using the mobile Femtocell technology. This study has stated that the spectral efficiency of mobile Femtocell's UE can be improved with the use of two resource partitioning schemes, orthogonal and non-orthogonal. However, this study has neglected the generated interference between the mobile Femtocells themselves and it assumed that the chosen environment is an ideal environment where there is no noise and interference, which may affect the accuracy and reliability of the achieved results since it does not match the real life scenarios. While [22], has discussed the problem of resource allocation in a cellular network with mobile Femtocells. This study has shown that the speed and path information of the mobile Femtocells have been used to determine the interference correlations between different Femtocells at different time instants and represent them as a time interval dependent interference graph.

More studies have been done by [12, 23, 24] on the resource management of Femtocell by introducing several mechanisms and techniques to share resources between the Macrocells and Femtocells. Whilst sharing the spectrum efficiently among the Macrocell and Femtocell's UEs with less interference has been discussed in [25-27]. Other studies concentrated more on investigating the OFDMA technique based on LTE network like [26, 28]. These studies have shown the correlation between the OFDMA and Physical Resource Blocks (PRBs) as well as the impact of this access scheme on sharing the resources between Macrocell and Femtocell UEs. While other studies like [29, 30] have observed the impact of PRBs distribution on the achieved cell capacity in LTE networks. Distributing the PRBs and dynamically allocating the resources among UEs can greatly improve the achieved capacity as well as the spectrum efficiency. Hence, all the previous studies have shown that the distribution of the PRBs depends on the UE's traffic that the cell is experiencing which might be a burden on the network without an efficiently resource allocation scheme. However, most of the previous studies were limited to the fact that they have considered the case of fixed BSs with mobile UEs but not many who considered the mobile BSs with mobile UEs which is considered a challenge issue in this case.

On the other hand, many researchers have considered the impact of UEs' mobility and speed on the network throughput as an area of interest. In [31], the authors have presented a system level simulation results for a cooperative moving Relay node system deployed on a High-Speed Train (HST) to provide enhanced cellular coverage to UEs in public transportations, particularly HSTs, of which the modern construction materials and techniques cause high VPL when signals propagate into the train. This study has shown that the mobile relay nodes utilising antenna arrays on the exterior and interior of the train are a promising solution to overcoming this VPL in order to provide on-board UEs with improved services. The achieved results showed a slight improvement in the achieved throughput of on-board UEs when compared to direct transmission of those vehicular UEs. Another study has considered the mobile Relays as a solution to improve the vehicular UEs throughput as in [32]. Here, the authors have considered the mobile Relay node to be deployed on public transportation to serve vehicular UEs in order to reduce the impact of the penetration loss and improve the UEs throughput. However, both of the previous studies were limited to the fact of the limited number of served UEs i.e. max five UEs, and coverage areas while another issue added to the second study, which is the random movement of the mobile Relay that agitates the interference problems.

In contrast, the mobility management procedures in LTE systems such as HO and cell reselection are becoming more complex due to the dense deployment of different type of cells. Many studies have examined the mobility robustness and HO issues in heterogeneous networks. Researchers have focused on different topics such as signal strength during HO, interference, SINR, cells' properties, UEs positioning and outage probability. However, this chapter is more focused on UEs mobility and its impact on the outage probability and connection quality between the vehicular UEs and the serving BS. This is because; having a high outage probability can drop the network performance and affect the UE's quality of service. Therefore, several studies have considered the vehicular UEs outage probability as an issue to be solved in LTE networks. In [33], the authors have investigated the power outage probability of vehicular UEs who are served by half-duplex decode-and-forward Relay Nodes (RN) under co-channel interference. The aim of the proposed framework is to optimise the HO parameters, as well as numerically optimise the fixed RN position, which minimises the average power outage probability at the vehicular UEs end. Fixed RN shows its advantage in serving its nearby vehicular UEs while the moving RN has shown a better quality of connection in terms of those vehicular UEs. On the other hand, in [34], the authors have proposed an efficient HO scheme, which contains two procedures in order to mitigate the signal outage probability of vehicular

UEs and improve their performance. The first procedure is an enhanced measurement procedure, which can accelerate the measurement procedure when the mobile Relay knows that the train is moving towards some neighbouring Donor eNBs (DeNBs). The second procedure is a group in-network HO procedure, which can occur similarly to network HO procedures in the Core Network. However, the limitation in the previous two studies is the occurrence of the interference among RN UEs and Macro UEs as RNs use the unlicensed spectrum unlike the Femtocells in LTE networks.

Where, the authors in [35] have proposed an analytical model for mobile UE connection probability in Femto-Macro cellular networks. The aim of this study is to improve the performance of the connectivity probability in terms of the communication range when Macro UEs at the boundary of the Macrocell and at the same time at the boundary of the Femtocell. The achieved results have shown an improvement in the performance of the connectivity probability after increasing the communication range by increasing the number of deployed Femtocells in the Macrocell. However, this study raises the cost issue and the deployment of unnecessary number of Femtocells in the Macrocell. Increasing the number of Femtocells will bring more issues regarding the interference and the unnecessary number of HOs with the highly UEs speed. While in [36], the authors have evaluated the performance advantage of using Femtocells as mobile Relays by communicating with the Macro BSs to improve and extend coverage for mobile UEs. The proposed approach enables cooperative strategies between Home BSs and Macrocell BSs in order to mitigate the signal outage probability of vehicular UEs. Whereas in [21], the authors have investigated the effects of using mobile Femtocells to serve vehicular UEs and the impact of that on the amount of signaling overhead with the Macrocells. The results show that there is a large saving in the volume of control signaling after using mobile Femtocells to communicate with the eNB on behalf of the on-board mobile UEs. This has its own positive impact on improving the signals inside vehicles. Other studies like [19] have concentrated on improving the signal quality inside vehicles. This study has proposed the deployment of Femtocells in vehicles to improve the uplink throughput for mobile UE. The mobile BS is connected to the Operator's Core Network through the Macrocell BS or a satellite where the results have shown that mobile Femtocells can improve service quality and maintain an acceptable level of SINR.

Hence, all the previous works together with [31] and [37] have shown that the increased demand for using the new multimedia services and features of today's Smart-Phones in vehicular environment have been considered as a drawback in nowadays networks. This is because, vehicular UEs may not be able to connect to the network directly without the use of an efficient technology to cover the network holes and improve the vehicular UEs performance. Also it was quite noticeable that very few studies have considered the Mobile-Femto as a research of interest unlike the Fixed-Femto and Relay nodes that have been discussed in many studies like [38] and [39]. This is due to several challenges come with the deployment of Mobile-Femto technology. These challenges are needed to be considered from different aspects like; resource distribution between Macrocells and Mobile-Femtos, UEs scheduling process and link capacity to accommodate the increased transmitted data rate. Additional to that, the main concern will be reducing the effect of path-loss, penetration loss, link outage probability, interference and improve the performance of vehicular UEs.

11.2.3. Small Cells Interference

The deployment of low power nodes in heterogeneous networks has a beneficial impact on enhancing the system's capacity by bringing the transmitter and the receiver closer to each other. However, the dense deployment of these small cells has created a serious interference issue. Thus, in order to reduce and mitigate the interference between those small BSs and between the Macrocell, many studies have been carried out based on the power control for Uplink (UL) and Downlink (DL) signals, radio resource scheduling, Femto/Macro local information, UE measurement, cooperative approach, time/frequency allocation, QoS requirements, frequency division, frequency reuse and network planning approaches. While other studies like [40-42] have discussed the impact of the interference issue on the network performance in LTE and LTE-A networks. The previous studies have shown that, the position of the Femtocells needs to be chosen wisely in order to have the smallest possible number of Femtocells but with the largest possible coverage area. Using this technique reduces the implementation cost and the interference issue as it is necessary to specify the cell threshold areas and place the Femtocells near to these areas.

Other studies have considered the transmission power as a way of controlling the interference issue like in [43]. In this study, the author has proposed a method for power control of randomly deployed Femtocells. The system model is based on the idea of a constant Femtocell coverage area by enabling the control of pilot power and data on the DL transmission. In the UL, the transmission power of the UE is limited to a predefined value in order to guarantee that interference is at its minimum level for the existing Macro-UEs. While in [44], a dynamic power control algorithm is proposed in order to reduce the interference level while maximising the indoor coverage. The aim of this scheme is considering the load balance of the Femtocells so that loaded Femtocell reduces its coverage area by adjusting the transmission power and vice versa, which helps to reduce the interference issue.

In contrast, in [45] the authors have described a communication mechanism over the X2 interface between the Macrocell and Femtocells for Inter-Cell Interference Coordination (ICIC). One example is that when a certain cell needs to transmit, it has to inform its neighbours about its transmission power level on the DL. Another approach utilises the UL by exchanging the measurements about the interference levels among different cells. While in [46], the authors have proposed a decentralised resource allocation scheme for hybrid networks. In this scheme, the available radio resources are divided into time and frequency domains. The Macrocell can select and use all the resources, while the Femtocell has to select only a subset from the available OFDMA frequency resources when it wants to transmit in a random manner. The aim is to minimise the probability of interference occurring for every resource block. While in [47] a Femtocell management model was proposed, where resources are allocated to Femtocells and Macrocells orthogonally in time and frequency domains. However, this approach requires a high level of synchronisation between different cells in addition to the overhead caused by the large amount of signaling. In [48], the cognitive approach has been proposed for Femtocells in LTE systems. The Femtocells share information among themselves about the path-loss, then the interference is estimated, and the Femtocell selects a component carrier according to mutual interference levels.

In [49], a dynamic technique for interference avoidance between the Macrocell and Femtocells was put forward. The proposed scheme is a combination of power control and HO process. When the Macro-UE is discovered to be under interference from a Femtocell, the Macrocell will perform intra HO so that the UE is assigned to a different channel with lower interference. This also applies to Femtocells for changing their allocated channels. While, in [50], the authors have presented a method for interference control based on clustering, BW division and power control. In this approach, the Femtocells are allocated different frequencies in a frequency-reuse manner, according to the cluster position interference level.

It is worth mentioning here that the Radio Resource Management (RRM) model used in LTE system is responsible for the spectrum resources, channel allocations, transmission powers and modulation schemes. Several studies have proposed different resource allocation schemes to allocate resources between the Macrocells and the Femtocells like [46, 51, 52]. In those studies, the Femtocells try to learn the resource usage pattern of Macrocells based on their synchronisation and adjust the resource block pattern based on the interference level. The Femtocell finds a free slot from Macrocell and allocates the free resource block to Femtocell's UE; this is applicable only when there is less traffic. While [53] has proposed the approaches like cognitive radio resource and dynamic Fractional Frequency Reuse (FFR) scheme to reduce the interference between eNB's via X2 interface and sub-carrier allocation. The FFR and Soft Frequency Reuse (SFR) are existing solutions for frequency reuse in LTE Femtocell based systems [54]. The FFR splits the given BW into an inner and an outer part. It allocates the inner part to the near users (located close to the BS) with reduced power applying a frequency reuse factor of one. For users closer to the cell edge (far users), a fraction of the outer part of BW is dedicated with the frequency reuse factor greater than one. In contrast, with the SFR, the overall BW is shared by all BSs (i.e. a reuse factor of one is applied), but for the transmission on each subcarrier, the BSs are restricted to a certain power bound. Another drawback in the SFR is that, if the user density is not high in a particular region, the spectrum band used in that region will be wasted.

Hence, study [55] has introduced a frequency planning technique to avoid the co-channel interference between the Femtocell and the Macrocell. This study has shown that the mobile Operator has three basic options (scenarios) for allocating available frequencies in Femtocell deployments. The first scenario represents a dedicated radio channel for Femtocell deployment that provides separate Macrocell and Femtocell radio channels. This has the advantage of minimising the interference between the two networks and simplifying initial deployment of Femtocells. This Scenario is typically more suitable in rural areas where the mobile Operators may have unused radio channels. The second scenario shares all available radio channels between the Macrocell and Femtocell networks. This has the advantage of providing more degrees of freedom to manage interference between Femtocells, especially in dense urban deployments, but also requires the greatest degree of interference management to ensure minimal impact on the Macrocell network from the co-channel Femtocells. While the third scenario represents a compromise between the first and the second scenarios in which some radio channels are shared between the Macrocell and Femtocell networks and other radio channels are reserved for the Macrocell network only. In this scenario, the Macrocell can redirect the mobile devices that are serving over the shared radio channel to a dedicated Macrocell radio channel when they approach a Femtocell. This approach may not work perfectly in the case of vehicular environment since the Mobile-Femtos and UEs keep changing their locations from one place to another.

Further studies have introduced the co-channel interference as a significant issue in LTE networks. In [56], the authors have stated that the traditional Femtocell shares the same licensed frequency band with Macrocell, which causes the co-channel interference. Therefore, a resource allocation scheme of using TV White Space (TVWS) for LTE Femtocell Network was introduced, showing the potential to solve the co-channel interference in traditional Femtocell deployment. While in [57], the authors deal with the problem of aggregate interference modelling and static resource allocation in OFDMA-based two-tier Femtocell networks for both close and open access Femtocells. It has been assumed that for Macrocell UEs, the spectrum allocation is accomplished through FFR scheme. The authors' objective is to maximise the Femtocell UEs' throughput while maintaining as low as it is possible the impact of interference on the Macrocell UEs' performance. Their proposed algorithm is decentralised and based on some measurements that need to be performed only when Femtocell access point is plugged in. While studies like [58] have shown that Femtocells are capable of achieving higher capacity and improving the indoor coverage where FFR is used to improve spectral efficiency in emerging OFDMA networks and avoid the interference issue.

There are many other related studies on the frequency reuse techniques that have been used to mitigate the interference problem in LTE-A deployments. The evaluation of various frequency reuse schemes including Integer Frequency Reuse (IFR), FFR and Two Level Power Control (TLPC) have been presented in the work of [59]. A brief comparison of these schemes has been provided and concluded that FFR and TLPC with appropriate settings of inner region radius and power ratio provide the best performance together with applying fair-share scheduler. While [60] has proposed a frequency planning mechanism, in which Femtocells choose the frequency sub-bands that will not be used in the sub-region of a Macrocell using FFR in an integrated Macrocell/Femtocell network. In [61], the authors have proposed a frequency partitioning method, in which both sub-channels for inner cell region and sub-channels for outer region are allowed to be used in the inner region of cells while sub-channels for outer region are defined differently from cell to cell to reduce co-channel Interference. Furthermore, FFR scheme is introduced by [62], which has stated that the inner and the outer regions can be served in a different way, not only in terms of frequency sub-bands but also in terms of time slots. This scheme has been extended further by [63] by employing the concept of cell sectors while an optimised FFR mechanism is presented in [64] to achieve better system performance based on dynamic cluster sizing and frequency allocation.

Hence, it is very noticeable that most of the used interference techniques have been applied on fixed BSs, while very few have been implemented for vehicular BSs. Therefore, this chapter pays more attention towards the vehicular BSs interference together with its impact on the achieved vehicular UEs throughput and SINR.

11.3. Mobile Femtocell Technology

In mobile and ubiquitous networks, it is desirable that UEs do not experience fluctuations in the services quality when they are moving from one place to another, i.e., they should not be aware of mobility. Nowadays, vehicular UEs demand for the quality in their own network rather than in mobile devices. In this sense, an efficient technology is needed to prevent the UEs from detecting changes in the QoS when they are continuously moving. Thus, Mobile-Femto architecture has been designed to improve the 3G and 4G connectivity inside buses environment to support mobility between the bus passengers and the Core Network in LTE network. The Mobile-Femto concept has been derived from combining the concept of two technologies; the Fixed-Femto technology and the Mobile Relay Node technology. The main advantage of implementing the Mobile-Femto technology is the ability of this small cell to move around and dynamically change its connection with the Operator's Core Network. This Mobile-Femto concept can be seen as a practical implementation of the moving networks that can be deployed in public transportations like buses to overcome the high penetration loss and path-loss issues. Therefore, Mobile-Femto is seen as the new paradigm of the Femto-cellular network technology that reduces the impact of vehicular environment on the UEs SINR, throughput and spectral efficiency as will be shown in this chapter. However, the approach of network configuration should be different for different vehicular environments due to vehicles speed variation, availability of wireless backhaul networks and Doppler Shift in the case of high-speed trains.

UEs inside public transportations may execute multiple HOs at the same time that may cause a significant increase in the signaling load and drop in the network connections. This has led to look at the Mobile-Femtos as a solution to minimise the signaling load, the number of dropped connections and the number of HOs [65]. Hence, Fig. 11.4 represents the Fixed and Mobile Femtocells that could be either installed inside buildings, on streets or public transportations like trams and buses.

As mentioned earlier, Mobile-Femto is a moving hotspot with multiple UEs who are requesting diverse data services, e.g. web browsing, Voice over IP (VoIP), e-mailing and video streaming. It adapts the LTE's standard radio interface to communicate with the serving eNB and the group of UEs who are within the coverage area of that particular Mobile-Femto. Fig. 11.5 shows that there are three types of links that have been utilised to differentiate between the eNB/Mobile-Femto link, the Mobile-Femto/UE link and the eNB/UE link. The previous three links are known as the backhaul link, the access link and the direct link respectively.

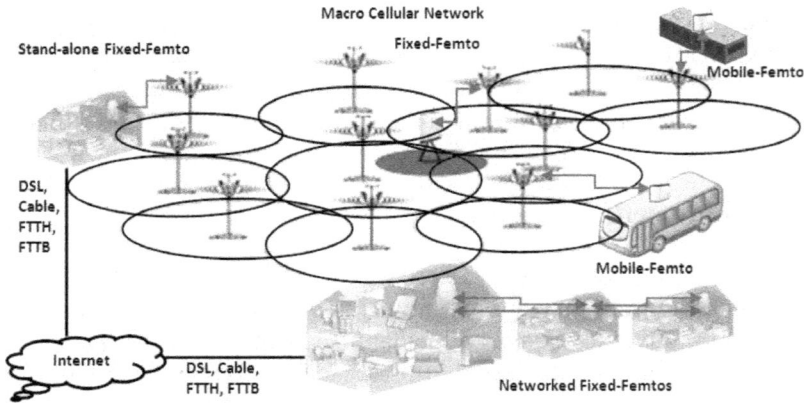

Fig. 11.4. Fixed and Mobile Femtocell Technologies.

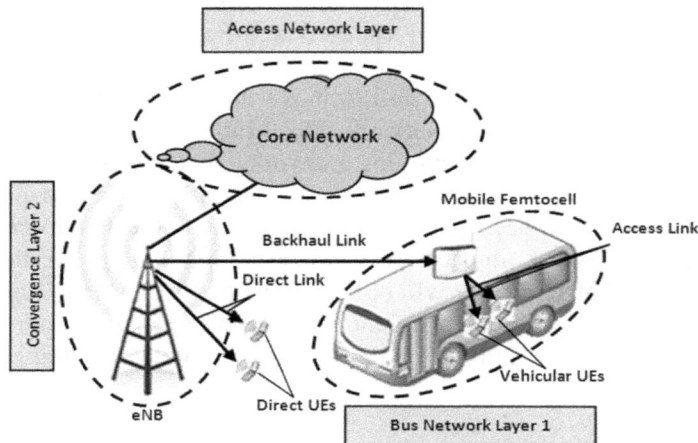

Fig. 11.5. Mobile Femtocell architecture with its layering system.

Moreover, the Mobile-Femto architecture relies on three different designed layers as the following:

- **The Bus Network Layer (BNL)** consists of the Mobile-Femto along the bus and all the vehicular UEs (passengers) attached to this Femtocell.

- **Convergence Layer (CL)** aggregates the traffic sent by the Mobile-Femtos in the BNLs via the backhaul links and forwards it to the Internet. The eNBs or the mother BSs enable connectivity for the Mobile-Femto technology that is installed in the bus with the outside environment.

- **The Access Network Layer (ANL)** is composed by the outdoor wireless technology that is available along the bus paths, e.g. LTE technology. Thus, the ANL is the LTE Core Network and it is the decision maker ahead of the eNB in the LTE systems.

Hence, using Mobile-Femto BS has three potential advantages:

a) It avoids the multiple HOs procedures since a single HO is required between the vehicular UE and the serving Femtocell inside the bus.

b) It improves the mobile devices battery life due to the short distance between those mobile devices and the serving Femtocell that is installed inside the bus.

c) It makes a better use of the coverage area because of the use of the omnidirectional antenna that gives equal signal quality distribution. This makes those vehicular UEs enjoy a better signal quality inside vehicles by mitigating the high LTE penetration loss. The result is a longer use of the wireless link, thus, less signal outage probability.

11.4. Vehicular UEs Performance Analysis in LTE Networks

The main aim of this section is to look into the Fixed-Femto and Mobile-Femto technologies and their impact on the vehicular UEs' performance in LTE and next generation networks. However, it is to be noticed here that the backhaul link between the eNB and the Mobile-Femto experiences fast fading with a Non Line of Sight (NLOS) channel while the access link between the Mobile-Femto and the UE is assumed to be Line of Sight (LOS) with slow fading channel. Also it is worth mentioning that the wireless backhaul link transmission does not interfere with the Small/ Macro-cells arrangement. This is because it uses a frequency band that has not been used before. This is required in order to avoid the interference between different wireless backhaul links as well as with different direct and access links. Hence, this chapter shows the potential of deploying the Mobile-Femtos in the Macrocell as they have been assumed to be installed across the roof of public transportations e.g. buses to improve the capacity, reliability and coverage for vehicular UEs in 5G and future wireless networks. To this end, the following will draw a mathematical comparison between the vehicular UEs performance before and after deploying the Fixed-Femto and Mobile-Femto in LTE Macrocell.

11.4.1. System Model

First it is important to clarify the communication process between the eNB and the Femtocell and between the Femtocell and the UE in LTE system. The eNB gathers the Channel State Information (CSI) from all UEs and Fixed-Femtos/Mobile-Femtos in the Macrocell. Likewise, the UEs within the Fixed-Femto/Mobile-Femto coverage will feedback this information only to the Fixed-Femto/Mobile-Femto BS. In the transmission process, the eNB transmits the data to the selected Fixed-Femto/Mobile-Femto via the backhaul link and then the Fixed-Femto/Mobile-Femto will fully decode the data, buffer it and retransmit it to its UE via the access link. However, as depicted in Fig. 11.6, the considered eNB has a fixed coverage of D meters depends on the chosen transmission power, while one vehicle (bus) is moving along the highway with a number of UEs inside it. It has been assumed that both the Fixed-Femto and the Mobile-Femto consider the dual-hop transmission where the eNB transmits to a vehicular UE via the Fixed-Femto/Mobile-

Femto and vice versa. Additionally, d meter is the distance between the eNB and the Fixed-Femto while x is the distance between the eNB and the vehicular UE.

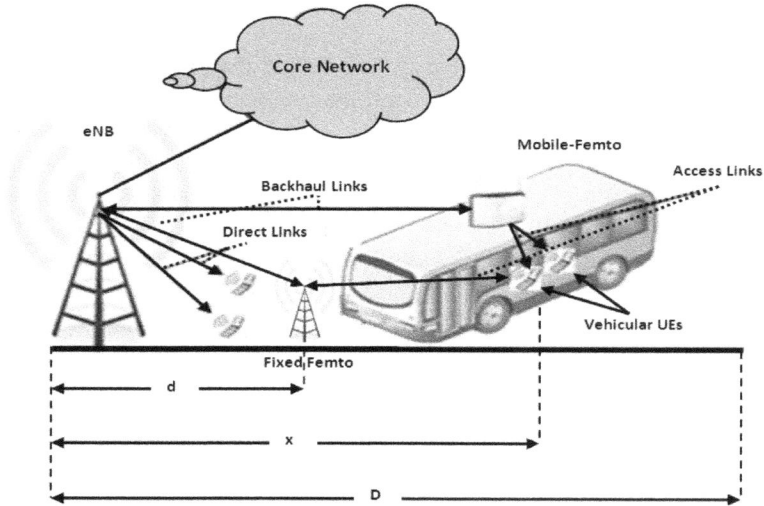

Fig. 11.6. eNB, Fixed-Femto and Mobile-Femto system model architecture.

However, before clarifying the communication links between the eNBs, Fixed-Femtos and Mobile-Femtos, it is significant first to discuss the UEs scheduling process and resource allocation scheme in these BSs. A multiuser scheduling scheme is assumed where the Macrocell UEs, the Mobile-Femtos and the Fixed-Femtos are served over k PRBs, indexed by $k=1,...,K$. The Fixed and Mobile Femtos are scheduled over a dedicated time-frequency zone in such a set of Fixed-Femtos and Mobile-Femtos that are selected according to scheduling criterion e.g. serving priority or channel condition. Fig. 11.7 shows the scheduling mechanism in terms of eNB, Mobile-Femtos and Fixed-Femtos UEs.

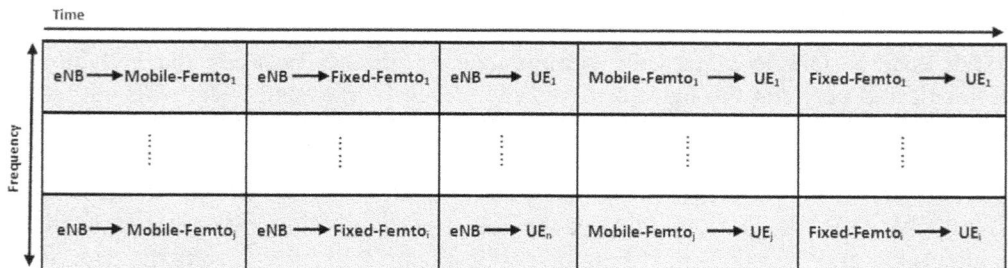

Fig. 11.7. Time sharing strategy for Fixed and Mobile Femtos in LTE system.

As it is known, the eNB is responsible for scheduling all the links of the network, Femtocells' links and UEs' links. The Femtocell nodes only forward the received data and signaling from/to the eNB without any scheduling. The scheduler in the eNB should take into account the limitation of the Control Channel Elements (CCEs) when allocating the PRBs to the UEs in both directions UL and DL. Therefore, the UEs scheduling has two successive scheduling decisions; the candidates selection followed by frequency domain resources allocation to assign the PRBs among the selected UEs. Here the candidates' selection could be either UEs or Femtocells who need to be scheduled in the Macrocell. The eNB will schedule the Mobile-Femtos like any other UEs but of course, more PRBs will be allocated to those access points than normal UEs need. Hence, the scheduling process can be summarised as the following:

a) The time domain scheduler will prioritise the UEs based on a given priority criterion e.g. Proportional Fair (PF) Scheduler.

b) It selects only Marco UEs or Mobile-Femtos/Fixed-Femtos with highest scheduling priority taking into account the total CCEs constraints as well as the number of available PRBs. This can be defined as (N UEs $n \leq N_{max}$), (J Mobile-Femtos $j \leq J_{max}$) or (I Fixed-Femtos $i \leq I_{max}$), where $n \in N = \{1,..., N\}$ denotes the set of UEs who communicate directly with the eNB (Macrocell UEs). While N_j refers to a group of UEs within a Mobile-Femto j and N_i denotes the group of UEs within a Fixed-Femto i, where Mobile-Femto $j \in J = \{1, ..., J\}$ and Fixed-Femto $i \in I = \{1,...,I\}$.

As mentioned earlier, this work considers the PF scheduling policies. This type of scheduling refers to the amount of resources allocated within a given time window to UEs with better channel quality in order to offer high cell throughput as well as fairness satisfactory. The PF scheduling mechanism has been presented by Fig. 11.8 [66].

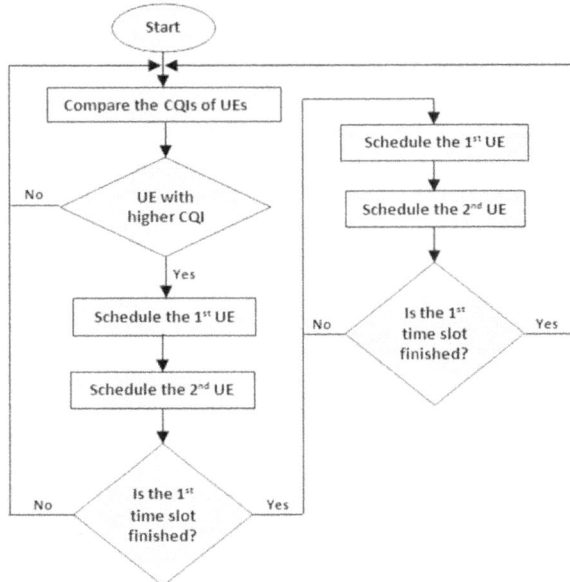

Fig. 11.8. Proportional fair scheduler for UEs in LTE network.

As the above figure shows, the scheduling technique works as the following; firstly, the scheduler sorts the UEs in descending order according to the proportional fair metric [67] and then it picks up only some of the UEs depending on the availability of the CCE, the PRBs and UE's CQI. Secondly, the scheduler allocates the PRB k to UE n, Mobile-Femto j or Fixed-Femto i according to the following criterion:

$$n_k = \arg \max_{n \in N} \frac{R_n(k,t)}{\bar{R}_n(t-1)}, \tag{11.1}$$

where $\bar{R}_n(t-1)$ denotes the average data rate of UE n before the current scheduling subframe t. Thus, $\arg max_{n \in N} R_n(k,t), k = 1, \ldots \ldots, K$ and $R_n(k,t) \propto SNR_n$ is the instantaneous achievable rate on PRB k for a user n, which can be calculated according to Shannon formula:

$$R_n(k,t) = \frac{BW}{k} \log_2(1 + SNR(k,t)). \tag{11.2}$$

Hence, the average data rate of UE n can be calculated using an exponential average filtering, which will be unpdated using the following formula:

$$\bar{R}_n(t) = \left(1 - \frac{1}{T}\right)\bar{R}_n(t-1) + \frac{1}{T}\sum_{k=1}^{K} R_n(k,t)d_n(k,t), \tag{11.3}$$

where T is the average window length, which is considered as an important element in the PF scheduler and $d_n(k,t)$ is a binary indicator that is set to 1 if the user n is scheduled on PRB k at time 1 and to 0 otherwise.

Since the main concern of this work is the vehicular environment, vehicular UEs and Femtocells scheduling process may occur differently from the traditional one. Therefore, the following algorithm represents the UEs and Femtocells PF scheduling scheme in the Macrocell under different traffic loads (low, medium and heavy) traffics.

In Algorithm 1, the availability of BW resulted in resource blocks plays important role in the scheduling process of both UEs and Femtocells. This is because; there is a positive correlation between the used BW and the transmitted data rate (R_n). In other words, whenever the used BW is large, the ability of allocating more PRBs to UEs and Femtocells increases. This has its positive influence on the transmitted bit rate and achieved throughput.

Hence, based on [68], the received signal-to-noise ratio (SNR) in (dB) at the receiver (R_x) side can be given by

$$SNR_{R_x} = \frac{P_x|G|^2 PL(L)\varepsilon}{P_{noise}}, \tag{11.4}$$

where G represents the channel gain and PL has been used to model the Path-Loss when the receiver R_x is at distance L away from the transmitter $T_{x,.}$. The P_x is the average transmission power at the transmitter T_x while ε is the VPL and P_{noise} represents the noise power.

Algorithm 1. Scheduling UEs and Femtocells.

1:      */* Bandwidth Scheduling to Macro UEs, Mobile-Femtos & Fixed-Femtos*
2:      *for N = {1,..., N}, J = {1,....J}, I = {1,....I}*
3:      *compute CCE*
4:      *compute CQI*
5:      *compute max_RBs*
6:              *if (Sch_ $BW_{eNB} \geq max\_R_n$) then*
7:      *($RBs_{eNB} \geq [R\ (t,k) = \frac{BW_{eNB}}{k} log_2(1 + SNR(t,k))])$ %thus, do the following*
8:              *sch_N*
9:              *sch_J*
10:             *sch_I*
11:             *accept_Transmission then*
12:     *for n=1 % for all UEs do the following calculations*
13:     *ergodiccapacity = calculate_ ergodiccapacity(n)*
14:     *spectraleffeciency = calculate_ spectraleffeciency(n)*
15:     *throughput = calculate_ throughput(n)*
16:     *sinr = calculate_sinr(n)*
17:     *end for*
18:                 *end if*
19:                 *else if (Sch_ $BW_{eNB} < max\_R_n$) then*
20:                 *($RB_{eNB} < [R\ (t,k_{eNB}) = \frac{BW_{eNB}}{k_{eNB}} log_2(1 + SNR(t,k_{eNB}))])$*
21:                 *rej_sch*
22:                 *rej_Transmission*
23:                 *end*
24:     *end for*

Thus, since the vehicular UEs is at distance x away from the eNB, the SNR of the eNB direct transmission can be given by

$$SNR_{eNB-UE} = \frac{P_x^{eNB}|G_1|^2\ PL(x)\varepsilon}{P_{noise}}. \tag{11.5}$$

While, for the Fixed-Femto assisted-transmission the equation is given as

$$SNR_{FFemto-UE} = \frac{P_x^{FFemto}|G_2|^2\ PL(x-d)\varepsilon}{P_{noise}}. \tag{11.6}$$

On the other hand, the distance between the transmitter T_x of the vehicular UE and the Femtocell that is allocated in the same bus (Mobile-Femto), is shorter than 5 meters at most. As a result, a LOS access link and a constant power loss C_{loss} = -84.55 dB have been assumed [69]. The SNR of the Mobile-Femto assisted-transmission can be given by

$$SNR_{MFemto-UE} = \frac{P_x^{MFemto}C_{loss}}{P_{noise}}, \tag{11.7}$$

where the P_x^{eNB}, P_x^{FFemto} and P_x^{MFemto} denote the average transmission power of the eNB, Fixed-Femto and Mobile-Femto while G_1 denotes the channel gain of the backhaul link and G_2 denotes the channel gain of the access link in the Fixed-Femto assisted transmission. Whereas the channel gain of the Mobile-Femto assisted-transmission has been assumed to be 1 (means unity) due to the short distance between the UE and the Femtocell as well as the LOS access link. After presenting the SNR of the eNB, Fixed-Femto and Mobile-Femto assisted transmissions, now it becomes necessary to state the Ergodic capacity of the direct, backhaul and access links, which are generally can be given by

$$C_{b,d,a} = [\min\{C_{backhaul}, C_{direct}, C_{access}\}]. \qquad (11.8)$$

The backhaul links between the eNB-Fixed Femtos and the eNB-Mobile Femtos are assumed to be NLOS outdoor links. As a result and based on Shannon equation capacity [70], the backhaul link capacity in (bits/s/Hz) between the eNB and the Fixed-Femto at distance d can be given by

$$C_{backhaul(eNB-FixedFemto)} = BW_{eNB-FixedFemto} \; \log_2\left(1 + \frac{P_x{}^{eNB}|G_1|^2 \, PL(d)}{P_{noise}}\right). \qquad (11.9)$$

While the backhaul link capacity between the eNB and the Mobile-Femto at distance x can be given by

$$C_{backhaul(eNB-MobileFemto)} = BW_{eNB-MobileFemto} \; \log_2\left(1 + \frac{P_x{}^{eNB}|G_1|^2 \, PL(x)}{P_{noise}}\right). \qquad (11.10)$$

Here the $BW_{eNB\text{-}FixedFemto}$ and $BW_{eNB\text{-}MobileFemto}$ represent the bandwidth of backhaul links between eNB-FixedFemto, eNB-MobileFemto while $BW_{FixedFemto\text{-}UE}$ and $BW_{MobileFemto\text{-}UE}$ represent the bandwidth of access links between FixedFemto-UE and MobileFemto-UE respectively.

It should be noticed that, in the backhaul link capacity between the eNB and the Mobile-Femto there is a small channel gain like the Fixed-Femto due to the distance gap between the Mobile-Femto and the eNB as well as the NLOS backhaul link. While, the $C_{direct(eNB\text{-}UE)}$ can be given as the same as the $C_{backhaul(eNB\text{-}Mobile\text{-}Femto)}$ in Eq. (11.10), since the direct link between the eNB and the vehicular UEs is a NLOS link and the distance between the eNB and the UE is the same as the distance between the eNB and the Mobile-Femto.

Hence, based on the above equations, the access link capacity between the Fixed-Femto and the vehicular UE at distance x-d can be derived and given as

$$C_{access(FFemto-UE)} = BW_{FixedFemto-UE} \; \log_2\left(1 + \frac{P_x{}^{FFemto}|G_2|^2 \, PL(x-d)\varepsilon}{P_{noise}}\right). \qquad (11.11)$$

While the access link capacity between the Mobile-Femto and the vehicular UE is a special case scenario as the penetration loss does not exist in this case. This is because there are no boundaries between the UEs and the serving small BS so nothing resists the signal from reaching the UEs without losses. Therefore, the link capacity can be given by

$$C_{access(MFemto-UE)} = BW_{MobileFemto-UE} \; \log_2\left(1 + \frac{P_x{}^{MFemto}C_{loss}}{P_{noise}}\right). \qquad (11.12)$$

However, it is important to state that the backhaul link between the eNB and the Mobile-Femto is the capacity bottleneck of the Mobile-Femto technology and many challenges can be faced with deploying this technology. One of these challenges is observed when the UE and Mobile-Femto speed goes up, the rapid variations of mobile channels combined with feedback delays reduce the accuracy of the CSI at the eNB side. The CSI feedback inaccuracy at the eNB limits the use of advanced MIMO transmission schemes that are needed in LTE and 5G systems to further increase the throughput of the backhaul link. To control the previous issues, the speed of the designed scenarios has been controlled. Nevertheless, new challenges regarding the interference management arise due to the use of Mobile-Femtos. This is because; the backhaul link becomes complicated as the interference is expected between different Fixed-Femtos and Mobile-Femtos backhaul links. The previous issue can be mitigated with an efficient coverage planning scheme as well as specifying the buses paths (buses zones) within the Macrocell to avoid the overlapping areas. However, there is always a worst-case scenario in terms of the interference problem, which will be further investigated in this chapter.

After discussing the scheduling process, PRBs distribution among Macro UEs and Femtocells in the Macrocell and clarifying the communication links between the eNBs, Fixed-Femtos and Mobile-Femtos, it is important now to discuss the effect of this on the achieved spectral efficiency. Once the Fixed-Femtos/Mobile-Femtos are deployed, the spectrum has to be allocated efficiently among the different links; the backhaul, direct and access links. It is essential therefore, to design an efficient method that improves the spectral efficiency among the previous three links. Here, the non-orthogonal resource allocation scheme has been applied in which the radio resources are reused by the direct and access links while the radio resources are orthogonally allocated between the backhaul and the direct links, and between the backhaul and the access links. The non-orthogonal resource allocation scheme indicates that there will be an Inter-Carrier Interference (ICI) to the access and direct UEs due to the simultaneous transmissions from the Mobile-Femto/Fixed-Femto and eNB on the same sub-channels. This scheme has several advantages over the orthogonal resource allocation scheme since it improves the resource utilisation as well as it gives the flexibility to implement the RRM at the eNB and the Mobile-Femto/Fixed-Femto independently.

Hence, in order to compute the UEs spectral efficiency, it is essential first to calculate the SINR of Macro and Femtocell UEs as the signal strength of vehicular UEs is the main concern of this work. Thus, the received SINR in (dB) for Direct vehicular UE (SINR$_D$) is given by

$$SINR_{m(D)} = \frac{P_x^{eNB}|G_1|^2 PL(x)\varepsilon}{(I_{MFemto}+I_{FFemto})+P_{noise}}, \quad (11.13)$$

where I_{FFemto} and I_{MFemto} is the ICI from the Fixed-Femto and Mobile-Femto respectively, P_{noise} is the noise power, and the G_1 is the channel gain over the direct link. On the other hand, the received SINR for the Access vehicular UE (SINR$_A$) in the case of the Fixed-Femto transmission can be calculated according to the following equation

$$SINR_{FFemto(A)UE} = \frac{P_x^{FFemto}|G_2|^2 PL(x-d)\varepsilon}{(I_{eNB}+I_{MFemto})+P_{noise}}, \quad (11.14)$$

267

where I_{eNB} is the ICI from the eNB and G_2 is the channel gain over the access link between the Fixed-Femto and the vehicular UE. Whilst, the received SINR for the Access vehicular UE (SINR$_A$) in the case of the Mobile-Femto can be calculated according to the following formula

$$SINR_{MFemto(A)UE} = \frac{P_x^{MFemto} C_{loss}}{(I_{eNB} + I_{FFemto}) + P_{noise}}. \tag{11.15}$$

As mentioned earlier, the channel gain over the Mobile-Femto access link is unity (equal to 1) because of the LOS link and the short distance between the Mobile-Femto and the UE while there is a constant loss C_{loss} between the two parties. However, the UEs might experience some interference from the eNB and the nearby Fixed-Femtos as it may affect the SINR value. This will be further investigated later in this chapter.

All the previous formulas have helped in calculating the spectral efficiency and throughput in order to evaluate the vehicular UEs performance. As explained, the spectral efficiency is defined for each location, as the ratio of throughput to the available BW for a UE under the assumption of one single subscriber in the cell as the following represents [71]

$$Spectral\ effciency = \frac{Throughput}{AvaliableBW}. \tag{11.16}$$

Moving on from the previous concept of the spectral efficiency to Shannon capacity formula ($C = BW \log_2 \left(1 + \frac{S}{N}\right)$), this will create the base to calculate the spectral efficiency [72]. Hence, the spectral efficiency of eNB$_{UE}$, Fixed-Femto$_{UE}$ and Mobile-Femto$_{UE}$ can be derived from the following

$$\frac{C}{BW} = \log_2 \left(1 + \frac{S}{N}\right), \tag{11.17}$$

where c is the achieved capacity that can be given by bits/sec/Hz. Based on Eq. (11.16) and (11.17) the spectral efficiency in (bits/cu) can be given by

$$Spectral\ effciency = \log_2 \left(1 + \frac{S}{N}\right). \tag{11.18}$$

Thus, the spectral efficiency of direct vehicular UEs can be given by

$$Spectral\ effciency\ of\ eNB_{UE} = E\left[\log_2 \left(1 + \frac{P_x^{eNB}|G_1|^2 PL(x)\varepsilon}{P_{noise}}\right)\right], \tag{11.19}$$

where E represents the minimum expected value of the Spectral Efficiency over time.

While the spectral efficiency of the Fixed-Femto vehicular UE at distance x-d can be calculated by the following

$$Spectral\ effciency\ of\ FFemto_{UE} = E\left[\log_2 \left(1 + \frac{P_x^{FFemto}|G_2|^2 PL(x-d)\varepsilon}{P_{noise}}\right)\right]. \tag{11.20}$$

On the other hand, the spectral efficiency of the Mobile-Femto UE can be given by the following formula

$$\text{Spectral effciency of MFemto}_{UE} = \log_2\left(1 + \frac{P_x{}^{\text{MFemto}}C_{\text{loss}}}{P_{\text{noise}}}\right). \quad (11.21)$$

In this case, there is no penetration loss due to the absence of walls between the vehicular UE and the Mobile-Femto that both share the same bus. That means there is no resistance against the signal, which can be transmitted smoothly without substantial power degradation.

Since this work considers the spectral efficiency of vehicular UEs, it is important to consider the throughput of those UEs as well. This is because of the positive correlation between the vehicular UEs spectral efficiency and the achieved throughput by those UEs as the following formula shows

$$\text{UEs\_Throughput} = \text{UEs\_Spectral Effeciency} * \text{BandWidth}. \quad (11.22)$$

So based on shannon equation, the throughput in (bits/s) of direct vehicular UEs at distance x from the eNB can be given by

$$\text{Throughput of eNB}_{UE} = E\left[\log_2\left(1 + \frac{P_x{}^{\text{eNB}}|G_1|^2 PL(x)\varepsilon}{P_{\text{noise}}}\right) * BW_{\text{eNB}}\right], \quad (11.23)$$

whereas the following represents the throughput of Fixed-Femto vehicular UEs at distance *x-d* where the penetration loss plays an important role in this case as below shows

$$\text{Throughput of FFemto}_{UE} = E\left[\log_2\left(1 + \frac{P_x{}^{\text{FFemto}}|G_2|^2 PL(x-d)\varepsilon}{P_{\text{noise}}}\right) * BW_{\text{FFemto}}\right]. (11.24)$$

Likewise the following equation represents the throughput of Mobile-Femto UEs with the absence of the penetration loss as explained earlier.

$$\text{Throughput of MFemto}_{UE} = \log_2\left(1 + \frac{P_x{}^{\text{MFemto}}C_{\text{loss}}}{P_{\text{noise}}}\right) * BW_{\text{MFemto}}, \quad (11.25)$$

where the BW_{eNB}, BW_{FFemto} and BW_{MFemto} is the avaliable bandwidth at the eNB, Fixed-Femto and Mobile-Femto respectvily to serve the vehicular UEs.

However, in the presence of fading, penetration loss and path-loss, there is always a probability that the received SINR at the receiver is below a given threshold to support a required transmission rate of R bits/sec. This is because all types of services have some minimum bit-error-rate requirements. These requirements can be translated to the minimum required average of the received SINR at the receiver. Based on what has been illustrated earlier, the probability that the received SINR falls below a given SINR threshold is referred to an outage probability. Nowadays, systems try always to keep this probability as low as possible to improve their network performance by improving their vehicular UEs connections. Therefore, the outage probability has been analysed in terms of the Fixed-Femto and Mobile-Femto deployments in the LTE Macrocell. This analysis has been done according to three transmission schemes; direct transmission from the eNB, Fixed-Femto assisted-transmission and Mobile-Femto assisted-transmission as shown earlier in Fig. 11.6. Thus, it is required here to calculate the outage probability of the vehicular UE in three cases; when this UE is connected directly with either the eNB, the

Fixed-Femto or the Mobile-Femto. Having high outage probability means that the vehicular UE has made the wrong choice of connection with one of surrounding BSs [73]. Hence, the outage probability of a single-hop system when there is a direct transmission from the eNB to the vehicular UE, can be given by

$$P_{out\_D}(SINR_{th\_D}) = P_r(SINR_{R_x} < SINR_{th\_D}), \qquad (11.26)$$

where the $SINR_{R_x}$ represents the instantaneous received SINR at the receiver R_x and P_r is the received power at the R_x [74]. The $SINR_{th\_D}$ is the required SINR threshold at the receiver R_x to support a given target rate over the direct link between the eNB and the vehicular UE. While the threshold here is a system designated parameter and depends on several factors such as the achievable target rate of an LTE system. It is worth noting that the signal will be outage when the received SINR at the receiver side is below the given SINR threshold. This has its negative impact on the achieved system performance especially when the UEs feel that they are not getting the best services they are paying for.

Thus, the SINR threshold in the case of direct transmission when a rate of R bits/sec is required at the vehicular UE's end, is given by

$$SINR_{th\_D} = SINR_{th\_femto} = 2^R - 1. \qquad (11.27)$$

Femtocells are using the full duplex transmission mode like the eNB and this is why in Femtocell-assisted system, both the backhaul and the access links are supporting a given end-to-end rate of R bits/sec at the receiver R_x like the direct transmission of the eNB, thus the same SINR threshold will be used.

However, if the transmitter T_x has an average transmission power of P_x and the receiver R_x is at distance y from the transmitter T_x, the received power (P_r) at R_x can be given by

$$P_r(y) = P_x L(y)\psi|G|^2, \qquad (11.28)$$

where $L(y)$ models the path-loss when the receiver R_x is at distance y from the transmitter T_x. ψ denotes the power loss caused by shadowing and G represents the channel gain. Here the threshold $SINR_{th\_D}$ or $SINR_{th\_femto}$ varies according to different QoS requirements like the achievable rate in bits/sec of an LTE system as mentioned earlier, which is in turn based on Shannon capacity that can be given by the following

$$R = BW_{eff} \log_2(1 + SINR), \qquad (11.29)$$

where BW_{eff} adjusts for the bandwidth efficiency and the SINR has been considered here instead of the SNR because of the existence of the interference issue in this case [75]. This is due to the multiple access method, since several transmitters can send information simultaneously over a single communication channel. This allows several UEs to share a band of frequencies, which may cause interference between UEs as will be shown later in this chapter.

Hence, the $\text{SINR}_{\text{th\_D}}$ and $\text{SINR}_{\text{th\_femto}}$ can be obtained from Eq. (11.27) with a required end-to-end rate of R bits/sec for direct, backhaul and access links transmission as the following formula shows

$$\text{SINR}_{\text{th\_D}} = \text{SINR}_{\text{th\_femto}} = (2^{R/BW_{\text{eff}}} - 1). \tag{11.30}$$

Thus, after calculating all the required parameters, it becomes now easier to find out the outage probability of Mobile-Femto and Fixed-Femto assisted transmissions. In Mobile-Femto assisted system, the access link is not corrupted or affected by the multipath fading due to the LOS link and the short distance between the UE and the serving Mobile-Femto. In this case, the outage probability can be obtained similarly to the single-hop system of direct transmission with the use of $\text{SINR}_{\text{th\_femto}}$ which is the same of the $\text{SINR}_{\text{th\_D}}$ as both are using the full duplex mode. However, as a result of the removed channel gain in the case of Mobile-Femto transmission, the P_r value will be different from the P_r of the direct transmission. This is due to the LOS access link, the short distance between the receiver and the transmitter and the absence of the penetration loss between the two parties. Hence, the Mobile-Femto assisted-transmission P_r can be obtained by

$$P_r(y) = P_x^{\text{MFemto}} PL(y)\psi, \tag{11.31}$$

where the $PL(y)\psi$ is equal to the constant loss (C) of the free space path-loss model.

On the other hand, the outage may occur if either the backhaul or the access link is outage in the case of the Fixed-Femto assisted system. Thus, the outage probability for Fixed-Femto assisted system can be given by

$$P_{\text{out\_F}}(\text{SINR}_{\text{th\_femto}}) = P_r(\min(\text{SINR}_{\text{backhaul}}, \text{SINR}_{\text{access}}) < \text{SINR}_{\text{th\_femto}}). \tag{11.32}$$

Consequently, it can be concluded that vehicular UEs are most likely to be exposed to performance degradation and dropped connections as they are in continuous movement under different speeds and more affected by path-loss and penetration loss. Besides that, their unpredictable movement creates a performance fluctuation whenever they are close or far from any BS. Therefore, the achieved results in the following section will show the impact of deploying Fixed/Mobile Femtocells on the achieved vehicular UEs performance and link outage probabilities.

11.4.2. Results and Discussion

In order to simulate the vehicular environment, Mobile-Femtos have been implemented by considering several criteria e.g. speed, direction and distance from the eNB. The Mobile-Femto speed has been set to be the same speed of the UEs inside the bus as they are moving as one unit at the same speed and direction. Moreover, the VPL has been added to all signals traversing the chassis of the vehicle from the eNBs and the Fixed-Femtos. Furthermore, the Microcell NLOS path-loss model has been used between the eNB's transceiver and UEs or between the Femtocells and UEs to model the propagation path-loss via using the distance and the antenna gain.

The performance of vehicular UEs in LTE network has been evaluated via using the dynamic System Level Simulator, which considers the LTE specification [76]. This simulator has been modified to use the Microcell NLOS path-loss model, which is based on the COST 231 Walfish-Ikegami NLOS model with urban environment that will be further explained in this chapter. This model is more appropriate when the distance between two BSs is less than 1 Km [77]. The vehicular UEs who have been served by the eNB, Fixed-Femtos and Mobile-Femtos were distributed randomly in the Macrocell, while the Femtocells' coverage has been distributed based on the Microcell NLOS path-loss model. The fast fading model [76] is generated according to the speed of the UEs/Mobile-Femtos and the used transmission mode. The environment uses the Proportional Fair scheduler, as it is more efficient in the case of vehicular environment in order to avoid interference and improve throughput. The directional TS36.942 antenna specification is used for the simulated eNBs with a gain of 15 dBi while omnidirectional antenna is used for the Fixed-Femtos and Mobile-Femtos with a gain of 0 dBi. Other options are also available, depending upon the environment and frequencies. The MIMO is used as a transmission mode in order to have a better throughput and serve more UEs. All the mathematical equations and algorithms have been integrated along the simulations in order to create the required environment.

The eNB and Fixed-Femto/Mobile-Femto UEs were assumed to be 40 and 10 respectively in each Macrocell. The LTE frame structure has been considered, which consists of blocks of 12 contiguous subcarriers in the frequency domain and 7 OFDM symbols in the time domain. The carrier BW is fixed at 10 MHz with 50 PRBs. A full eNB buffer is considered where there are always buffered data ready for transmission for each node. The transmission power of the eNB and Fixed-Femtos/Mobile-Femtos were assumed to be 46 dBm and 24 dBm respectivly. Furthermore, the speed of the Mobile-Femto and the vehicular UEs in the bus were assumed to range from 3 km/h to 160 km/h where VPL scales have been considered in the simulated environment.

To begin with, the Ergodic capacity of vehicular UEs' links plays an important role in evaluating their performance as it is significantly affected by both penetration loss and path-loss. Fig. 11.9 shows that when there is no VPL, the direct transmission always achieves the highest Ergodic capacity. Low penetration loss means low resistance against the transmitted signal and the signal can pass through easily without facing a dramatic reduction in the signal's power. Even though the Mobile-Femto is seen as a better option rather than using the Fixed-Femto for vehicular UEs, however, at 500 m to 1000 m distance from the serving eNB, the Fixed-Femto shows a flat capacity improvement. This is because; when the penetration loss is equal to 0dB, being served by Fixed-Femtos at high distances from the eNB, sounds a better option than using the Mobile-Femto. That is due to the backhaul link variation between the eNB and Mobile-Femto in high path-loss areas, which in turn limits the communication between the two and becomes more obvious in the absence of the VPL. Subsequently, this limits the achieved Ergodic capacity of Mobile-Femto UEs' access links.

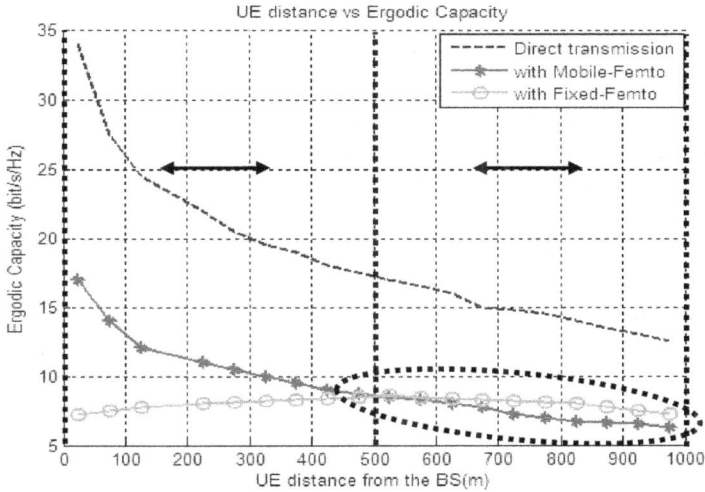

Fig. 11.9. The Ergodic Capacity when VPL = 0 dB.

On the other hand, as the VPL increases, both the Mobile-Femto and Fixed-Femto assisted-transmission schemes achieve higher capacities especially when the vehicular UE moves away from the eNB as shown in Figs. 11.10 and 11.11. Hence, Fig. 11.10 shows the Ergodic capacity when the penetration loss is equal to 25 dB. At 500 m distance from the eNB, the Mobile-Femto starts to achieve higher capacity in the case of vehicular UEs who are facing the high penetration loss and signal variation. This is because with the increased penetration loss and path-loss due to the distance from the eNB, the Mobile-Femto in the bus is seeing as a better option for the vehicular UEs to be connected to, and improve their throughput, subsequently their performance.

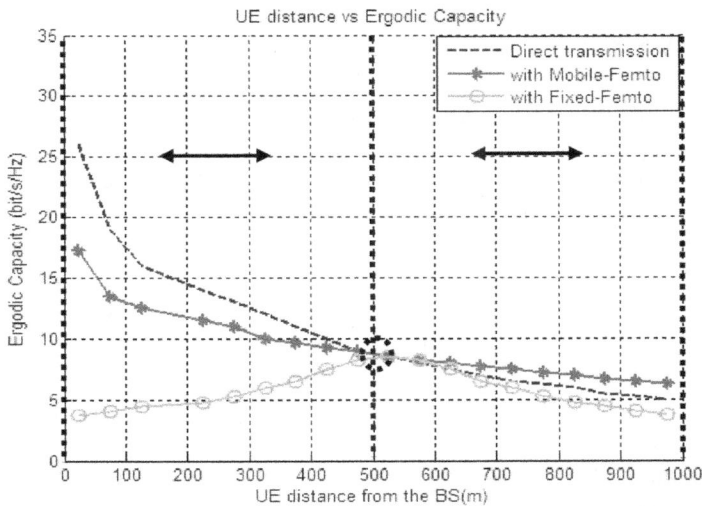

Fig. 11.10. The Ergodic Capacity when VPL = 25 dB.

273

While in Fig. 11.11, it is important to notice that at a certain stage both of the direct and the Fixed-Femto transmission Ergodic Capacity will be poor as the penetration loss and the path-loss increase due to the distance gap between the vehicular UE and the eNB. At this stage, deploying Mobile-Femtos inside buses will be a promising solution to overcome the signal reduction with the increased in both distance and penetration loss for vehicular UEs. Moreover, at almost 440 m distance apart from the eNB, the Fixed-Femto starts to achieve higher capacity than the eNB as those vehicular UEs experience very high penetration loss, distance gap and weak signal from the eNB. Therefore, the best option for those vehicular UEs is to connect to any nearby Fixed-Femto even for few moments just to maintain the signal connection.

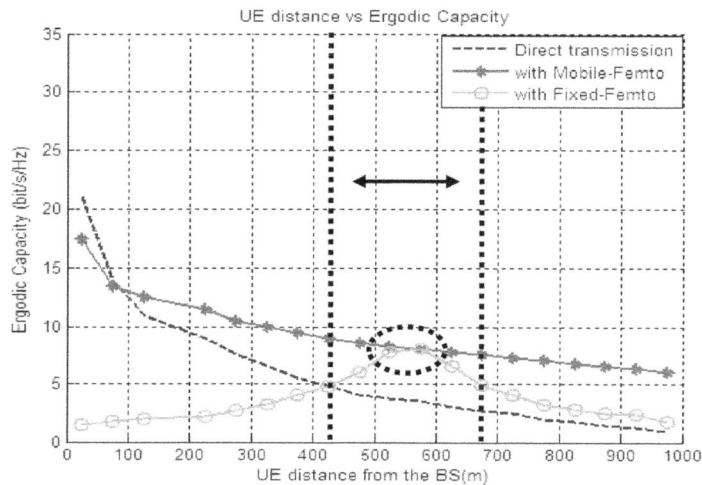

Fig. 11.11. The Ergodic Capacity when VPL = 40 dB.

After discussing the results of the vehicular UEs links Ergodic capacity, it is important now to consider the other performance evaluation elements like the spectral efficiency, throughput and SINR. Fig. 11.12 represents a comparison between the spectral efficiency in respect to the *Empirical Cumulative Distribution Function* (ECDF) of vehicular UE in the case of direct transmission from the eNB and in the case of implementing the Fixed-Femto and Mobile-Femto in the Macrocell. There is a noticeable improvement in the average spectral efficiency of the vehicular UE after implementing the Mobile-Femto rather than the Fixed-Femto. This is because; firstly, the UEs are vehicular and moving from one place to another in public transportations like buses, which makes it hard for them to establish a long connection with the nearby Fixed-Femto unless those vehicular UEs have stopped for few minutes close to a Fixed-Femto that has been deployed in a nearby bus station. This is why the figure shows a slight improvement in the case of vehicular UEs spectral efficiency when they are attached to the Fixed-Femto over the direct transmission UEs. As the distance between the vehicular UEs and the eNB increases, the Fixed-Femto starts to look as a better option than relying on the eNB to provide the connection for them. Secondly, the high penetration loss that is 25 dB in the case of vehicular UEs plays an important role in the poor spectral efficiency of direct transmission as well as the Fixed-Femto UEs transmission. Therefore, 90 % of the vehicular UEs have enjoyed a spectral efficiency around 3.7 bit/cu when they are connected to the Mobile-Femto, while 2.5 bit/cu in the case of direct and Fixed-Femto transmissions.

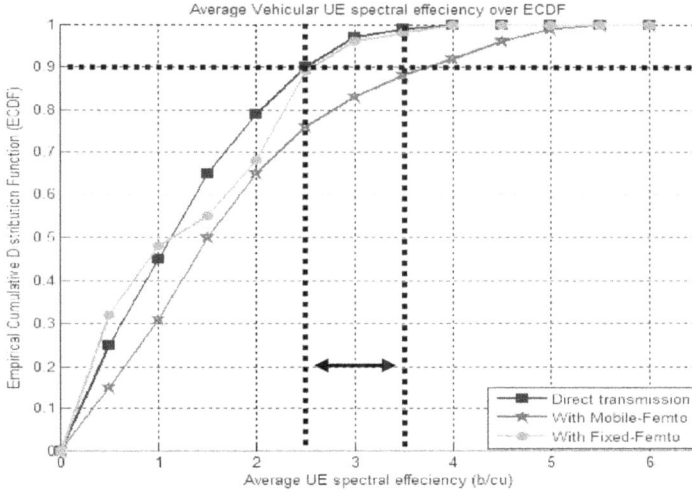

Fig. 11.12. Average spectral efficiency of vehicular UEs at VPL=25 dB.

Since the spectral efficiency has a positive correlation with the throughput, the achieved throughput of vehicular UEs needs to be demonstrated before and after deploying Femtocells in the Macrocell. Hence, Fig. 11.13 shows that at a certain distance, the Fixed-Femto vehicular UEs start to achieve a slight higher throughput than the eNB vehicular UEs. This is because the vehicular UEs are moving close and far from a nearby Fixed-Femto especially when the distance gap increases between the vehicular UEs and the eNB, but then the throughput drops again as the penetration loss and distance gap (path-loss) increase between the Fixed-Femto and vehicular UE. Therefore, deploying the Mobile-Femto in the Macrocell shows an improvement in the vehicular UEs average throughput as 80 % of the vehicular UEs throughput increased by 300 Kbps. Furthermore, the Mobile-Femtos themselves have a higher throughput around 500 Kbps due to the additional gain in the received SINR on the backhaul link which has improved the transmitted signal between the eNB and the Mobile-Femto, thus the achieved throughput [78]. This gain can be achieved by using a highly directional antenna pattern in the eNB and directing it towards the positioned Mobile-Femto antenna.

In addition, the SINR of vehicular UEs plays an important role in measuring the UEs performance. It indicates the signal strength especially for those UEs who suffer from high penetration loss and path-loss. Fig. 11.14 presents the vehicular UEs SINR before and after implementing Femtocells into the Macrocell. The results show that 80 % of the vehicular UEs have increased their SINR with 4 dB and that means implementing the Mobile-Femto into the Macrocell has overcome the signal degradation. While, the vehicular UEs SINR who have been served by the Fixed-Femtos showed slight improvement as the distance between the vehicular UEs and the eNB is more than 500 m. Therefore, those vehicular UEs will try to establish a connection to maintain their signal with any nearby Fixed-Femto even for a short period of time. As a result, deploying Mobile-Femtos as well as Fixed-Femtos in the Macrocell can be seen as the future of next generation networks to provide internet along the buses and buses' routes when the penetration and path losses are very high.

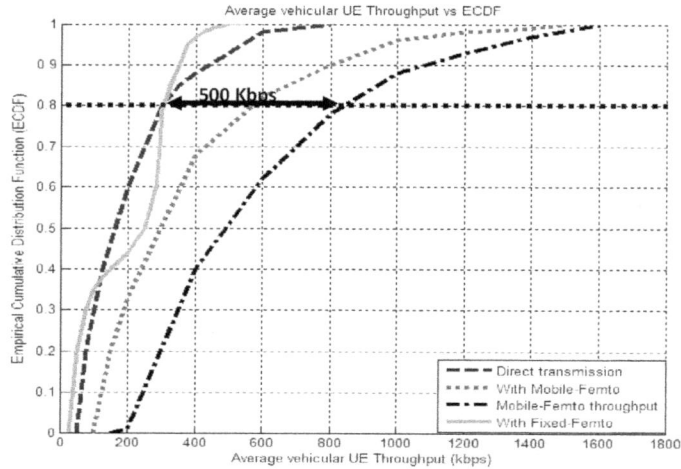

Fig. 11.13. Average vehicular UEs throughput at VPL=25 dB.

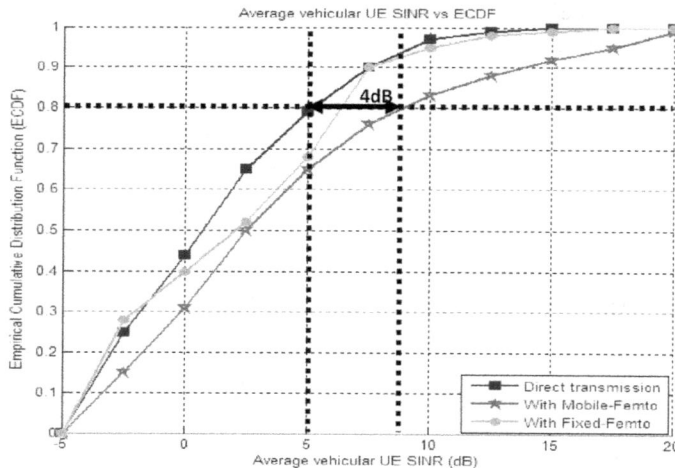

Fig. 11.14. SINR of vehicular UEs at VPL=25 dB.

On the other hand, the following discusses the achieved results in terms of the vehicular UEs outage probability. These results create a comprehensive comparison between the signal quality of the direct transmission from the eNB, the transmission from the Fixed-Femto and finally the transmission from the Mobile-Femto. In these results, two parameters play an important role in the achieved performance, which are the path-loss and the penetration loss. Table 11.1 gives detailed parameters of the simulated scenarios.

Table 11.1. Detailed simulated parameters of the Outage Probability scenarios.

Parameters			Values		
Schemes	Duplex Mode	Path-loss Model	UE Position Distribution	VPL (dB)	Transmission Power
Direct Transmission	Full Duplex	3GPP SCM Urban NLOS Microcell model	Uniform distribution	0, 25, 40	$P_x^{eNB} = 46$ dBm
Fixed-Femto assisted Transmission	Full Duplex	Urban SCM NLOS Microcell model for both backhaul and access links	Uniform distribution	0, 25, 40	$P_x^{eNB} = 46$ dBm and $P_x^{FFemto} = 24$ dBm
Mobile-Femto assisted Transmission	Full Duplex	Backhaul link: SCM urban NLOS Microcell model. Access link: Constant path-loss	Uniform distribution	0	$P_x^{eNB} = 46$ dBm and $P_x^{MFemto} = 24$ dBm
Path-loss Model in dB			$PL(L) = 34.53 + 38\log_{10}(L)$		
Cell Radius D			1000 m		
Receiver noise figure for both Femtocell and UE			9 dB		
Minimum required rate R at the UE			2 bits/sec		
Fading			Small-scale fading		

The results in Fig. 11.15 show that, when the VPL is equal to 0 dB, the direct transmission from the eNB always achieves the lowest outage probability since there is no resistance against the transmitted signal. While at 400 m distance from the BS, the vehicular UEs who are being served by the Fixed-Femto achieve lower outage probability than the Mobile-Femto UEs. This is because when there is no penetration loss, the Fixed-Femto signal can travel smoothly without massive reduction in the received power (P_r) at the vehicular UE's end. Besides that, the mobility aspect of the Mobile-Femto creates link variation in the backhaul link between the eNB and the serving Mobile-Femto, which becomes obvious, compared to the other BSs transmissions with the absence of the VPL.

While in Fig. 11.16, the outage probability of the Mobile-Femto assisted-transmission at a VPL of 25 dB is always lower than the outage probability of the direct transmission form the eNB. It is also lower than the outage probability of the Fixed-Femto assisted-transmission. This indicates that the Mobile-Femto assisted system is better at maintaining a given rate of R bits/sec for a vehicular UE, which can be translated to a good QoS compared to the other two transmissions. However, with the increased distance -over

300 m- the Fixed-Femto assisted-transmission starts to achieve lower outage probability than the direct transmission from the eNB and this is because of the high path-loss of the eNB signal over distance when it tries to reach the vehicular UEs inside vehicles. Here it raises the case of being able to be attached to any nearby Fixed-Femto when the Mobile-Femto channels and PRBs are fully occupied by other UEs. This might be a temporary solution to improve the vehicular UEs signal quality unless the required PRBs are available in the Mobile-Femto to accommodate those UEs. Otherwise, the connection will be dropped as soon the distance increases between the vehicular UE and the serving Fixed-Femto.

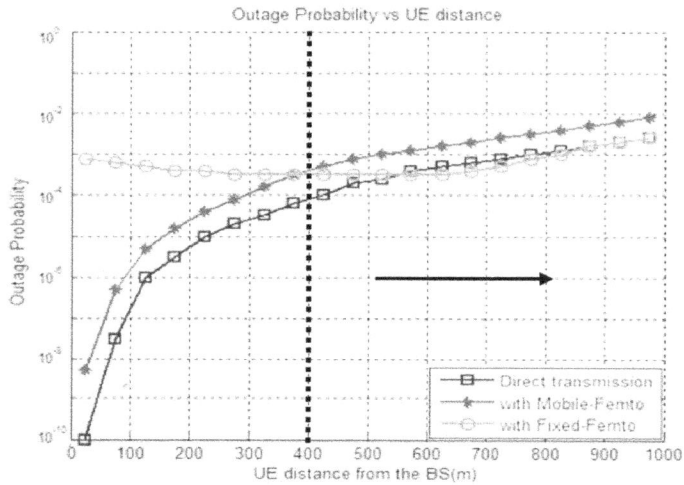

Fig. 11.15. Outage Probability when VPL= 0 dB.

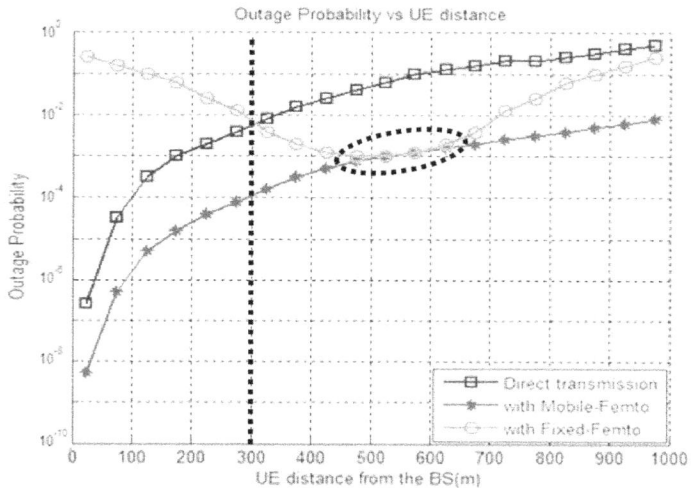

Fig. 11.16. Outage Probability when VPL= 25 dB.

As the VPL value goes up to 40 dB, the Mobile-Femto assisted-transmission outperforms the direct transmission when the vehicular UE is fairly near the eNB (around 50 m). The 40 dB vehicular penetration loss is presented as an extreme case, which can rarely be observed in the current wireless systems. However, higher vehicular penetration losses are foreseeable if higher frequency bands are used [79], e.g., the 3.6 GHz band is allocated to next generation mobile communication systems. Hence, Fig. 11.17 represents the outage probability of vehicular UEs when the penetration loss is equal to 40 dB. From the figure below, it is obvious that vehicular UEs are enjoying better quality of connection and signal strength when they are served by the Mobile-Femto rather being served directly by the eNB or the Fixed-Femto. However, in terms of Fixed-Femtos at 500 m the results again show less outage probability compared to the eNB transmission, although the outage probability increases again as the vehicular UEs move away from the serving Fixed-Femto.

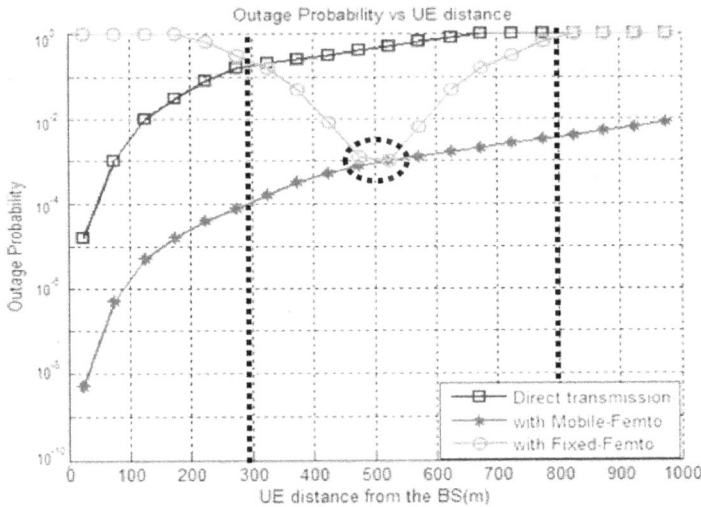

Fig. 11.17. Outage Probability when VPL= 40 dB.

Thus, it can be concluded that under high VPLs, the vehicular UEs have enjoyed better connection and less outage probability when they are connected directly to the Mobile-Femto. On the other hand, the Fixed-Femto outage probability was changing between lows and highs as the path-loss and penetration loss increase between the BS and the vehicular UEs.

11.5. Interference Management for Co-Channel and Co-Tire Femtocells Technology

The dense deployment of Fixed-Femtos and Mobile-Femtos within the Macrocell's coverage is expected to dominate the future of LTE networks. Many studies have focused on the Femtocell planning and management scheme when thousands of these small cells are deployed in the LTE networks. This is because; the densely deployment of those low power nodes is one of the main features of the future of 4G networks. In general, a feature of Femtocells is that they are deployed without planning; therefore, a Macro-UE operating close to a Femtocell may experience severe interference, as well as the other neighbouring Femto-UEs [80]. Subsequently, the interference management is essential in Fixed and Mobile Femtocells dense deployment schemes, as they have to operate in a coordinated way in order to avoid the generated interference. Here, the interference management is even more important when there is a Macro BS that may share all or part of the BW with the deployed Femtocells. As in this case, the already existing Macrocells and UEs may experience interference from the deployed Fixed-Femtos/Mobile-Femtos as well as the Fixed-Femtos/Mobile-Femtos may experience interference from each other. However, the presence of small BSs within the coverage of larger ones changes the architecture of the cellular system. As the Fixed and Mobile Femtocells are deployed within the coverage of the Macrocell, this can cause interference within the same tier or on different tiers. In another words, the co-channel interference happens when the Femtocell causes interference on the DL signal of the Macrocell to the Macro UE, and also the Macro UE causes interference on the UL signal to the Femtocell. Where, the co-tier interference happens when the Femtocell causes interference on the DL signal to another Femtocell UE, and also when the Femtocell UE causes interference on the UL signal to another nearby Femtocell. Consequently, the following two scenarios are considered as the worst-case scenarios in the case of the Mobile-Femto deployment. This is because; in the OFDM system, the interference level varies depending on the subcarriers' allocation as shown in [81, 82]. In the first scenario, the interference occurs when a cell-edge Macro UE inside the bus has to transmit with high power due to its far position from the Macrocell to overcome the path-loss issue and vice versa. Thus, if the Mobile-Femto is installed in the same bus, it will receive a massive amount of interference from the Macro UE. While in the second scenario, the interference occurs when two buses are close to each other and a UE attached to Mobile-Femto$_B$ is being interfered by a DL signal from Mobile-Femto$_A$ and vice versa. Hence, optimising the coverage planning scheme and controlling the transmission power have been considered to overcome the generated interference with the highly deployed dense of Femtocells in LTE and 5G networks. While an efficient frequency reuse scheme, has been proposed for the worst-case interference scenarios.

11.5.1. Coverage Optimisation

An important challenge for Femtocells is optimising their radio coverage area dynamically. The goal is to achieve a desired level of performance for mobile transmission, avoid the undesired interference and reduce power consumption. Providing optimal Femtocell signal coverage is important to improve UEs' mobile usage experience as well as reducing service cost expenditure [83]. In outdoor enterprise environment, a number of Femtocells may be deployed together to achieve joint coverage. This is also done to cover a large area while balancing the UEs load, minimising coverage gaps and overlaps between multiple Femtocells. Hence, researchers have begun to consider the problem of coverage optimisation of Femtocells in LTE networks as an important aspect that influences the amount of generated interference on UEs performance [84]. However, most of the previous works focused on the coverage optimisation of Fixed-Femtos rather than on the coverage optimisation of Mobile-Femtos as the mobility of Mobile-Femtos is considered a challenge issue in this case.

Hence, the purpose of this section is to find an efficient Femtocells coverage planning solution that involves identifying locations for installing Femtocells that are either Fixed or Mobile in LTE and 5G Networks. Femtocells locations play a vital role in achieving a good trade-off between coverage and interference. Poor choices of Fixed or Mobile Femtocells locations may result in coverage holes. Coverage hole occurs when Femtocells are either placed or moving distant apart from each other, while the maximum transmission power-level is not sufficient to cover certain regions in the outdoor environment. Moreover, incorrect locations such as placing or moving Femtocells too close to each other may result in excessive interference to Mobile or Fixed Femtocells' UEs and Macro UEs. Therefore, in the case of Mobile-Femto, identifying the bus path that the Mobile-Femto will be installed in will be a reasonable solution to avoid the interference issue between any nearby Mobile-Femtos. Thus, the goal of the coverage planning scheme is to have good radio conditions everywhere within the outdoor area or in the vehicular environment so UEs can acquire, initiate and sustain voice calls, as well as overcome the penetration loss and path-loss issues. Good coverage performance is addressed by limiting the maximum path-loss, which can be mitigated by deploying Fixed-Femtos and Mobile-Femtos in the right positions within the Macrocell, and added to this a proper Fixed-Femtos and Mobile-Femtos power calibration is needed.

For that reason, it is important to state that the used path-loss model in this work is the Microcell NLOS path-loss, which is based on the COST 231 Walfish-Ikegami NLOS model. The COST 231 Walfish-Ikegami model is an evolution of the Ikegami model [85]. It is developed for urban areas and it takes into consideration obstructing building height and street width, as well as other factors related to the urban environment. Therefore, in order to calculate the Microcell NLOS path-loss, the following parameters are needed: BS antenna height 12.5 m, building height 12 m, building to building distance 50 m, street width 25 m, Mobile Station (MS) antenna height 1.5 m, orientation 30 deg for all paths, and selection of metropolitan center. And the distance between two BSs is less than 1 Km [86]. Hence, based on the previous parameters, the NLOS path-loss equation simplifies to

$$PL(L) = -55.9 + 38 \log_{10}(L) + \left(24.5 + 1.5 * \frac{f_c}{925}\right) * \log_{10}(f_c). \quad (11.33)$$

This is resulting that the path-loss at 1900 MHz is given by

$$PL(L) = -55.9 + 38 \log_{10}(L) + \left(24.5 + 1.5 * \frac{1900}{925}\right) * \log_{10}(1900). \quad (11.34)$$

Thus, it is equal to

$$PL(L) = 34.53 + 38 \log_{10}(L), \quad (11.35)$$

where L is the distance in meters and it is at least 20 m.

On the other hand, the Microcell LOS path-loss is based on the COST 231 Walfish-Ikegami street canyon model with the same parameters as in the NLOS case [86]. The LOS path-loss is given by

$$PL(L) = -35.4 + 26\log_{10}(L) + 20\log_{10}(f_c). \qquad (11.36)$$

This is resulting that the path-loss at 1900 MHz is equal to:

$$PL(L) = -35.4 + 26\log_{10}(L) + 20\log_{10}(1900). \qquad (11.37)$$

Thus, it is equal to

$$PL(L) = 30.18 + 26\log_{10}(L). \qquad (11.38)$$

Hence, it becomes easier to classify the achieved path-loss equations among the direct and access links of eNB and Femtocells in LTE and next generation networks. The used path-loss on the direct transmission between the eNB and the vehicular UEs is the urban NLOS Microcells model as shown in Eq. (11.35). The same NLOS model has been used for the Fixed-Femto for both; the backhaul link between the eNB and the Fixed-Femto, and the access links between the Fixed-Femtos and vehicular UEs. Therefore, based on Eq. (11.4) the received SNR at the NLOS receiver R_x can be modified to $SNR_{R_x} = \frac{P_x|G|^2(34.53 + 38\log_{10}(L))\varepsilon}{P_{noise}}$. Based on Fig. 11.6, L here is changeable according to the distance between the transmitter and the receiver, which could be either x in the case of the direct transmission from the eNB or $x\text{-}d$ in the case of the Fixed-Femto transmission.

On the other hand, in the case of the Mobile-Femto, the backhaul link between the Mobile-Femto and the eNB has used the same previous NLOS model. Whereas, for the LOS access link between the Mobile-Femto and the vehicular UEs in the same bus a Constant Path-Loss has used, which is a free space loss when there is no obstacle against the transmitted and received signals. The free space loss can be very small or sometimes unity when the distance between the receiver and the transmitter is very small and does not exceed few meters. However, there is always a constant power loss of -84.55 [69] and based on Eq. (11.7), the received SNR at the LOS receiver R_x is given by $SNR_{R_x} = \frac{P_x{}^{MFemto}C_{loss}}{P_{noise}}$.

Thus, identifying the used path-loss model between the links of two BSs or between the links of BSs and vehicular UEs has helped in reducing the interference issue by allowing the Operator to know where the Fixed and Mobile Femtocells are needed to be installed based on the used path-loss model. In the case of the Mobile-Femto, identifying the path-loss is an important factor for the network Operator because it helps the Operator to indicate the Mobile-Femto paths, which are the bus paths. This is due to the fact that, as much as it is important to reduce the path and penetration losses between the vehicular UEs and the serving Femtocell, it is more important to maintain the backhaul path-loss between the eNB and Mobile-Femto, so it can serve those vehicular UEs without losing the connection with the mother BS (eNB). Having strong connection between the Mobile-Femto and the eNB as well as between the Mobile-Femto and the vehicular UEs in the same bus has helped in mitigating the generated interference. This strong connection can be achieved by having an efficient coverage planning technique together with choosing the right path-loss model. Hence, after discussing the chosen path-loss model in the case of vehicular environment and its impact on mitigating the interference, it is important now to discuss the effect of the chosen transmission power on the generated interference as the following section shows.

11.5.2. Transmission Power Control

Controlling the BS transmission power is a very important factor in 4G and next generation systems since it indicates the transmission range of BSs e.g. eNBs and Femtocells. However, the power may leak between neighbouring Femtocells due to the dense deployment, and this creates dead zones, which means areas without coverage (holes). While, high transmission power with high Femtocells dense deployment creates excessive interference issue. Therefore, the organised installation of Fixed-Femto makes it easier to manage the Femtocells planning and the distribution dense, thus, the interference issue. In contrast, the environment and location changes of the Mobile-Femto without any prior knowledge available of other Mobile-Femtos within the network, makes it very important to equip the Femtocell with some kind of interference management schemes. In the Mobile-Femto scenario, the distance between the Femto UEs and the serving Femtocell does not require high transmission power due to the close distance between the transmitter and the receiver. Added to this, the channel between the Femtocell and its UEs is considered in a good state which is less affected by the interfered signals from the eNB. Besides that, the high penetration loss in the case of vehicular UEs due to the vehicle chassis can be counted as an advantage as it prevents the outside signals from passing through to the vehicular UEs. This means that the bus chassis can isolate the vehicular UEs from the outside signals as they can be served only by the installed Femtocell (Mobile-Femto) in the same bus, which can help to mitigate the interference issue.

Hence, indicating the Mobile-Femtos paths as well as the Fixed-Femtos positions based on the used path-loss model has helped in mitigating the interference and expanding the coverage area. Additionally, the transmission power of Fixed-Femtos and Mobile-Femtos plays an important role in mitigating the interference as well as in filling the coverage holes. Therefore, the chosen transmission power in the case of Fixed-Femtos and Mobile-Femtos has been assumed to be 24 dBm while 46 dBm in the case of eNB transmission. The previous transmission power values have been chosen after running the simulation several times with different transmission powers and taking into consideration the impact of that on the achieved throughput and SINR. This has led to the following conclusion, choosing lower transmission powers has its negative impact on the achieved throughput and number of scheduled UEs even when the interference is mitigated. This is because limiting the transmission power is not only about the interference but it is also about the network performance and UEs throughput in general.

However, there is always a worst-case scenario especially in the Mobile-Femto case, as the mobility aspect plays an important role in increasing the interference percentage. This is because with the vehicular environment, unexpected scenarios can occur at any time. This means a bus might change its route and pass close to another Mobile-Femto in another bus or a Mobile-Femto is interfering with the eNB signal; therefore, an efficient interference management scheme is required. In fact, the Fixed-Femtos interference is easier to be mitigated with a good coverage planning and good transmission power control as mentioned earlier, while the Mobile-Femto technology requires an efficient interference management scheme due to its high mobility and movement from one place to another.

11.5.3. The Proposed Interference Management Scheme

The mobility of Mobile-Femto positioning in public transportations like buses makes it difficult for the cellular Operator to control the interference of these new small BSs. Therefore, the aim here is to enable the system to avoid the generated interference between the eNB/Mobile-Femtos and between the Mobile-Femtos themselves. Fig. 11.18 represents three possible interference scenarios that might occur in the DL signal of each the eNB and the Mobile-Femto. The first interference scenario occurs when a vehicular UE inside Bus$_A$ who is being served by the Femtocell that is installed in the same bus interferes with the DL signal of the eNB. This scenario occurs when Bus$_A$ gets close to the eNB, which makes the eNB DL signal strength equal to the DL signal strength of the Mobile-Femto in Bus$_A$. On the other hand, the second interference scenario may occur when two nearby Mobile-Femtos interfere with one another e.g. a UE in Bus$_B$ can be interfered by the DL signal of Mobile-Femto in Bus$_A$ especially when the used transmission power of the Mobile-Femtos is quite high and vice versa. While the third interference scenario occurs when a Macro UE (primary UE) is close to Bus$_A$ and interferes by the DL signal of the Femtocell in Bus$_A$.

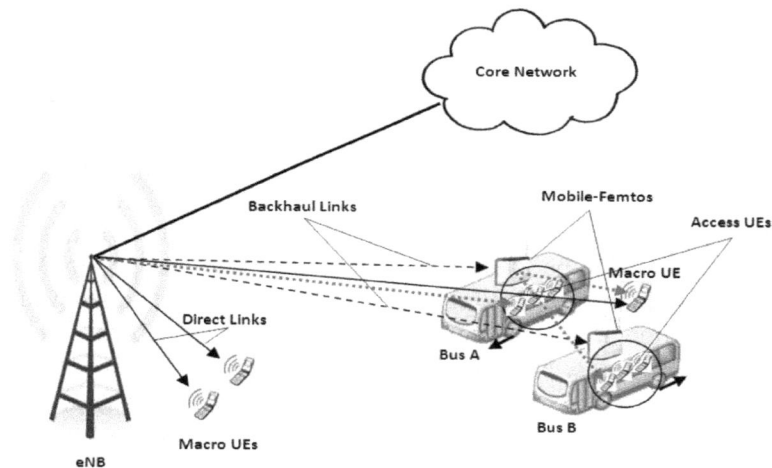

Fig. 11.18. Summarisation of three interference scenarios.

To overcome the previous interference issues as well as to improve the vehicular UEs throughput and save the used frequency, the FFR scheme has been used in the Mobile-Femtos deployment. The FFR is one of the solutions to reduce the ICI in the Macrocell system, especially for the cell-edge UEs. In addition, it is helpful to achieve the reuse factor of one for the centre zone of the Macrocell. Under this condition, the interference from the Femtocell deployment should be minimised for the Macro UEs as well as for Femto UEs. Therefore, the focus here is to mitigate the interference between the Macrocell and the Mobile-Femtos as well as between the Mobile-Femtos themselves via using the FFR scheme.

The FFR scheme has been discussed in the OFDMA based network, such as the LTE, to overcome the co-channel interference problems [57]. In the FFR, the whole frequency band is divided into several sub-bands, and each sub-band is differently assigned to centre zones and edge zones. The reuse factor of one is used for the centre zone while the edge zone adopts bigger reuse factor. Subsequently, the ICI is substantially reduced and the system throughput is improved. On this basis, this work proposes an efficient interference management scheme in the LTE Mobile-Femto systems via using the FFR management scheme. When the Macrocell allocates frequency band using the FFR, the Mobile-Femto chooses sub-bands which have not been used by the Macrocell sub-area and other Mobile-Femtos to avoid the generated interference.

Thus, in order to understand the proposed scheme, Fig. 11.19 represents the allocated frequency sub-bands to the Macrocells and Femtocells in details. The Macrocell coverage is divided into centre zone and edge zone including the frequency sub-band F_0 for the Macrocell centre zone while the edge zone has three frequency sub-bands denoted by, F_1, F_2 and F_3. The reuse factor of one is applied in the centre zone, while the edge zone adopts the reuse factor of three. The sub-band F_0 is used in the centre zone of Macrocell$_1$, Macrocell$_2$ and Macrocell$_3$, and sub-band F_1, F_2 and F_3 is applied in the edges of Macrocell$_1$, Macrocell$_2$ and Macrocell$_3$ respectively. Under this circumstance, the Mobile-Femto chooses sub-bands which have not been used by the Macrocell sub-area and other Mobile-Femtos. It is to be mentioned that, the used frequency for the backhaul link transmission between the eNB and the Mobile-Femto is different from the below frequency set. This is because it uses a frequency band which has not been used before. This is required in order to avoid the interference between different backhaul links as well as with different direct and access links.

For instance, when a Mobile-Femto is moving in the cell-edge of Macrocell$_2$, it can reuse the centre zone's frequency sub-band F_0 or F_1+F_3 to serve the Femto UEs, while the Macrocell uses sub-band F_0 for its centre zone. On the other hand, if a Mobile-Femto is moving in the central zone of Macrocell$_2$ (eNB$_2$), sub-band F_1+F_3 is applied. The Mobile-Femto cannot reuse sub-band F_0 which is already been used by the Macrocell in the central zone. Besides that, it cannot reuse sub-band F_2 which is used by the Macrocell to serve the cell-edge Macro UEs of the Macrocell$_2$. This is because the transmission power of the eNB in each case is different. The centre UEs who are close to the eNB require low transmission power whereas, the eNB should transmit in maximum power in order to satisfy the cell-edge UEs. Following the previous approach has helped in avoiding the interference between the transmitted signals. In this study, more subcarriers are allocated to the Mobile-Femto that is located in the cell-edge rather than the centre zone in order to improve the performance of the vehicular cell-edge UEs. While in order to mitigate the interference between two nearby Mobile-Femtos that belong to the same Macrocell, the proposed scheme has used the same frequency reuse approach.

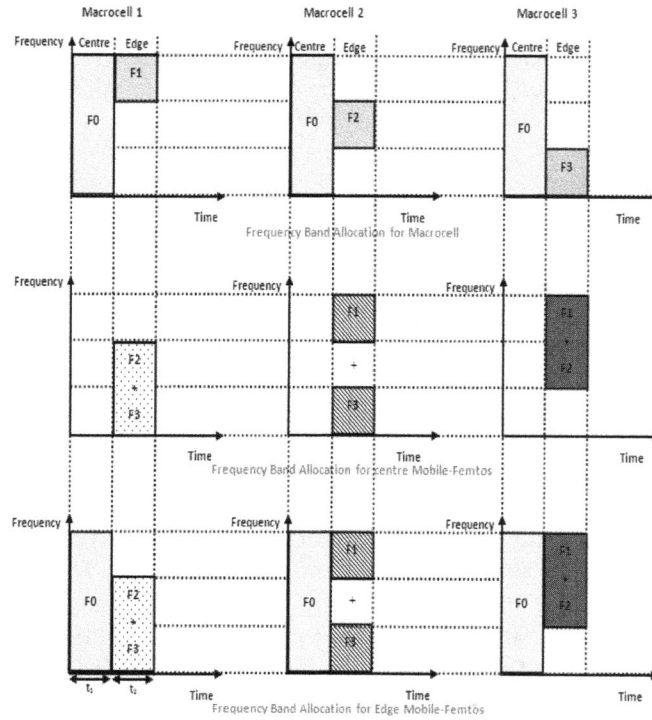

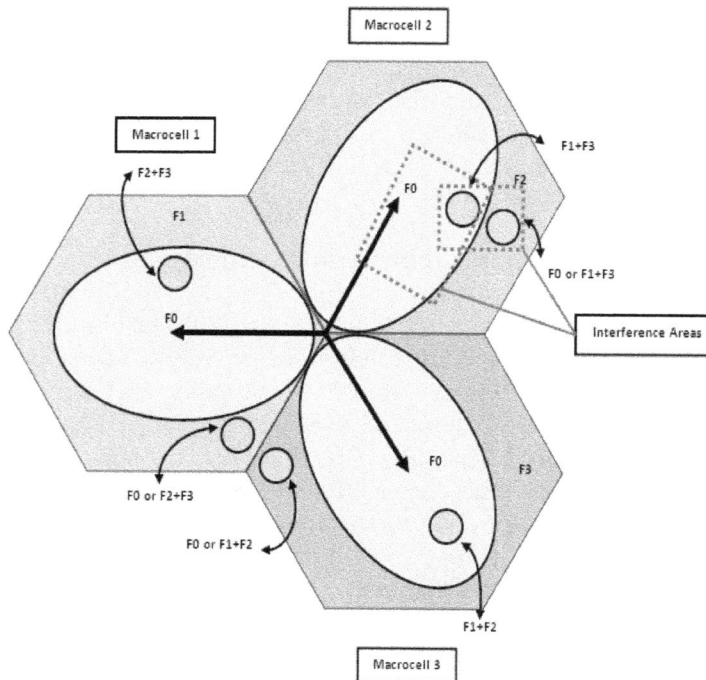

Fig. 11.19. The proposed interference management scheme based on the FFR scheme.

For example, when a Mobile-Femto is moving in the cell-edge of Macrocell$_2$, it uses sub-band F_0 or $F_1 + F_3$ as mentioned earlier, while the other Mobile-Femto that is moving in the centre zone of Macrocell$_2$ (eNB$_2$), will be using sub-band $F_1 + F_3$ to avoid the signals conflict between the two Mobile-Femtos. In this manner, each of the Mobile-Femtos will not be using the frequency sub-band that has already been used by the other Mobile-Femtos that belong to the same Macrocell.

It is quite important then to state how this interference management scheme works in real life scenarios. The proposed scheme receives as an input the Macrocell dimensions, the number of Mobile-Femtos and their positions as well as the Macrocell and Mobile-Femtos transmission powers. Then, it calculates the received power from the serving cells as well as from the interfering cells. Based on the pervious values, the scheme is able to make estimation for the SINR and throughput at any given position of the examined LTE network. Thus, the proposed scheme follows the approach presented below.

a) Calculate the inner cell radius

Based on the Macrocells' characteristics and the used transmission power, the inner cell radius can be calculated which will help later to indicate the centre zone and cell-edge zone as Fig. 11.20 shows. Indicating the best dimensions of the inner cell optimises the UEs' throughput, thus achieves better performance.

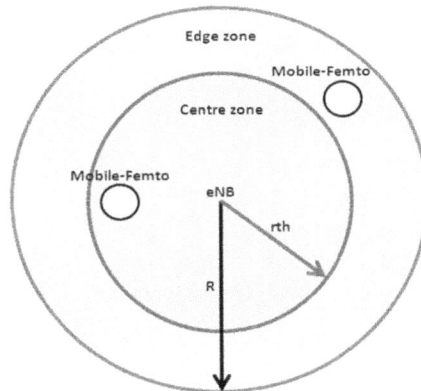

Fig. 11.20. Inner Cell radius.

b) Find the optimum frequency band division

The aim of this step is to calculate the UEs throughput as well as the UEs SINR for every possible combination of the spectrum division. As mentioned earlier, the radio spectrum is divided into frequency sub-bands reserved for a single use or a range of compatible uses. Within each band, individual transmitters often use separate frequencies or channels, so they do not interfere with each other. Hence, the available spectrum will be allocated to UEs according to the combination that maximises their throughput. Thus, there are two disjoint sets of subcarriers: the subcarriers in the centre zone and the subcarriers in the cell-edge zone. Every time the frequency band that is allocated to the cell-edge zone is equally divided between F_1, F_2 and F_3 as stated earlier. Initially, let it be assumed that the set of the centre zone is an empty set and all the subcarriers are contained in the set of cell-edge zone, that means all the subcarriers that are allocated to the cell-edge zone, and each one of F_1, F_2 and F_3 equals to $Total_{subcarriers}/3$. Each time one subcarrier is removed from the set of cell-edge zone and added to the set of centre zone. Finally, the set of the centre zone will consist of $Total_{subcarriers}$ and the set of the cell-edge zone will be an empty set.

c) Allocate frequency band to the Mobile-Femtos

In this step, the frequency is allocated to the Mobile-Femtos with the process presented in Fig. 11.20. Hence, the proposed scheme can be summarised in the following Algorithm 2.

Algorithm 2. Frequency Reuse Scheme for Interference Avoidance
for Macrocell/Mobile-Femtos

1:	*create_Networktopolgy();*	*/*defines Macrocells, Mobile-Femtos and UEs*
2:	*for r = 0: R*	*/*inner cell radius*
3:	*for f = 0: Total-subcarriers*	*/*frequency band division*
4:	*allocate frequency band for macrocells;*	
5:	*if r > = distance_mfemto*	*/*Mobile-Femto belongs to Macro centre zone*
6:	*allocate frequency band for mfemto;*	
7:	*else if r < distance_mfemto*	*/*Mobile-Femto belongs to Macro cell-edge zone*
8:	*allocate frequency band for mfemto;*	
9:	*End*	
10:	*End*	
11:	*for u = 1: U /*For all UEs calculate*	
12:	*sinr = calculate_sinr(u)*	
13:	*capacity = calculate_capacity(u)*	
14:	*throughput = calculate_ throughput(u)*	
15:	*outageprobability = calculate _Outageprobability(u)*	
16:	*End*	
17:	*End*	
18:	*End*	
19:	*define_FFR(max_ user_ throughput)*	

11.5.4. The Proposed Scheme (System Performance Analysis)

It is important to formulate the DL SINR, throughput and capacity to have clear comparison between the UEs performance before and after implementing the proposed scheme. Needless to say that, the only considered interference scenarios in this section are the interference between Mobile-Femtos and Macrocells (Co-Channel Interference) and interference among Mobile-Femtos themselves (Co-Tier Interference). The Macro UE can be interfered by the DL signal of any adjacent Mobile-Femto and the same may happen for the Mobile-Femto UE with the Macrocell or any adjacent Mobile-Femto. Thus, based

on Eq. (11.13), the received SINR for an outdoor direct Macro UE (m) on subcarrier n
can be expressed as

$$SINR_{m(D),n} = \frac{P_x^{eNB}|G_{1,eNB,n}|^2 PL(x)\varepsilon}{\sum_{eNB'=1}^{neNB'} P_x^{eNB'}|G_{1,eNB',n}|^2 + \sum_{MFemto=1}^{nMFemto} P_x^{MFemto}|G_{2,MFemto,n}|^2 + P_{noise}}, \quad (11.39)$$

where P_x^{eNB} and $P_x^{eNB'}$ are the transmission power of the serving Macrocell (eNB) and the
neighbouring Macrocell (eNB') on subcarrier n respectively. $G_{1,eNB,n}$ is the channel gain
between the Macro UE m and serving Macrocell (eNB) on subcarrier n. Channel gain
from neighbouring Macrocells is denoted by $G_{1,eNB',n}$. Similarly, P_x^{MFemto} is the
transmission power of neighbouring Mobile-Femto (MFemto) on sub-carrier n.
$G_{2,MFemto,n}$ is the channel gain between the Macro UE m and neighbouring Mobile-Femto
(MFemto) on sub-carrier n. Whilst, *PL* has been modelled to express the Path-Loss model
and x is the distance between the Macro UE and the serving eNB. ε is the VPL and P_{noise}
is the white noise power. Hence, Eq. (11.39) has made it clear that outdoor Macro UEs
can be interfered by two interference sources, one is the interference from neighbouring
Macrocells signals which has been given by $\sum_{eNB'=1}^{neNB'} P_x^{eNB'}|G_{1,eNB',n}|^2$ and the second is
the interference from Mobile-Femtos in the same Macrocell which has been given by
$\sum_{MFemto=1}^{nMFemto} P_x^{MFemto}|G_{2,MFemto,n}|^2$.

In contrast, the Macrocell or any adjacent Mobile-Femto can cause interference to a
Mobile-Femto's UE based on Eq. (11.15), thus, the received SINR of a Mobile-Femto UE
(mf) on subcarrier n can be formulated as

$$SINR_{mf(A),n} = \frac{P_x^{MFemto} C_{loss}}{\sum_{eNB=1}^{neNB} P_x^{eNB}|G_{1,eNB,n}|^2 + \sum_{MFemto'=1}^{nMFemto'} P_x^{MFemto'}|G_{2,MFemto',n}|^2 + P_{noise}}. \quad (11.40)$$

There is a system constant loss but no channel gain over the Mobile-Femto LOS access
link between the UE and the serving Mobile-Femto. The only existing channel gain
($G_{2,MFemto',n}$) is the channel gain between the Mobile-Femto UE with other adjacent
Mobile-Femtos that may cause interference to the considered UE.

Hence, the Macro-UE capacity on a specific subcarrier n can be estimated via the SINR
from the following equation [54]

$$C_{m,n} = BW. \log_2(1 + \alpha SINR_{m(D),n}), \quad (11.40)$$

where BW is the available bandwidth for subcarrier *n* divided by the number of UEs that
share the specific subcarrier and α is the coding margin and in this equation it is a constant
for target Bit Error Rate (BER) that defined by $\alpha = -1.5/\ln(5BER)$ [87]. Here, the BER
has been set to 10^{-6}. So the overall throughput of the serving Macrocell M can be expressed
as

$$Throughput_M = \sum_m \sum_n \beta_{m(D),n} C_{m(D),n}, \quad (11.41)$$

where $\beta_{m,n}$ here notifies the subcarrier assignment for Macro UEs. When $\beta_{m(D),n} = 1$, the
subcarrier n is assigned to Macro UE m and otherwise $\beta_{m(D),n} = 0$. From the characteristics

of the OFDMA system, each subcarrier is allocated to only one Macro UE in a Macrocell in every time slot. This implies that $\sum_{m=1}^{N_m} \beta_{m(D),n} = 1$ for $\forall k$, where N_m is the number of Macro UEs in a Macrocell and k is the available PRBs. While similar expression for Mobile-Femto UEs related to the practical capacity and the overall throughput is possible except $\sum_{mf=1}^{N_{mf}} \beta_{mf(A),n} = 3$ for $k \in F_{\text{Mobile-Femto}}$. N_{mf} is the number of Mobile-Femto UEs in a Macrocell and $F_{\text{Mobile-Femto}}$ is the available sub-bands allocated to Mobile-Femtos in Macrocells. This implies that the proposed scheme reuses the full frequency band three times in the considered Macrocells.

Thus, the Mobile-Femto UE capacity on a specific subcarrier n can be estimated via the SINR as the following equation shows

$$C_{mf,n} = BW. \log_2(1 + \alpha SINR_{mf(A),n}), \qquad (11.43)$$

where the overall throughput of the serving Mobile-Femto (MFemto) can be expressed as

$$\text{Throughput}_{\text{MFemto}} = \sum_{mf} \sum_n \beta_{mf(A),n} C_{mf(A),n}. \qquad (11.44)$$

After discussing the SINR, throughput and capacity, here it is important to state that the outage probability plays an essential role in the proposed scheme as it affects the network performance by affecting the data rate and throughput of UEs. If the outage probability is small, the throughput increases, and then when the throughput increases the data rate increases, thus, the performance improves and the interference decreases. However, the outage probability affects the performance of cell-edge UEs more than the Macro UEs due to their high path-loss. To find out the outage probability it is required first to identify the SINR threshold value in the range of 0 dB to 30 dB. The outage probability (P_{out}) here is determined when the SINR level of a subcarrier is below the designated threshold and it can be given by

$$P_{out} = \frac{\sum_u \sum_n \delta_{u,n}.SINR_{u,n}}{\sum_u \sum_n \beta_{u,n}.SINR_{u,n}}, \qquad (11.45)$$

where $\delta_{u,n}$ indicates the failed subcarrier of UE u on subcarrier n. This can occur when the penetration loss and path-loss issues are severe especially in the vehicular environment, which may affect the received SINR at the receiver side. Thus, if $\delta_{u,n}=1$, the SINR of that subcarrier is under the SINR threshold ($SINR_{u,n} < SINR_{\text{threshold}}$). As a result of that, the ratio between the number of subcarriers under the SINR threshold and the number of the total subcarriers is the outage probability.

The subsequent results will show that the proposed scheme is greatly capable of preventing the interference among the Macro UEs as well as the interference among the Mobile-Femto UEs. The Mobile-Femto UEs effect on Macro UEs and vice versa is less in the proposed scheme in comparison with the previous schemes. For example, the FFR-3 (frequency factor reuse 3) and NoFFR (frequency factor reuse 1) techniques assign random subcarriers to the Femto UEs, regardless of the subcarriers that have been used by the Macro UE. Therefore, the Macro UEs and the Femto UEs may use subcarriers very nearer to one another that cause interference. Due to this fact, the interference between the Macro UEs and the Femto UEs is higher than the proposed scheme. Nevertheless, the proposed scheme avoids this interference at minimal degradation and the total amount of available subcarriers for Mobile-Femto UEs is three times of the full band in this scheme. The above comparison has been summarised in Table 11.2. Also, the table shows the fact that Macro UEs have the option to choose one frequency band at a time in each Macrocell and their choice is limited on whether those UEs are in the centre zone or edge zone. In contrast, the Mobile-Femto UEs have the choice to choose between three different frequency bands at a time and their choice is again based on UEs locations and whether they are in the centre zone or edge zone. (Refer to Fig. 11.19)

Table 11.2. Comparison between the proposed and previous interference management schemes.

Schemes	Macro-user		Femto-user	
	Frequency	Amount	Frequency	Amount
Proposed for the Mobile-Femto/ Macrocell	FFR	**1** Centre zone: F_0 Edge zone: F_1, F_2 or F_3 depends on the Macro UE position in each Macrocell $1 + 1 = 2$ $\frac{2}{2} = 1$	Divide centre and edge zones	**3** Centre zone has **2** options Edge zone has 4 potions: F_0, F_1, F_2 or F_3 depends on the Mobile-Femto UE position in each Macrocell $2 + 4 = 6$ $\frac{6}{2} = 3$
FFR-3 (reuse 3)	FFR		Random	
NoFFR-3 (reuse 1)	Random	1	Random	1
Note: Amount column implies value of $\sum_{m=1}^{N_m} \beta_{m,n}$ and $\sum_{mf=1}^{N_{mf}} \beta_{mf,n}$ for Macro and Mobile-Femto UEs respectively.				

11.5.5. Results and Discussion

The following section represents the simulated results of the interference avoidance mechanisms, e.g. the coverage planning technique and the transmission power controlling technique. As mentioned earlier, those mechanisms can work efficiently in the case of fixed BSs in close areas. While in the case of mobile BSs the path of the Mobile-Femto (bus) has been specified based on the chosen path-loss model, however the street traffic is unpredictable and Mobile-Femtos can change their directions, which might create interference problems in some scenarios. These interference scenarios can occur when; Macro UE interferes with a DL signal from the nearby Mobile-Femto, a Mobile-Femto UE interferes with a DL signal from the eNB and finally a Mobile-Femto UE interferes by a DL signal of neighbouring Mobile-Femto. Hence, the proposed scheme has been compared with previous reuse frequency interference mitigation schemes like FFR-3 scheme (frequency reuse factor 3) and NoFFR-3 scheme (frequency reuse factor 1) in order to have a comprehensive comparison between the three schemes.

It is obvious that, the signal loses its strength along the distance, which affects the UEs ability of being attached to the Macrocell. This issue is worst in the case of cell-edge vehicular UEs because besides the high path-loss, they are more affected by the mobility and the VPL. For this reason, implementing Femtocells in Macrocells is one of the solutions to improve the UEs performance. Since the concern here is about vehicular UEs who suffer from high path-loss, high penetration loss and shadow fading, implementing Femtocells (Fixed and Mobile) is a reasonable solution for the previous issues. However, due to vehicular UEs mobility, the coverage distribution and signal transmission power of each Fixed-Femto and Mobile-Femto play important roles in improving the UEs performance as well as reducing the interference issue as much as possible. Therefore, the following two scenarios represent the coverage distribution as well as the assumed transmission power of Fixed-Femtos and Mobile-Femtos. These two scenarios were simulated separately in order to see clearly the impact of Fixed-Femtos and Mobile-Femtos coverage distribution and transmission power on the generated interference and coverage overlapping.

a) Scenario 1: Fixed-Femtos with vehicular UEs

The first simulated scenario represents the Fixed-Femtos deployment to serve vehicular UEs. The Fixed-Femtos have been distributed in the Macrocell based on the used path-loss model as shown in Fig. 11.20.

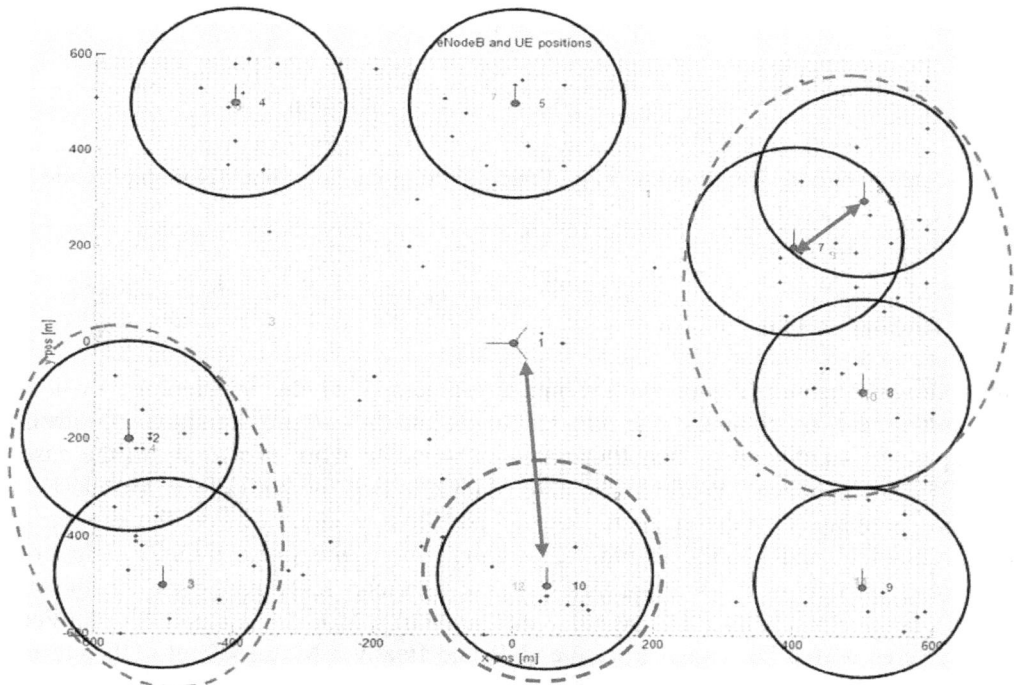

Fig. 11.20. Fixed-Femtos coverage distribution when the transmission power is 24 dBm and VPL is 25 dB.

Added to this, the use of high transmission power, which is 24 dBm, was required to reach the vehicular UEs and overcome the high VPL. However, this has created overlapping areas in the three considered Macrocells coverages since those Femtocells are in fixed positions when trying to serve vehicular UEs. The occurred interference is between the Femtocells themselves and the Femtocells and the eNB as the figure shows.

The previous interference issue can be mitigated by reducing the used transmission power, although, this will have negative impact on the achieved throughput as using lower transmission power makes the signal too weak to penetrate the chassis of the vehicle and makes it unable to reach the distant vehicular UEs.

While Fig. 11.21 represents the Fixed-Femtos distribution in the Region of Interest (ROI) in terms of path-loss with and without the shadow fading. In this figure, the dark red areas represent the coverage area with good signal strength as the path-loss is at its lowest value. In this case, the highest SINR value that represents the signal strength is 20 dB. In contrast, the dark blue areas represent the areas with low signal strength as the path-loss is at its highest value. In this case, the SINR reaches its lowest at -5 dB. It is noticeable that in this figure there are many overlapped areas and the highlighted area represents the worst overlapped area due to the high transmission power of those Fixed-Femtos in the coverage area of the Macrocell. Microcell path-loss model together with 25 dB VPL and shadow fading played important roles in the simulated plots. These three parameters were the base behind designing the required environment. They helped also in illustrating the impact of interference between BSs and UEs on the achieved SINR and spectral efficiency.

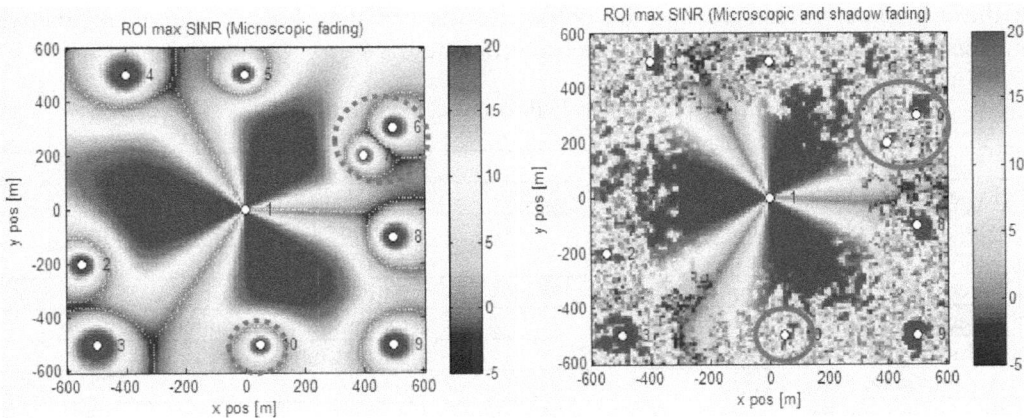

Fig. 11.21. Fixed-Femtos Sector SINR, calculated with distance dependent Microcell path-loss (left) while (right) represents the distributed space correlation with shadow fading.

The overlapped areas shown in the previous figure do not only have a negative impact on the SINR but also have a negative impact on the spectral efficiency as Fig. 11.22 shows. The figure shows that in the highlighted areas, the spectral efficiency is low as the yellow colour represents the low spectral efficiency of 3 bits/cu. While other Fixed-Femtos have more of the red and dark red areas that represent the areas with less interference as the spectral efficiency is high. Added to this, the right side of the figure represents the spectral efficiency with the effect of the shadow fading. It is obvious that the signal strength of the interfered highlighted areas fades faster than other Fixed-Femtos because of the interference effect on the signal strength as well as the achieved spectral efficiency.

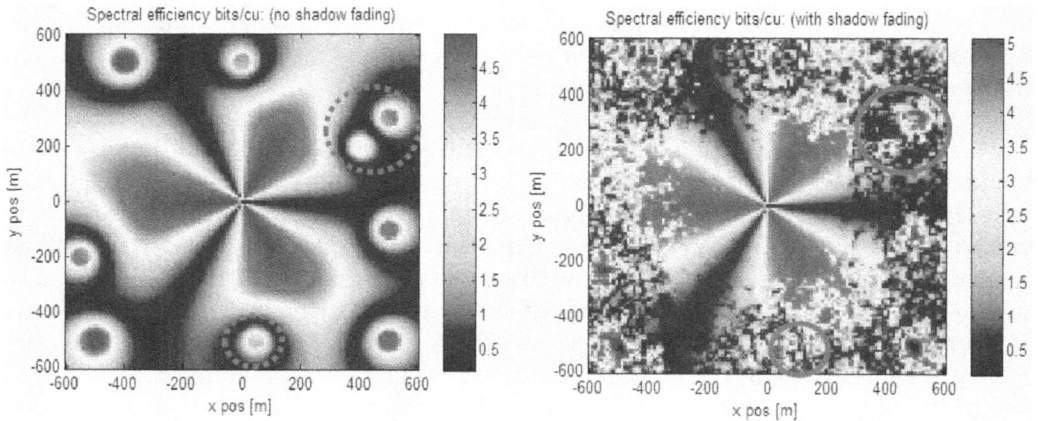

Fig. 11.22. Fixed-Femtos Sector Spectral efficiency with distance dependent Microcell path-loss (left) while (right) represents the distributed spectral efficiency correlation with shadow fading.

Hence, Table 11.3 represents the used parameters in the simulated scenario as any changes in these parameters can change the output results and may have negative or positive impacts on the achieved SINR and spectral efficiency.

Table 11.3. Fixed-Femto simulated parameters.

Parameters	Value
Type of Femtocell	Fixed
Number of Fixed-Femtos	3 Fixed-Femto per each 1 Km^2
Number of Fixed-Femto's server UEs	10 UEs
Fixed-Femto transmission power	24 dBm
Fixed-Femto antenna type	Omnidirectional antenna
Vehicular UEs speed	3-160 km/h
Antenna gain Fixed-Femto	0
Shadowing standard deviation eNB-Fixed-Femto	8
Path-loss model	SCM urban NLOS Microcell model for both backhaul and access links $PL(dB) = 34.53 + 38\log_{10}(L)$

In order to mitigate the interference issue between the Macro and Fixed-Femto UEs, the transmission power of these Fixed-Femtos can be reduced to avoid the overlapped areas. Reducing the transmission power can limit the coverage area of the Fixed-Femto, which may increase the cost since more Fixed-Femtos are required to be deployed to cover all the high path-loss areas. Moreover, reducing the transmission power can increase the number of unnecessary HOs especially in the case of vehicular UEs who are in continuous movement and keep changing their locations and connections with the serving BS. To sum up, all the previous results had shown that having Fixed-Femtos to serve vehicular UEs even in bus stops or along the bus paths is not always the best solution, and hereby the next scenario comes.

b) Scenario 2: Mobile-Femtos with vehicular UEs

The second scenario represents the case when there is Mobile-Femto installed in a bus while the UEs are inside the bus. The aim of this scenario is to improve the signal connection of those vehicular UEs with the serving network. Those UEs are suffering from high penetration loss and high path-loss as they are at the edges of the Macrocell besides of the vehicle chassis that works as a signal barrier against in and out signals.

The distribution of the Mobile-Femtos is based on the used path-loss model that helps in creating the bus paths of the Mobile-Femtos. Installing the Mobile-Femtos efficiently in public transportations like buses has helped in mitigating the penetration loss, path-loss and cost to serve vehicular UEs. A constant loss has been assumed between the Mobile-Femtos and the vehicular UEs due to the LOS access link as well as the short distance between the two, which is less than 5 m. In addition, the used transmission power in the Mobile-Femto is 24 dBm while the path of the Mobile-Femto -which is the bus path- has been specified as mentioned earlier, and two Mobile-Femtos have been assumed to be deployed in every 1 km^2. Using this technique has its advantages in mitigating the interference and the path-loss. This has given the Mobile-Femto the ability to cover wider areas and reach the remote areas that are hard for the eNB to reach due to the high path-loss and penetration loss. Hence, Fig. 11.23 shows the movement of the Mobile-Femto, which is a uniform movement in the areas that suffer from high path-loss as the arrows show.

The max SINR of the ROI has been improved after deploying the Mobile-Femtos to serve vehicular UEs. This is because Mobile-Femtos are able to overcome the high penetration loss, path-loss and interference issues that vehicular UEs are suffering from by improving their connection with the serving network compared to the Fixed-Femto deployment as Fig. 11.24 shows. This figure represents the Mobile-Femtos distribution in the ROI in terms of the path-loss with and without the shadow fading. As mentioned earlier, the dark red areas represent the coverage area with good signal strength as the path-loss is at its lowest value. The dark blue areas, however, represent areas with low signal strength as the path-loss is at its highest value. It has become quite obvious to the reader that the Mobile-Femtos have achieved better coverage area as the paths of these Mobile-Femtos were specified based on the path-loss model so it is rare that the Mobile-Femtos coverage is overlapped with each other. In addition, the figure shows that each Mobile-Femto has strong signal strength at the centre then the signal fades gradually as the highlighted area shows in both cases with and without the shadow fading.

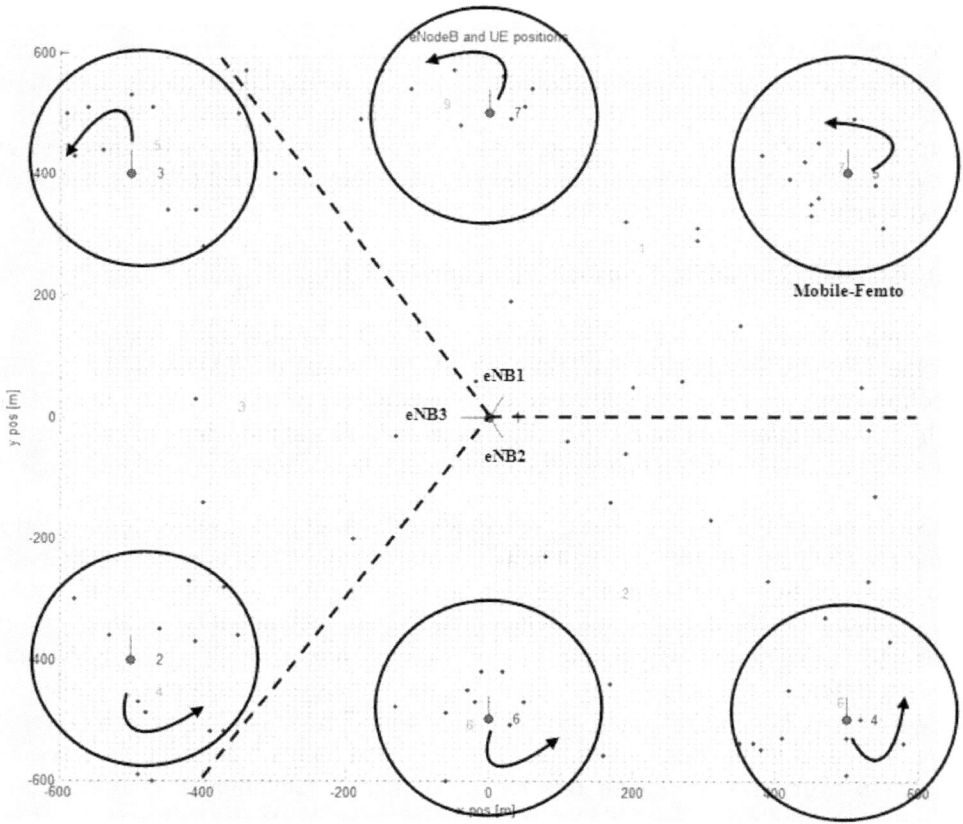

Fig. 11.23. Mobile-Femtos coverage distribution when the transmission power is 24 dBm and VPL is 25 dB.

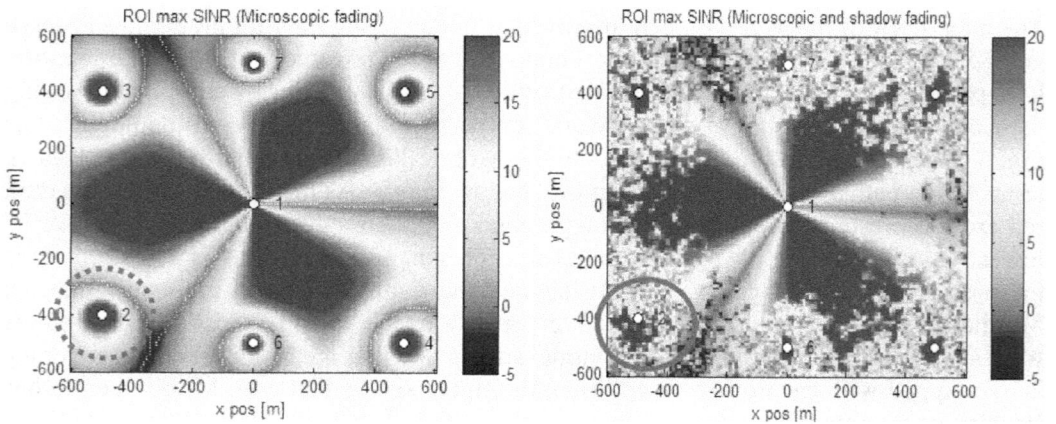

Fig. 11.24. Mobile-Femtos Sector SINR, calculated with distance dependent Microcell path-loss (left) while (right) represents the distributed space correlation with shadow fading.

The SINR improvement is accompanied with an improvement in the achieved spectral efficiency as Fig. 11.25 shows. The figure shows that in the highlighted area the spectral efficiency of the Mobile-Femto is high as the dark red colour in the centre represents a spectral efficiency of higher than 4.5 bits/cu. This high spectral efficiency reflects that the Mobile-Femto coverage is less susceptible to interference compared to the Fixed-Femto. This is due to the organised distribution and the specified paths in the case of the Mobile-Femto, which makes it less exposed to the interference issue. Thus, the Mobile-Femto signal fades gradually since the interference is low as the right side of the figure shows.

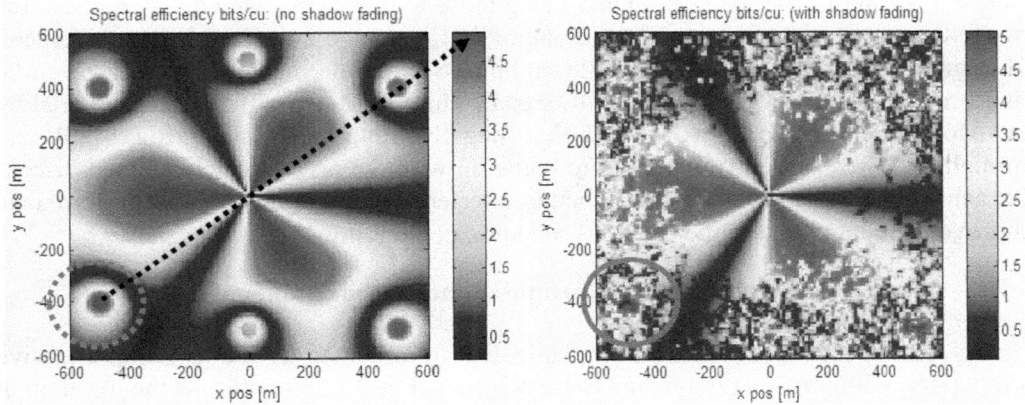

Fig. 11.25. Mobile-Femtos Sector Spectral efficiency with distance dependent Microcell path-loss (left) while (right) represents the distributed spectral efficiency correlation with shadow fading.

The following table represents the used parameters in the Mobile-Femto simulated scenario in order to create the required environment.

Table 11.4. Mobile-Femto simulated parameters.

Parameters	Value
Type of femtocell	Mobile
Number of Mobile-Femtos	2 MFemto per each 1 km^2
Number of Mobile-Femto's server UEs	10 UEs
Mobile- Femto transmission power	24 dBm
MFemto antenna type	Omnidirectional antenna
Speed of Mobile-Femto and passengers in the bus	3-160 km/h
Antenna gain Mobile-Femto	0
Shadowing standard deviation eNB-Mobile-Femto	8
Path-loss model	3GPP SCM urban NLOS Microcells model for backhaul link while constant path-loss for the LOS access links NLOS PL(dB) = $34.53 + 38\log_{10}(L)$

Hence, it can be concluded that the Mobile-Femto shows a better coverage distribution over the Fixed-Femto as the SINR and spectral efficiency have been improved in the case of vehicular UEs. Additionally, less interference occurred in the case of the Mobile-Femto and this is because of the uniform movement of these mobile BSs that improved the vehicular UEs signal strength and mitigated the penetration loss and path-loss. Thus, deploying the Mobile-Femtos is seen as a promising solution for the next generation technology as it is able to improve the vehicular UEs performance by mitigating the penetration loss, path-loss and interference. However, it is to be emphasised that there is always worst-case interference scenarios especially when the Mobile-Femto changes its route e.g. facing traffic congestion and being close to other BSs, which may cause three interference scenarios. First, the Mobile-Femto's UE can be interfered by the Macrocell DL signal, which can affect the UE throughput. Second, the Mobile-Femto UE can be interfered by another Mobile-Femto DL signal. Third, a Macro UE can be interfered by the Mobile-Femto DL signal, which has negative impact on the UE's SINR and throughput. Therefore, it was needed to come up with the proposed scheme and discuss the achieved results that showed a better interference management and performance improvement compared to the previous schemes as the following section shows.

c) **The simulated results of the proposed interference scheme**

The proposed scheme has been implemented in order to create a comprehensive comparison with previous frequency reuse schemes. Fig. 11.26 represents the throughput of Macro UEs located in the Macrocell coverage as the number of Mobile-Femtos varies.

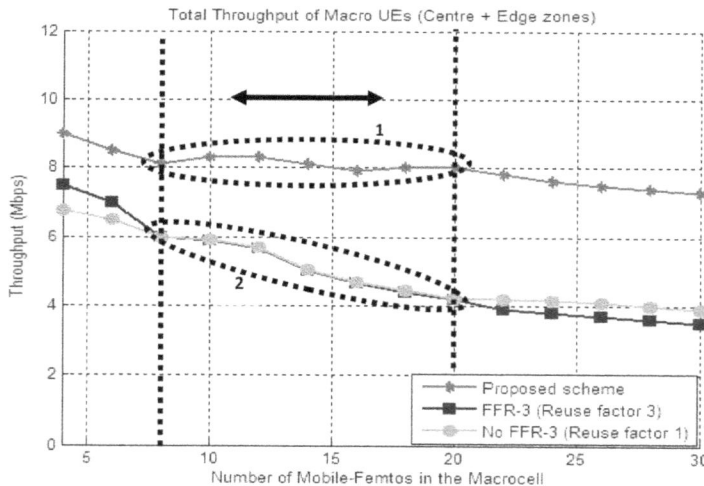

Fig. 11.26. Average Throughput of Macro UEs located in centre and edge zones vs. no. of Mobile-Femtos.

In the proposed frequency reuse scheme, the Mobile-Femto UEs can allocate the subcarriers, which are not used by the Macro UEs at each location. Thus, the interference between the Macro UEs and Mobile-Femtos is greatly avoided. The Mobile-Femtos effect to Macro UEs who are located in the same zone is less in the proposed scheme compared to the previous schemes. This is because; in FFR-3 and NoFFR-3 schemes, the subcarriers are randomly assigned to the Mobile-Femto UEs regardless of the subcarriers used by the Macro UEs. This means that the same subcarriers are used by the Macro UEs and the Mobile-Femto UEs who are very close to each other, which can create severe interference for the two. As a result, the throughput gap between the proposed and previous schemes is bigger when more Mobile-Femtos are deployed in the Macrocell. This is because with highly deployed number of Mobile-Femtos, the chance of Macro UE to be interfered with the increased number of deployed Femtocells is greater. Hence, the proposed scheme has achieved an acceptable, almost stable, throughput around 8 Mbps in the range between 8 to 20 deployed numbers of Mobile-Femtos in the Macrocell, while for the same numbers of Mobile-Femtos, the FFR-3 and NoFFR-3 schemes have achieved lower throughput that starts with 6 Mbps and then decreases to 4 Mbps. This is due to the increased unavoidable interference between those Mobile-Femtos and Macro UEs with the increase number of Mobile-Femtos in the Macrocell.

On the other hand, Fig. 11.27 illustrates the throughput of Macro and Mobile-Femto UEs located at the edge zone only. In the OFDMA cellular network, the performance of the edge zone is always poor due to the ICI that increases with the increased path-loss. However, in the proposed scheme, allocating more subcarriers to cell-edge Mobile-Femtos can greatly improve the throughput of cell-edge vehicular UEs. While, the compared schemes assign the subcarriers irrespective of the UEs location, centre zone or edge zones and this is why those schemes achieved less throughput compared to the proposed interference scheme. However, it is to be mentioned that the FFR-3 scheme shows higher throughput than the NoFFR-3 scheme and this is because the FFR-3 scheme reduces interference by applying FFR to the Macro UEs but still the interference from other Mobile-Femtos is unavoidable. Since cell-edge vehicular UEs with the proposed interference scheme has enjoyed higher throughput than the FFR-3 and NoFFR-3 interference mitigations schemes; that means the proposed scheme is able to mitigate the UEs' interference not only between the Mobile-Femtos and the Macro UEs but also between the Mobile-Femtos UEs themselves. The figure shows that with 30 deployed number of Mobile-Femtos in the Macrocell, the cell-edge UEs have achieved a max throughput of almost 29 Mbps while 20 Mbps for the FFR-3 scheme and 14 Mbps for the NoFFR-3 scheme. This is due to the increased interference with the increased number of deployed Mobile-Femtos in the Macrocell and the adaptability of each scheme with the generated interference issue.

While Fig. 11.28 represents the total throughput of centre and cell-edge zones after deploying the Mobile-Femto into the Macrocell. The achieved simulated results show that the throughput increases with the increased number of deployed Mobile-Femtos in the Macrocell. This is due to the ability of this Mobile BS to reach vehicular UEs and improve their performance fast and efficiently compared to other technologies. However, deploying many Mobile-Femtos to improve the vehicular UEs performance by reducing the vehicular penetration loss in LTE network can increase the interference issue between Macro and Mobile-Femto UEs. Therefore, implementing the proposed interference mitigation scheme has shown better throughput compared to the FFR-3 and No-FFR schemes as the max-achieved throughput is almost 58 Mbps in the proposed scheme while it is 40 Mbps in the FFR-3 and No-FFR schemes. Implementing the proposed scheme has shown an improvement of 18 Mbps over the previous schemes. Actually, with the increased number of deployed Mobile-Femtos in the Macrocell and exactly 30 Mobile-Femtos, the FFR-3 and the NoFFR-3 achieved the same throughput as the generated interference between the Macro and Mobile-Femto UEs is unavoidable in this case. In other words, at a certain number of deployed Femtocells in the Macrocell, the FFR-3 scheme will allocate the same subcarriers to the Macro UEs and the Mobile-Femto UEs who are very close to each other.

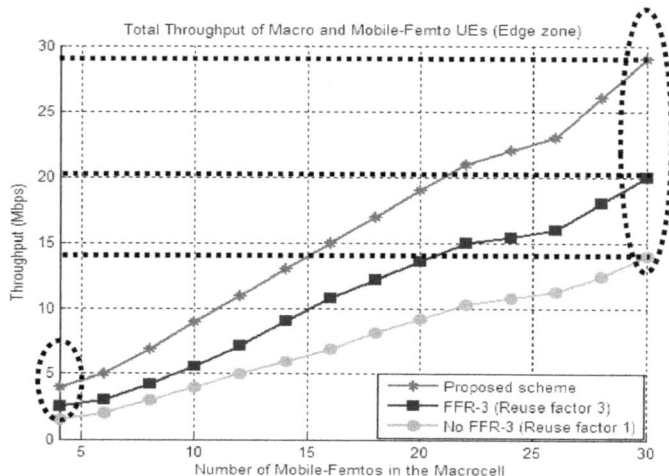

Fig. 11.27. Total Throughput of Macro and Mobile-Femto UEs located only in the edge zone vs. no. of Mobile-Femtos.

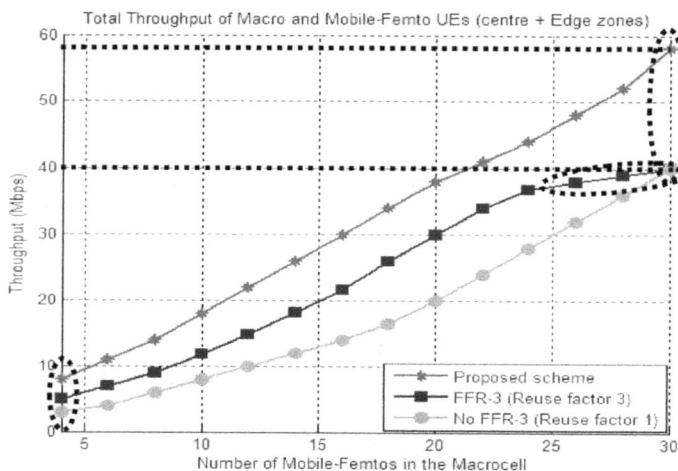

Fig. 11.28. Total throughput of Macrocell and Mobile-Femto UEs in the entire Macrocell vs. no. of Mobile-Femtos.

Now, after discussing the impact of the proposed scheme on the achieved throughput of Macro and Mobile-Femto UEs, it is necessary to consider the impact of this scheme on the achieved signal outage probability with the increased number of Mobile-Femtos in the Macrocell. It is to be mentioned that with the increased number of deployed small BSs in the Macrocell, the generated interference issue increases, thus the signal outage increases as well. However, this does not mean that less number of Mobile-Femtos should be deployed in the Macrocell to mitigate the generated interference and signal outage. The signal outage can be reduced with the proposed scheme as Fig. 11.29 shows. The figure depicts the outage probability of UEs according to the SINR threshold, when 30 Mobile-Femtos are deployed in the Macrocell. In a given SINR threshold, the proposed scheme indicates lower outage probability than the FFR-3 and NoFFR-3 schemes. Added to this, it decreases the outage probability more at low SINR threshold while it reaches its max at 0.96 with high SINR threshold. This implies that the proposed scheme effectively supports more UEs, even though the interference is excessive. While in the case of the NoFFR-3, UEs have achieved the highest outage probability, which has started with 0.4 under low SINR threshold and continuously increased with the increased SINR threshold. However, at 15 dB SINR the outage probability of the proposed scheme has been reduced by 0.07 compared to the FFR-3 scheme while reduced by 0.14 compared to the NoFFR-3 scheme.

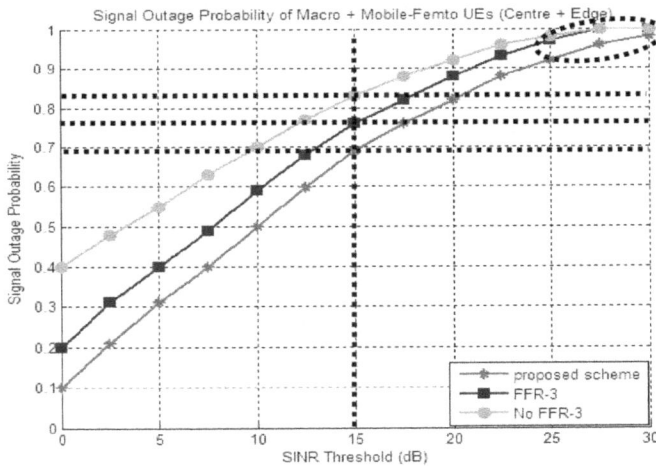

Fig. 11.29. Signal Outage Probability of Macro and Mobile-Femto UEs vs. the SINR threshold when 30 Mobile-Femtos are deployed in the Macrocell.

Thus, it is obvious that using the proposed frequency reuse scheme in the presented Femtocell technology has improved the UEs throughput in the cell centre and edge zones for both the Macro and Mobile-Femto UEs. Improving the UEs throughput and reducing the signal outage are very important factors to improve the vehicular UEs performance in LTE and 5G Networks. However, this solution can be generalised on different types of BSs like fixed and mobile BSs, as well as different type of UEs e.g. fixed, mobile and vehicular UEs.

11.6. Conclusions and Further Work

The on-going increasing demand for wireless based services nowadays makes it necessary to be available everywhere and at any time by introducing new technologies, standards and polices. This can be implemented in a way that makes the wireless environment match with this high evolution in wireless devices and technology. Hence, the overall goal of this work is to design and implement a technology that solves the drawbacks of vehicular users' performance for the future of wireless technologies e.g. 5G Networks. The Mobile-Femto technology allows higher throughput, signal strength and capacity for the served vehicular users by the 4G networks. Upon proposing the Mobile-Femto with the required methods to avoid interference and to improve the vehicular users' performance, it is strongly felt that there is scope for further potential developments related to the work done especially in the interference management area.

11.6.1. Conclusion

11.6.1.1. Performance Evaluation

The performance evaluation has been done for three technologies in order to generate a comprehensive comparison between the vehicular UEs who have been served by the eNB direct transmission, Fixed-Femto assisted-transmission and finally the Mobile-Femto assisted-transmission. The achieved results through MATLAB simulator showed that Mobile-Femto UEs have achieved higher throughput and performance in comparison with other transmissions. This is because; Mobile-Femto is more effective in mitigating the penetration loss inside public transportations which has been noticed through the significant improvement in the SINR value. Moreover, the Ergodic capacity in terms of vehicular UEs' links has improved under high penetration losses e.g. 25 dB and 40 dB, due to the implementation of Mobile-Femto inside public transportations. In addition, Mobile-Femto UEs have achieved the highest spectral efficiency in comparison with the spectral efficiency of eNB and Fixed-Femto UEs. This has been noticed as, 90 % of the vehicular UEs have enjoyed better spectral efficiency up to 3.7 bit/cu when they were connected to the Mobile-Femto, while 2.5 bit/cu in the case of direct and Fixed-Femto transmissions. On the other hand, when the VPL is equal to 25 dB, the Mobile-Femto assisted-transmission achieved lower outage probability than the eNB direct transmission. It is also lower than the outage probability of the Fixed-Femto assisted-transmission. This indicates that the Mobile-Femto assisted system is better at maintaining a given rate of R bits/sec for vehicular UE, which can be translated to a better QoS in comparison with the other two transmissions. However, as the VPL value goes up to 40 dB, the Mobile-Femto assisted-transmission outperforms the direct transmission when the vehicular UE is fairly near the eNB (around 50 m). From the achieved results, it was obvious that vehicular UEs enjoy better quality of connection and signal strength when they are served by the Mobile-Femto rather than being served directly by the eNB or the Fixed-Femto BSs.

11.6.1.2. Interference Management Evaluation

Deploying Fixed-Femtos and Mobile-Femtos in Macrocells has created an excessive interference issue between the served vehicular UEs and the deployed BSs. Therefore, the coverage planning solution based on the path-loss model has been examined together with controlling the transmission power scheme. The achieved results showed that the previous solutions can be regarded as reasonable solutions to solve the interference issue in the Fixed-Femtos scenario while it is challengeable in the vehicular environment scenario especially when here the BSs are vehicular. Therefore, an efficient frequency reuse scheme has been proposed to mitigate the interference between the Mobile-Femto UEs, Macro UEs and BSs. In this scheme, each cell has been divided into two zones; the centre zone and the edge zone. The Mobile-Femtos in the cell centre zone will be using the frequencies allocated for the edge zones and likewise for the cell-edge zone where the Mobile-Femtos will be using the frequencies that have been allocated for the centre zones. The achieved results via MATLAB simulator showed a noticeable improvement in the achieved throughput after implementing the proposed scheme compared to previous frequency reuse schemes. This improvement has been noticed for Macro and Mobile-Femto UEs when they are at the centre and edge zones. It is to be emphasised, that the cell-edge UEs have achieved high throughput because the proposed scheme allocates more subcarriers to cell-edge Mobile-Femtos. While, the compared schemes assign the subcarriers irrespective of the UEs location, centre zones or edge zones and this is why those schemes have achieved less throughput compared to the proposed interference scheme. Additionally, the achieved results showed that the proposed scheme is able to mitigate the UEs' interference not only between the Mobile-Femtos and the Macro UEs but also between the Mobile-Femtos UEs themselves. This has been noticed as the number of deployed Mobile-Femtos increased in the Macrocell up to 30, the cell-edge UEs have achieved a max throughput of almost 29 Mbps while 20 Mbps for the FFR-3 scheme and 14 Mbps for the NoFFR-3 scheme.

Added to this, implementing the proposed interference mitigation scheme has reduced the outage probability of the signals. The achieved results showed that at a given SINR threshold, the proposed scheme indicated lower outage probability than the FFR-3 and NoFFR-3 schemes. Hence, this solution can be generalised on different types of BSs like fixed and mobile BSs, as well as different type of UEs e.g. fixed, mobile and vehicular UEs.

For all the above discussed reasons, Mobile-Femto is seen as the future of next generation networks, which can effectively improve the vehicular UEs performance in public transportations like buses. All the achieved results showed an improvement in the vehicular UEs performance after deploying the Mobile-Femtos in the LTE Macrocell with the appropriate distribution and interference mitigation methods. However, deploying Mobile-Femtos in high-speed trains to improve the train passengers' performance and provide them with the internet is seen as the future of next generation networks as many challenges are raised here that will be discussed briefly in the future work section.

11.6.2. Future Work

The aim of the mobile communications Enablers for Twenty-twenty (2020) Information Society (METIS), which is an EU co-funded research project, is to lay the foundation for the next generation of the mobile and wireless communications system [4]. The project is examining the possibility of improving the cellular network by supplementing the infrastructure with simple, low power small BSs. It also has introduced the proposal of placing mobile BSs in vehicles, such as cars, trucks, buses and trains to improve the vehicular UEs performance. That is why, considering the train Mobile-Femtos technology as the future work of this research is seen as a complement of what has been proposed earlier in regard to the deployment of Mobile-Femto technologies in public transportations (i.e. buses).

The cooperative and coordinated Mobile-Femto systems in train carriages need to be further investigated as it consists of a set of interconnected and coordinated Mobile-Femtos as shown in Fig. 11.30. The coordinated Mobile-Femtos would be implemented on trains or on other vehicles with large spatial dimensions where those coordinated Mobile-Femtos are connected to one another through the CrX2 coordination interface. This allows several optimisation methods for group HO and backhaul link optimisation.

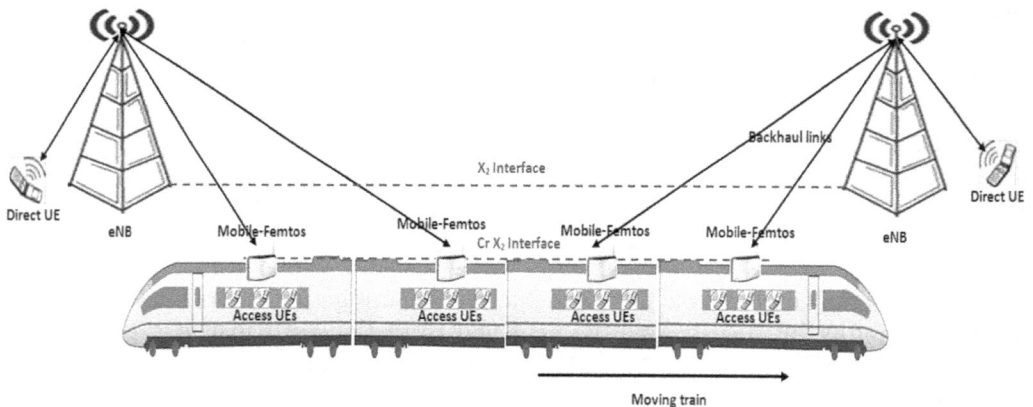

Fig. 11.30. Coordinated Mobile-Femtos system.

Actually, there are four challenging reasons that affect communication services on high-speed trains. These challenges are listed as the following:

a) Large penetration loss via the shield of the train, which is expected to be 40 dB.

b) Large number of HOs in very short time. This is due to the hundreds or thousands of UEs who need HO procedures from one site to another concurrently/sequentially. This phenomenon affects system stability and eats-up capacity.

c) High power consumption for UEs transmission and this is because UEs on the train need higher transmission power to overcome the large penetration loss in uplink as well.

d) High-speeds of trains are considered one of the challenges as some trains speeds range between 300 km/h to 350 km/h, which makes the impact of Doppler Shift is even more severe.

Thus, a common adopted solution for high-speed trains is to condense the network along the railway to combat the large penetration loss. However, this will make the second issue more severe, as the number of HO procedures is increased due to smaller site-to-site distance. Another way is to increase the transmission power of the BSs, which helps to solve the issue of the large penetration lose but, this can create excessive ICI for the DL and UL signals. Besides that, neither of these solutions is cost-effective. Therefore, thinking about an effective solution to solve the above challenges was the motive behind proposing the coordinated Mobile-Femtos system in trains as the future work of this research. The interference between the coordinated Mobile-Femtos and eNB's and interference between Backhaul links need to be further investigated. Consequently, the train coordinated Mobile-Femtos is seen as the future of high speed trains to overcome the previous issues and connect the vehicular UEs to the network anywhere and at any time regardless of the obstacles that might face the transmitted signals.

References

[1]. A. M. Sadekar, LTE-a enhanced inter-cell interference coordination (eICIC) with pico cell adaptive antenna, PhD Thesis, Electrical and Computer Engineering, *Carleton University*, Ottawa, Ontario, 2015, pp. 1-6.

[2]. M. Ahmad, 4G and 5G wireless: How they are alike and how they differ, https://www.androidauthority.com/4g-and-5g-wireless-how-they-are-alike-and-how-they-differ-615709/

[3]. Mundy, J. What is 5G? 5G vs 4G and the future of mobile networks, accessed on: 18/08/2016, available at: http://www.trustedreviews.com/opinion/what-is-5g-a-rough-guide-to-the-next-generation-of-mobile-networks-2911748

[4]. METIS, Scenarios, requirements and KPIs for 5G mobile and wireless system, accessed on: 30/11/2017, available at: https://www.metis2020.com/wp-content/uploads/deliverables/METIS_D1.1_v1.pdf

[5]. Enhancing LTE Cell-Edge Performance via PDCCH ICIC, White Paper, *Network Communications Inc.*, Fujitsu, 2011.

[6]. J. Kokkoniemi, J. Ylitalo, P. Luoto, S. Scott, J. Leinonen, M. Latva-aho, Performance evaluation of vehicular LTE mobile relay nodes, in *Proceedings of the IEEE 24th International Symposium on Personal, Indoor and Mobile Radio Communications: Mobile Wireless Networks Conference (PIMRC'13)*, 2013, pp. 1972-1976.

[7]. O. Anjum, C. Wijting, M. A. Uusitalo, K. Valkealahti, On the performance of decentralized cell edge coordinated scheduling in small cell clusters with difference densities, in *Proceedings of the European Conference of Networks and Communications (EuCNC'14)*, 2014, pp. 1-5.

[8]. Y. Sui, Z. Ren, W. Sun, T. Svensson, P. Fertl, Performance study of fixed and moving relays for vehicular users with multi-cell handover Co-channel interference, in *Proceedings of the Connected Vehicles and Expo International Conference (ICCVE'13)*, 2013, pp. 514-520.

[9]. X. An, F. Pianese, Understanding Co-channel interference in LTE-based multi-tier cellular networks, in *Proceeding of the 9th ACM Symposium on Performance Evaluation of Wireless Ad Hoc, Sensor, and Ubiquitous Networks (PE-WASUN'12)*, 2012, pp. 107-112.

[10]. Y. Sui, I. Guvenc, T. Svensson, Interference management for moving networks in ultra-dense urban scenarios, *EURASIP Journal on Wireless Communication Networking*, 2015, Art. 111.

[11]. IMT-2020 (5G) Promotion Group, 5G Wireless Technology Architecture, IMT-2020, White Paper, 2015, available at: www.imt-2020.org.cn/zh/documents/download/8.

[12]. J. Yoon, M. Y. Arslan, K. Sundaresan, V. Srikanth, S. Banerjee, A distributed resource management framework for interference mitigation in OFDMA femtocell networks, in *Proceedings of the 13th ACM international symposium on Mobile Ad Hoc Networking and Computing (MobiHoc'12)*, University of Wisconsin, USA, 2012, pp. 233-242.

[13]. B. A. Martin, C. P. Callender, S. I. Geary, A. G. Charles, S. Franklin and J. Howard, Fast Reselection Between Different Radio Access Technology Networks, Networks, Broadcom Corporation, White Paper, 2012, available at: https://ipv4.google.com/patents/EP2852224A2?cl=en

[14]. Vulnerabilities of LTE and LTEAdvanced Communications, White Paper, *Rohde and Schwarz*, 2014.

[15]. P. N. Vivas, C. Campo, C. G. Rubio, A. R. Carrion, Communication technology for vehicles, in *Proceedings of the 4th International Workshop, Nets4Cars/Nets4Trains*, Vilnius, 2012, pp. 174-185.

[16]. Signalling the Future, White Paper, *Vilicom*, http://people.ucsc.edu

[17]. I. Kostanic, C. Hall, J. McCarthy, Measurements of the vehicle penetration loss characteristics at 800 MHz, in *Proceedings of the IEEE Vehicular Technology Conference (VTC'98)*, 1998, pp. 1-4.

[18]. B. Ai, R. He, Z. Zhong, K. Guan, B. Chen, P. Liu, Y. Li, Radio wave propagation scene partitioning for high-speed rails, *International Journal of Antennas and Propagation*, Hindawi, 2012.

[19]. M. Chowdhury, S. Lee, B. Ru, N, Park, W. Jang, Service quality improvement of mobile users in vehicular environment by mobile femtocell network deployment, in *Proceedings of the Int. Conference on ICT Convergence (ICTC'11)*, Sept. 2011, pp. 194-198.

[20]. T. Beniero, S. Redana, J. Hamalainen, B. Raaf, Effect of relaying on coverage in 3GPP LTE-advanced, in *Proceedings of the IEEE Vehicular Technology Conference (VTC'09)*, 2009, pp. 1-5.

[21]. F. Haider, M. Dianati, R. Tafazolli, A simulation based study of mobile femtocell assisted LTE networks, in *Proceedings of the International Wireless Communication and Mobile Computing Conference (IWCMC'11)*, UK, July 2011, pp. 2198-2203.

[22]. S. Jangsher, V. O. K. Li, Resource allocation in cellular networks employing mobile femtocells with deterministic mobility, in *Proceedings of the IEEE Wireless Communications and Networking Conference (WCNC'13)*, 2013, pp. 819-824.

[23]. K. Sethom, A. B. Salem, F. Mhiri, R. Bouallegue, Resource management mechanism for femtocell enterprise networks, in *Proceedings of the IEEE Wireless Advanced Conference (WiAd'12)*, University of Carthage, Tunisia, 2012, pp. 74-78.

[24]. H. Tang, P. Hong, K. Xue, J. Peng, Cluster-based resource allocation for interference mitigation in LTE heterogeneous networks, in *Proceedings of the IEEE Vehicular Technology Conference (VTC Fall'12)*, 2012, pp. 1-5.

[25]. Y. Bai, J. Zhou, L. Chen, Hybrid spectrum usage for over laying LTE macrocell and femtocell, in *Proceedings of the IEEE Global Telecommunications Conference (GLOBECOM'09)*, 2009, pp. 1-6.

[26]. S. Al-Rubaye, Radio network management in cognitive LTE/femtocell systems, PhD Thesis, Electronic and Computer Engineering, School of Engineering and Design, *Brunel University*, September 2012.

[27]. L. Yang, L. Zhange, T. Yang, W. Fang, Location-based hybrid spectrum allocation and reuse for tiered LTE-a networks, in *Proceedings of the IEEE Vehicular Technology Conference (VTC Spring'11)*, 2011, pp. 1-5.

[28]. A. Hatoum, N. Aitsaadi, R. Langar, R. Boutaba, FCRA: femtocell cluster-based resource allocation scheme for OFDMA networks, in *Proceedings of the IEEE International Conference on Communication (ICC'11)*, 2011, pp. 1-6.

[29]. S. Ali, M. Zeeshan, A. Naveed, A capacity minimum quarantee-based service class-oriented scheduler for LTE networks, *EURASIP Journal on Wireless Communication and Networking*, 2013.

[30]. D. Karvounas, A. Georgakopoulos, V. Stavroulaki, N. Koutsouris, K. a Tsagkaris, P. Demestichas, Resource allocation to femtocells for coordinated capacity expansion of wireless access infrastructures, *EURASIP Journal on Wireless Communication and Networking*, 2012.

[31]. S. Scott, J. Leinonen, P. Pirinen, J. Vihriala, V. V. Phan, M. Latva-aho, A cooperative moving relay node system deployment in a high speed train, in *Proceedings of the IEEE 77th Vehicular Technology Conference (VTC Spring'13)*, Centre for Wireless Commun., Univ. of Oulu, Oulu, Finland, 2013, pp. 1-5.

[32]. Y. Sui, J. Vihriala, A. Papadogiannis, M. Sternad, Moving cells: a promising solution to boost performance for vehicular users, *IEEE Communications Magazine*, Vol. 51, Issue 6, 2013, pp. 62-68.

[33]. Y. Sui, A. Papadogiannis, W. Yang, T. Svensson, Performance comparison of fixed and moving relays under Co-channel interference, in *Proceedings of the 4th IEEE International Workshop on Heterogeneous and Small Cell Networks (HetSNets'12)*, 2012, pp. 574-579.

[34]. M. S. Pan, M. Lin, W. T. Chen, An enhanced handover scheme for mobile relays in LTE-a high-speed rail networks, *IEEE Transactions on Vehicular Technology*, Vol. 64, No. 2, February 2015, pp. 743-756.

[35]. M. Saied, A. El-atty, Z. M. Gharsseldien, Analytical model for mobile user connectivity in coexisting femtocell/macrocell networks, *International Journal of Wireless & Mobile Networks*, Vol. 4, No. 6, December 2012, pp. 61-72.

[36]. T. Elkourdi, O. Simeone, Femtocell as a relay: an outage analysis, *IEEE Transactions on Wireless Communications*, Vol. 10, Issue 12, 2011, pp. 4204-4213.

[37]. T. Mimura, K. Yamamoto, A. Iwata, A. Nishio, M. Morikura, Multi-operator mobile-relaying: effect of shared spectrum allocation, in *Proceedings of the IEEE 23th international conference on Personal Indoor and Mobile Radio Communications (PIMRC'12)*, 2012, pp. 1268-1272.

[38]. T. Badri, S. Rachid, M. Wahbi, M. Samir, Handover management schemes in LTE femtocell networks, *International Journal of Computer Sciences & Information technology*, Vol. 5, Issue 3, 2013, pp.90-100.

[39]. D. Xenakis, N. Passas, C. Verikoukis, An Energy- Centric Handover Decision Algorithm for the Integrated LTE Macrocell-Femtocell Network, Special Issue: Wireless Green Communication and Networking, *Elsevier*, 2012, pp. 1684-1694.

[40]. M. Bennis, L. Giupponi, E.M. Diaz, M. M. Lalam, Interference management in self-organized femtocell networks: the BeFEMTO approach, in *Proceedings of the 2nd International Conference Wireless Communication, Vehicular Technology, Information Theory and Aerospace & Electronics Systems Technology (Wireless VITAE'11)*, 2011, pp. 1-6.

[41]. A. Dalal, L. Hailong, D. P. Agrawal, Fractional frequency reuse to mitigate interference in self-configuration LTE-femtocell network, in *Proceedings of the IEEE 8th International Conference Mobile Adhoc and Sensor Systems (MASS'11)*, 2011, pp. 49-54.

[42]. K. Zheng, H. Fanglong, L. Lei, W. Wenbo, Interference coordination between femtocells in LTE-advanced networks with carrier aggregation, in *Proceedings of the 5th International ICST Conference Communication and Networking in China (CHINACOM'10)*, 2010, China, pp. 1-5.

[43]. H. Claussen, Performance of macro and co-channel femtocells in a hierarchical cell structure, in *Proceedings of the IEEE 18th international Symposium Indoor and Mobile Radio Communications (PIMRC'07)*, 2007, pp. 1-5.

[44]. D. Shin, S. Choi, Dynamic power control for balanced data traffic with coverage in femtocell networks, in *Proceedings of the 8th international Wireless Communications and Mobile Computing Conference (IWCMC'12)*, 2012, pp. 648-653.

[45]. M. Xiao, N. B. Shroff, E. K. P. Chong, Utility-based power control in cellular wireless systems, in *Proceedings of the 20th Annual Joint Conference of the IEEE Computer and Communication Societies (INFOCOM'01)*, 2001, pp. 412-421.

[46]. X. Chu, Y. Wu, L. Benmesbah, W. K. Ling, Resource allocation in hybrid macro/femto networks, in *Proceedings of the IEEE Wireless Communications and Networking Conference Workshops (WCNCW'10)*, 2010, pp. 1-5.

[47]. K. Sundaresan, S. Rangarajan, Efficient resource management in OFDMA femto cells, in *Proceeding of the 10th ACM International Symposium on Mobile Ad Hoc Networking and Computing (ACM MobiHoc'09)*, 2009, pp. 33-42.

[48]. L. Zhang, L. Yang, T. Yang, Cognitive interference management for LTE-a femtocells with distributed carrier selection, in *Proceedings of the IEEE 72nd Vehicular Technology Conference Fall (VTC-Fall'10)*, 2010, pp. 1-5.

[49]. D. Lopez-Perez, A. Ladanyi, A. Juttner, J. Zhang, OFDMA femtocells: intracell handover for interference and handover mitigation in two-tier networks, in *Proceedings of the IEEE Wireless Communication and Networking Conference (WCNC'10)*, 2010, pp. 1-6.

[50]. H. Li, X. Xu, D. Hu, X. Tao, P. Zhang, S. Ci, H. Tang, Clustering strategy based on graph method and power control for frequency resource management in femtocell and macrocell overlaid system, *Journal of Communications and Networks*, Vol. 13, 2011, pp. 664-677.

[51]. H. Zhang, X. Chu, X. Wen, 4G Femtocells: Resource Allocation and Interference Management, *Springer*, 2013, pp. 4-5.

[52]. A. U. Ahmed, M. T. Islam, A. Ismail, M. Ghanbarisabagh, Dynamic resource allocation in hybrid access femtocell network, *The Scientific World Journal,* Hindawi, 2014, pp. 1-7.

[53]. S. Hussain, Dynamic radio resource management in 3GPP LTE, PhD Thesis, *Blekinge Institute of Technology*, 2009.

[54]. R. Ghaffar, R. Knopp, Fractional frequency reuse and interference suppression for OFDMA networks, in *Proceedings of the 8th International Symposium Modelling and Optimization in Mobile, Ad Hoc and Wireless Networks (WiOpt'10)*, 2010, pp. 273-277.

[55]. R. Seetharaman and R. Vijayaragavan, Femtocells: Smart Approach Analysis to Macrocells, *International Journal of Computer Applications,* Vol. 49, No.13, 2012, available at: http://research.ijcaonline.org/volume49/number13/pxc3880996.pdf

[56]. F. Peng, Y. Gao, Y. Chen, K. K. Chai, L. Cuthbert, Using TV white space for interference mitigation in LTE femtocell networks, in *Proceedings of the IET International Conference Communication Technology and Application (ICCTA'11)*, 2011, Beijing, pp. 5-9.

[57]. S. Azam, H. Salati, M. Nasiri-Kenari, Aggregate interference modelling and static resource allocation in closed and open access femtocells, *IET Communications*, Vol. 8, Issue 7, May 2014, pp. 1007-1016.

[58]. A. Mahmud, K. A. Hamdi, On the Co-channel femtocells exclusion region in fractional frequency reuse macrocells, in *Proceedings of the Wireless Communications and Networking Conference (WCNC'14)*, Istanbul, 2014, pp. 2306-2311.

[59]. P. Godlewski, M. Maqbool, M. Coupechoux, J. M. Kelif, Analytical evaluation of various frequency reuse schemes in cellular OFDMA networks, in *Proceedings of the 3rd*

International Conference on Performance Evaluation Methodologies and Tools (ValueTools'08), Brussels, Belgium, 2008, pp. 32:1-32:10.

[60]. P. Lee, T. Lee, J. Jeong, J. Shin, Interference management in LTE femtocell systems using fractional frequency reuse, in *Proceedings of the International Conference on Advanced Communication Technology (ICACT'10)*, Phoenix Park, Korea, 2010, pp. 1047-1051.

[61]. S. S. Han, J. Park, T. J. Lee, H. G. Ahn, K. Jang, A new frequency partitioning and allocation of subcarriers for fractional frequency reuse in mobile communication systems, *IEICE Transactions on Communications*, 2008, pp. 2748-2751.

[62]. R. Giuliano, C. Monti, P. Loreti, WiMAX fractional frequency reuse for rural environments, *IEEE Wireless Communications*, Vol. 15, 2008, pp. 60-65.

[63]. S. Hamoudal, C. Yeh, J. Kim, S. Wooram, D. S. Kwon, Dynamic hard fractional frequency reuse for mobile WiMAX, in *Proceedings of the IEEE International Conference on Pervasive Computing and Communications (PerCom'09)*, Galveston, TX, USA, 2009, pp. 1-6.

[64]. D. Bilios, C. Bouras, V. Kokkinos, A. Papazois, G. Tseliou, Optimization of fractional frequency reuse in long term evolution networks, in *Proceedings of the IEEE Wireless Communications and Networking Conference (WCNC'12)*, Shanghai, Republic of China, 2012, pp. 1853-1857.

[65]. R. Raheem, A. Lasebae, J. Loo, Mobility management in LTE/mobile femtocell networks: outage probability and drop/block calls probability, in *Proceedings of the IEEE 3rd International Conference on Emerging Research Paradigms in Business and Social Sciences (ERPBSS'15)*, Dubai, United Arab Emirates, 2015.

[66]. M. H. Habaebi, J. Chebil, A. G. Sakkaf, T. H. Dahawi, Comparison between scheduling techniques in Long Term Evolution, *IIUM Engineering Journal*, Vol. 14, 2013, pp. 66-75.

[67]. E. Liu, K. K. Leung, Proportional fair scheduling: analytical insight under Rayleing fading environment, in *Proceedings of the IEEE Wireless Communications and Networking Conference (WCNC'08)*, 2008, pp. 1883-1888.

[68]. Audio precision, Signal-to-Noise Ratio, Introduction to the Basic Six Audio Testes, http://www.ap.com

[69]. H. Masui, T. Kobayashi, M. Akaike, Microwave path-loss modelling in urban line-of-sight environments, *IEEE Journal on Selected Areas in Communications*, Vol. 20, August 2002, pp. 1151-1155.

[70]. T. M. Cover, J. A. Thomas, Elements of Information Theory, *Wiley*, September 2006.

[71]. A. Patankar, What is Spectral Efficiency?, http://wireless-cafe.com/2009/02/05/what-is-spectral-efficiency/

[72]. S. Tombaz, M. Usman, J. Zander, Energy efficiency improvements through heterogeneous networks in diverse traffic distribution scenarios, in *Proceedings of the 6th IEEE International Conference on Communications and Networking in China (CHINACOM'11)*, Harbin, China, 2011, pp. 708-713.

[73]. R. Raheem, A. Lasebae, M. Aiash, J. Loo, Mobility management for vehicular user equipment in LTE/mobile femtocell networks, *International Journal of Information Systems in the Service Sector*, Vol. 9, 2017, pp. 60-87.

[74]. A. Goldsmith, Wireless Communications, *Cambridge University Press*, 2005.

[75]. Shannon Capacity CDMA vs OFDMA, http://www.raymaps.com/index.php/tag/shannon-capacity/

[76]. J. C. Ikuno, M. Wrulich, Dimensioning Vienna LTE simulators system level simulator documentation, v1.8r1375, Instit. of Comm. and Radio-Frequency Engg., *Vienna Univ. of Tech.*, Gusshausstrasse, Vienna, Austria, 2014, pp. 1-34.

[77]. 3GPP Specification, Physical layer aspect for evolved Universal Terrestrial Radio Access (UTRA), Technical Report, http://www.3gpp.org/ftp/Specs/html-info/25814.htm

[78]. O. Bulakci, On backhauling of relay enhanced networks in LTE-Advanced, Licentiate Seminar, Department of Communication and Networking, Alto University, Finland, 2010, pp. 1-4.

[79]. S. Stavrou, S. R. Saunder, Factors influencing outdoor to indoor radio wave propagation, in *Proceedings of the 12th International Conference of Antenna and Propagation (ICAP'03)*, 2003, pp. 581-585.

[80]. K. Doppler, M. Moisio, K. Valkealahti, On interference management for uncoordinated LTE-femto cell deployments, in *Proceedings of the 17th European Wireless Conference-Sustainable Wireless technologies (European Wireless'11)*, 2011, pp. 1-6.

[81]. A. Adhikary, V. Ntranos, G. Caire, Cognitive femtocells: breaking the spatial reuse barrier of cellular systems, in *Proceedings of the information Theory and Applications Workshop (ITA'11)*, 2011, pp. 1-10.

[82]. G. Gur, S. Bayhan, F. Alagoz, Cognitive femtocell networks: an overlay architecture for localized dynamic spectrum access [Dynamic Spectrum Management], *IEEE Wireless Communications*, Vol. 17, 2010, pp. 62-70.

[83]. T. Ma, P. Pietzuch, Femtocell Coverage Optimisation Using Statistical Verification, https://lsds.doc.ic.ac.uk/content/femtocell-coverage-optimisation-using-statistical-verification

[84]. Z. Lu, T. Bansal, P. Sinha, Achieving user-level fairness in open-access femtocell based architecture, *IEEE Transactions on Mobile Computing*, Vol. 12, Issue 10, 2013, pp. 1943-1954.

[85]. Y. A. Alqudah, On the performance of cost 231 Walfish Ikegami model in deployed 3.5 GHz network, in *Proceedings of the Technological Advances in Electrical, Electronics and Computer Engineering International Conference (TAEECE'13)*, Konya, 2013, pp. 524-527.

[86]. A. Naguib, Channel Models for IEEE 802.20 MBWA System Simulations, Rev 08r1, *IEEE 802.20 Working Group on Mobile Broadband Wireless Access*, 2007.

[87]. A. Nungu, B. Pehrson, J. Sansa-Otim, E-infrastructure and e-services for developing countries, in *Proceedings of the 6th International Conference on e-Infrastructure and e-Services for Developing Countries (AFRICOMM'14)*, Kampala, Uganda, 2014, pp. 3-6.

Chapter 12
The Buffer Delay Correction Algorithm for VoIP Communication

Fábio Sakuray, Robinson S. V. Hoto, Gean D. Breda and Leonardo S. Mendes

12.1. Background

Fig. 12.1 shows packets sent between two remote VoIP applications in a regular call, where talkspurts are periods with packets transmission and silence is periods without transmission. In a talkspurt k with n^k packets, a packet i is sent at instant t^k_i, received at instant a^k_i and executed in p^k_i (playout time) [3].

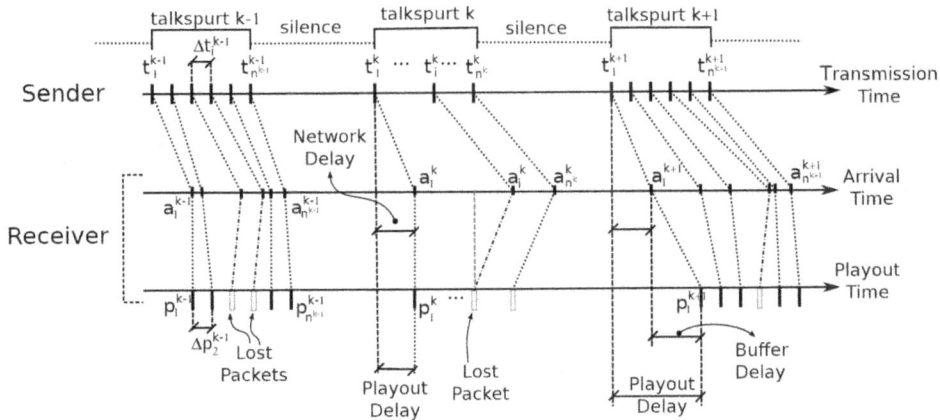

Fig. 12.1. Packet Audio Transmission.

In the receiver side of VoIP applications, the audio packets must be scheduled to playout with the same temporal spacing used in transmission ($\Delta t^k_i = \Delta p^k_i$). However, jitter makes

Fábio Sakuray
Computer Science Department, University of Londrina, Brazil

packets arrive after playout time ($p^k_i < d^k_i$), these packets are considered lost. To avoid this, most applications use a buffer delay that can be inserted at beginning of each talkspurt (see talkspurt k in Fig. 12.1), which is referred to as "inter-talkspurt" technique, or inserted inside a talkspurt, which is referred to as "intra-talkspurt". This work analyses only algorithms that act in silence periods, since they represent the most of BDA solutions in literature [4].

Lobina in [5], present an important classification of BDAs, as:

1. Packet Loss Intolerance: Algorithms that use high buffer delay values, avoiding packet loss. The simplicity of implementation is the main advantage of these algorithms;

2. Packet Loss Tolerance: audio applications can lose a certain number of packets without affecting audio quality. This class of algorithms adjusts buffer delay to control the packet loss rate;

3. Quality Based: this algorithm class monitors the call quality parameters to adjust the buffer delay.

Another element of voice call is the phenomenon called spike [6], defined as a sudden and large increase in the end-to-end delay. As a result the receiver has an interval without packets followed by a series of packets arriving almost simultaneously. Delay spikes represent a serious problem for VoIP applications, since they lead BDAs to overrated buffer delay values. A BDA must react adequately to the spike by changing its behavior.

Several BDAs have been developed with most of them trying to foresee network delay to set the buffer delay. Now let us consider some examples. The next two algorithms are packet loss intolerant. Ramjee in [7] presents four algorithms to measure the delay variance and estimate the average end-to-end delay, the fourth can detect spike and change the algorithm behavior. Barreto and Aragão in [8] present an algorithm based on the standard (Box-Jenkins) linear autoregressive (AR) model. The playout delay estimated ($\hat{d}^k$) of talkspurt k can be written by:

$$\hat{d}^k = \theta^\mu_1 \mu(A^{k-1}) + \theta^\sigma_1 \sigma(A^{k-1}) + \theta^\mu_1 \mu(A^{k-2}) + \cdots + \theta^\sigma_n \sigma(A^{k-n}), \quad (12.1)$$

where A^k is the network delay of k-th talkspurt; θ^μ_i and θ^σ_i are weights associated with mean ($\mu(A^k)$) and standard deviation ($\sigma(A^k)$), n is the sliding window size with last talkspurts received.

In a call with M talkspurts, (12.1) can rewrite by $\mathbf{d} = \mathbf{X}\theta$, where matrix $\mathbf{X} \in \mathbf{R}^{M \times 2n}$ is defined as:

$$X = \begin{bmatrix} \mu(A^n) & \sigma(A^n) & .. & \mu(A^1) & \sigma(A^1) \\ \mu(A^{n-1}) & \sigma(A^{n-1}) & .. & \mu(A^2) & \sigma(A^2) \\ \vdots & \vdots & \vdots & \vdots & \vdots \\ \mu(A^{M-2}) & \sigma(A^{M-2}) & .. & \mu(A^{M-n-1}) & \sigma(A^{M-n-1}) \\ \mu(A^{M-1}) & \sigma(A^{M-1}) & .. & \mu(A^{M-n}) & \sigma(A^{M-n}) \end{bmatrix}.$$

The vectors $\theta \in R^{2n}$ and $\mathbf{d} \in R^M$ are: $\theta = [\theta_1^{\mu} \theta_1^{\sigma} \dots \theta_n^{\mu} \theta_n^{\sigma}]^T$, $d = [d_{n+1} d_{n+2} \dots d_{M-1} d_M]^T$ with the superscript T denoting matrix transposition.

The estimate of θ is given by $\hat{\theta} = [X^T X]^{-1} X^T d$. However, the matrix $[X^T X]$ may be non-invertible, in which case Barreto and Aragao replace it by its regularized version:

$$\hat{\theta} = [X^T X + \lambda I]^{-1} X^T \mathbf{d}, \qquad (12.2)$$

where $I \in R^{2n \times 2n}$ is the identity matrix and $0 < \lambda \ll 1$. The values used by the authors are $\lambda = 0.01$ or $\lambda = 0.001$.

The next three algorithms are packet loss tolerant. Moon *et al.* [3] use the network delay distribution in the last w received packets and a desired packet loss rate. This algorithm can detect spike. Fujimoto *et al.* [9] uses the same idea, but focused on tail of the network delay probability distribution function. Assuming Pareto distribution for the tail, this approach presents better results when compared with algorithms that use a complete network delay distribution. In [10], Ramos *et al.* present the Move Average Algorithm (MA) to adjust the playout delay at each new talkspurt given a desired target of average loss percentage (ρ). The authors compute the optimal playout delay (D_k) at the beginning of talkspurt k as:

$$D_k = SORT\{Z_i^k\} with i = round(1 - \rho)N_k, \qquad (12.3)$$

with N^k the number of audio packets received during k-th talkspurt and Z_i^k the changing part of the end-to-end delay of i-th packet.

The predicted value of D_{k+1}, denoted by $\hat{D}_{k+1}$, is given by

$$\hat{D}_{k+1} = \sum_{l=1}^{M} a_l D_{k-l+1} \qquad (12.4)$$

The coefficients a_l must minimize the mean square error between D_{k+1} and $\hat{D}_{k+1}$. They can be found from solving the equation:

$$\sum_{m=0}^{M} a_{m+1} r_D(m - l) = r_D(l + 1) with l = 0, 1, \dots, M - 1. \qquad (12.5)$$

Suppose that it is known the last K values of r_D,

$$r_D(r) \cong \frac{1}{K - |r|} \sum_{k=1}^{K - |r|} D_k D_{k+|r|}, \qquad (12.6)$$

with $r = 0, \pm 1, \pm 2, \dots, \pm(K - 1)$. The model's order M is computed as follow: starting with $M = 1$, compute all values of $\hat{D}_k$, and estimate $E[(D_k - \hat{D}_k)^2]$, increase M and repeat the process. The model's order is taken equal to the lowest value of M preceding an increase in mean square error.

The next algorithms are quality based. Fujimoto *et al.* [11] shows that jitter, packet loss rate, codec and others parameters can affect call quality. Most solutions only allow packet loss rate setup. The algorithm presented in [11], called E-MOS, utilizes Mean Opinion

Score (MOS [1, 12]) classification as input to buffer delay adjust. MOS values are 1 to 5, where 1 is the worst and 5 is the best.

Valle *et al.* in [13] presents the Dynamic Management of Dejitter Buffer Algorithm (DMDB), which uses MOS rating as input to control the followings algorithms:

1. OpenH323: an open source and packet loss intolerant algorithm, used in CallGen323 application;

2. Window: histogram based algorithm with spike detection, presented in [3];

3. Adaptive: algorithm proposed by [14], which is also reactive and quality based, that tries to maximize the end-user notable quality.

12.2. Theatrical Aspects of De-Jitter Buffer

In the next sections, consider n^k the set of packets belonging to k-th talkspurt and p_i^k, a_i^k and t_i^k, respectively, the playout, receiver and transmission time. Using de-jitter buffer (or buffer delay - BD) in receiver side, with dynamic adjustment to each talkspurt, the playout time of i-th packet is:

$$p_i^k = a_1^k + BD^k + (i-1)\Delta t_i^k, \tag{12.7}$$

where $\Delta t_i^k = t_i^k - t_{(i-1)}^k$.

An audio packet i will be lost when does not match the jitter or latency restrictions [15, 16]:

- Latency: the end-to-end delay must be smaller than a delay value (L) defined by an application or Recommendation as G.114 [17]:

$$p_i^k - (t_i^k - C) \le L, \tag{12.8}$$

where C is the clock shift between sender and receiver clock.

- Jitter: the packet must be received before its playout time, i.e., $a_i^k \le p_i^k$, note that first packet always attend jitter restriction:

$$a_i^k \le p_i^k \Rightarrow a_i^k \le a_1^k + BD^k + (i-1)\Delta t_i^k \Rightarrow a_i^k \le$$
$$\begin{cases} a_i^k + BD^k with i = 1 \\ a_1^k + BD^k + (i-1)\Delta t_i^k if i > 1 \end{cases}. \tag{12.9}$$

The latency restriction produce the first property.

Property 1. *If i-th packet match the jitter restriction ($a_i^k \le p_i^k$), but it doesn't obey the latency restriction, i.e. $a_i^k - (p_i^k - C) > L$, the latency restriction will not be obeyed, no longer mattering BD^k value used.*

Proof. With $a_i^k \le p_i^k$ and the jitter restriction is matched by i-th packet ($a_i^k \le p_i^k$), then the latency restriction is broken, i.e. $p_i^k - (t_i^k - C) > L$ for any BD^k value used.

For the next results consider: (i) packets which attend the Property 1 will be unconsidered, in other words, they show an inevitable loss; (ii) the extra delay used in latency restriction are greater than delays used to jitter remove, therefore the buffer analysis presented are directed to Jitter Restriction and (iii) a packet will be lost when it don't meet jitter restriction [15] [16], i.e., BD^k is not enough to jitter removal in packet i, then:

$$a_i^k \le \begin{cases} a_i^k + BD^k & with \, i = 1 \\ a_1^k + BD^k + (i-1)\Delta t_i^k & if \, i > 1 \end{cases} . \tag{12.10}$$

The Theorem 1 presents a buffer delay value to prevent packet loss by jitter.

Theorem 1. *In a talkspurt k, with buffer delay BD^k, no packet is lost by jitter restriction violation if and only if*

$$BD^k \ge max_{i \in \{1,2,...,n^k\}}\{\delta_i^k - (i-1)\Delta t_i^k\}, \tag{12.11}$$

where $\delta_i^k = a_i^k - a_1^k$ for every $i \in \{1,2,...,n^k\}$.

Proof. Since there is no packets loss in talkspurt, this is equivalent to: $p_i^k - a_i^k \ge 0$ for every

$$i \in \{1,2,...,n\} \Leftrightarrow a_i^k \le p_i^k \Leftrightarrow a_i^k \le a_1^k + BD^k + (i-1)\Delta t_i^k \Leftrightarrow$$

$$\Leftrightarrow a_i^k - a_1^k \le BD^k + (i-1)\Delta t_i^k \Leftrightarrow BD^k \ge max_{i \in \{1,2,...,n^k\}}\{\delta_i^k - (i-1)\Delta t_i^k, \text{ for every}$$

$i \in \{1,2,...,n^k\}$. Notice that:

$$BD^k \ge \max_{i \in \{1,2,...,n^k\}}\{\delta_i^k - (i-1)\Delta t_i^k\},$$

where BD_{npl}^k is the buffer delay which does not present packet loss.

Wherefore we introduce the notion of limiting due to jitter. In the next definitions consider $\{1,2,...,n^k\}$ all packet indexes of the k-th talkspurt.

Definition 1. *The BD_c^k c-th limiting due to jitter, i.e., the value that remove jitter in a set Ω_c of packets in a talkspurt k is defined by*

$$BD_c^k = \max_{i \in \Omega_c}\{\delta_i^k - (i-1)\Delta t_i^k\},$$

where $\Omega_0 = N$, and $\Omega_c = N - (u_0 \cup u_1 \cup ... \cup u_{c-2} \cup u_{c-1})$ for $c > 0$, and $u_c = \{r_c^1, ..., r_c^{w_c}\}$ are the w_c packets where $p_i^k = a_1^k + BD_c^k + (i-1)\Delta t_i^k$ with $i \in u_c$.

Lemma 1. *There is a finite number of jitter limiting values in a talkspurt.*

Proof. The first jitter limiting value is: $BD_0^k = max_{\{i \in \Omega_0 = N\}}\{\delta_i^k - (i-1)\Delta t_i^k\}$, used by set of packets $u_0 \square \Omega_0 = N$. Consider $\Omega_1 = N - u_0 \square \Omega_0$, if $\Omega_1 = \square$, the proof is completed, otherwise it is possible to calculate other jitter limiting value: $BD_1^k = max_{\{i \in \Omega_1 = N\}}\{\delta_i^k - (i-1)\Delta t_i^k\}$ for which there is a non-empty set $u_1 \ \square \ \Omega_1 \ \square \ \Omega_0$ of packets. This reasoning is applied until one is found $\Omega_{m+1} = \square$, then the last jitter limiting value is $BD_m^k = max_{\{i \in \Omega_m = N\}}\{\delta_i^k - (i-1)\Delta t_i^k\}$, where $m \leq n$ and, there is $\square \neq u_m \ \square \ \Omega_m \ \square \ \Omega_{m-1} \ \square \ ... \ \square$ Ω_0 of packets that use that value to remove jitter. Therefore so, we obtain a finite number of jitter limiting value.

Lemma 2. *The jitter limiting values are presented in the format $BD_j^k < BD_{j-1}^k$ to j = 1,...,m.*

Proof. Considering $D_j^k < BD_{j-1}^k$, then:

$$BD_j^k = \max_{\{i \in \Omega_j\}}\{\delta_i^k - (i-1)\Delta t_i^k\},$$

$$BD_{j-1}^k = \max_{\{i \in \Omega_{j-1}\}}\{\delta_i^k - (i-1)\Delta t_i^k\},$$

where $\Omega_j = N - (u_0 \cup ... \cup u_{j-2} \cup u_{j-1})$ and $\Omega_{j-1} = N - (u_0 \ \square \ \cup ... \cup u_{j-2})$, so $\Omega_j \ \square \ \Omega_{j-1}$,

then $BD_j^k < BD_{j-1}^k$. If $BD_j^k < BD_{j-1}^k$ then $u_j \cap \ < u_{j-1} \neq 0$.

In a talkspurt, we have the following running order $BD_m^k < BD_{m-1}^k < ... < BD_0^k$.

Definition 2. *Intervals of type $P_m = [0, BD_m^k), P_{m-1} = [BD_m^k, BD_{m-1}^k), ..., P_0 = [BD_1^k, BD_0^k)$ will be referenced as steps due to jitter.*

Definition 3. *At each step, due to jitter we have associated a number named degree, given by:*

$$degree(P_j) = \sum_{c=0}^{j} w_c, \qquad (12.12)$$

where $BD_{m+1}^k = 0$, and j = 0,...,m.

Each degree is unique by definition. Besides $w_c \geq 1$ for each c, from what we can conclude:

$$0 < degree(P_0) < ... < degree(P_m).$$

The lemma 3 allows monitoring the packet loss behavior with each BD^k value used to jitter remove.

Lemma 3. *Using BD^k in a talkspurt, then the number of lost packets is equal to the degree of its step.*

Proof. In a talkspurt, we have the degrees P_j, with $j = 0,...,m$, due to Lemma 2 we have that $0 = BD_{m+1}^k < BD_m^k < \cdots < BD_{j+1}^k \leq BD^k < BD_j^k < \cdots < BD_1^k < BD_0^k$. Then, $max_{\{i \in \Omega_{j+1}\}}\{\delta_i^k - (i-1)\Delta t_i^k\} = BD_{j+1}^k \leq BD^k < BD_j^k < \cdots < BD_0^k$ and:

$$BD_j^k = \delta_i^k - (r-1)\Delta t_i^k, r \in u_j,$$

$$BD_{j-1}^k = \delta_i^k - (r-1)\Delta t_i^k, r \in u_{j-1},$$

$$\vdots$$

$$BD_0^k = \delta_i^k - (r-1)\Delta t_i^k, r \in u_0.$$

Assuming that $BD^k < (a_r - a_1) - (r-1)\Delta t_r^k$ for all r $\square$ $u_0 \cup \dots \cup u_j$, thus we have, $a_1 + BD^k + (r-1)\Delta t_i^k < a_r$, for $r \square u_0 \cup \dots \cup u_j$, i.e., the jitter restriction is broken for all r $\square$ $u_0 \cup \dots \cup u_j$, then packets $r_j, \dots, r_0$ are lost. On the other hand, with $BD^k \geq (a_r - a_1) - (r-1)\Delta t_r^k$, for all r $\square$ Ω_{j+1} $\square$ Ω_{j+2} $\square$ $\dots$ $\square$ Ω_m $\square$ Ω_{m+1}. Then $p_r \geq a_r$ for all r $\square$ Ω_j, and r $\square$ Ω_{j+2}, so on for all r $\square$ Ω_m. With u_{j+1} $\square$ $\Omega_{j+1}, \dots, u_m$ $\square$ Ω_m, the packets $r_{j+1}, \dots, r_m$ are not lost, and $\{u_0, \dots, u_m\}$ a subset of N, the total number of packets is $w_0 + \dots + w_j = \sum_{c=0}^{j} w_c = degree(P_j)$.

12.3. Buffer Delay Correction Algorithm (BDCA)

In the previous Section, we can see that there is a certain limit to Buffer Delay (BD^k), and above this level there is no packet loss. On the other hand, the good quality of voice communication admits a certain limit of packet loss. Therefore, let us suppose a $\lambda \in (0,1)$ of packets loss in a talkspurt, i.e., at most $\lfloor \lambda n^k \rfloor$ packets can be lost (see Fig. 12.2) where $\lfloor x \rfloor$ the floor function is (greater integer smaller than or equal to x). In this case, we are interested in solving (12.13) bellow.

$$min\{f(BD^k) = BD^k \vee \Psi(BD^k) \leq \lfloor \lambda n^k \rfloor, BD^k \in [0, +\infty)\}. \qquad (1.13)$$

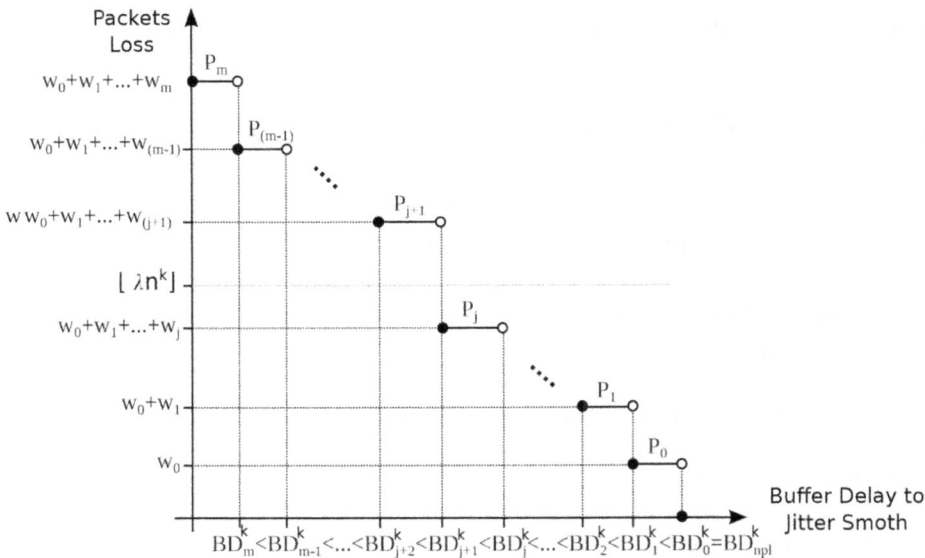

Fig. 12.2. Steps due to jitter in a talkspurt.

The value of equation 12.13 represents a minimum delay value inserted in a talkspurt k, with target loss factor λ or the Optimum Buffer Delay OBD_λ^k.

Theorem 2. *In a talkspurt that use BD^k, n' packets will be lost, if and only if, BD^k belongs to degree of with step n'.*

Proof. When $n' = 0$, i.e., no packet is lost, the Theorem 1 assure this proof. If $n' > 0$, consider $W = \{w_0, w_0 + w_1, \ldots, w_0 + \ldots + w_m\}$ a set of all packets lost by jitter, if $n' \in W$ with $n' = w_0 + w_1 + \ldots + w_j$ for any j, then $BD^k \in P_j$. If $BD^k \in P_h$ for $0 \le h < j$, less than n' packets would be lost, on other hand, if $j \le h < m$, more than n' would be lost. With $degree(P_j) = w_0 + \ldots + w_j$, the BD^k belongs to a degree, with step n'.

Looking for Theorem 2 and $BD^k \subset \{P_m, \ldots, P_0\}$ with $P_i \in [0, +\infty)$ we can write (12.13) as follow:

$$min\{min\{f(BD^k) = BD^k \vee \Psi(BD^k) \le \lfloor \lambda n^k \rfloor, BD^k \in I\}\}, \qquad (12.14)$$

where $I \in \{P_m, \ldots, P_0\}$ and $min\{f(BD^k) = BD^k \vee \Psi(BD^k) \le \lfloor \lambda n^k \rfloor, BD^k \in I\}$ can be solved by Weierstrass Theorem, because in this case, I is compact and f is continuous.

The BDCA adjust the value presented by one BDA (Buffer Delay Algorithm), i.e., approaching BD_{BDA}^k to OBD_λ^k, with packet loss rate in λ. To apply BDCA over talkspurt k, the Adjust Factor (AF) is calculate as:

$$AF(k) = \frac{1}{Z} * \sum_{i=(k-1-Z)}^{i=(k-1)} \frac{OBD_\lambda^i}{BD_{BDA}^i}. \qquad (12.15)$$

The window size (Z) has the last 40 received talkspurts to reduce computational costs, values greater than 40 do not change significantly the results. To find OBD_λ^i, the following elements are needed:

- N_i: Packets transmitted until talkspurt $(i - 1)$;

$$N_i = \sum_{j=1}^{i-1} n^j, \qquad (12.16)$$

- $PLost^{i-1}$: Packets lost from talkspurts 1 to $(i - 1)$;

- $\{(a_1^i, t_1^i), (a_2^i, t_2^i), \ldots, (a_{n^i}^i, t_{n^i}^i)\}$: set with transmission and reception time of all packets in talkspurt i;

- n^i: quantity of received packets during talkspurt i;

- λ: target packet loss rate.

The OBD_λ^i should be used in i-th talkspurt to bring the packet loss closer to λ. The BD_{BDA}^i is the value computed by selected BDA. Equation (4.17) shows Buffer Delay adjusted:

$$BD_{BDCA}^k = BD_{BDA}^k * AF(k). \tag{12.17}$$

Then, the adjusted playout time ($\hat{p}_i^k$) is defined by:

$$\hat{p}_i^k = a_1^k + BD_{BDCA}^k + (i-1)(\Delta t_i). \tag{12.18}$$

Consider the talkspurt $(k-1)$ received, the steps of BDCA to playout time adjust of talkspurt k is showed in Fig. 12.3.

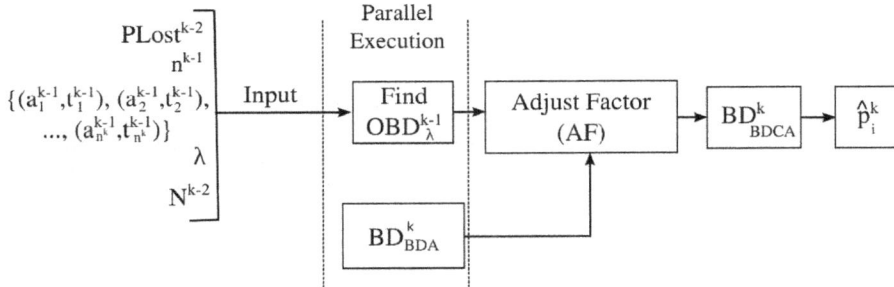

Fig. 12.3. Steps of Correction Algorithm.

The OBD algorithm can run in parallel with BDA selected, so the BDCA does not add extra time to the receiver application.

12.4. Implementation and Results

In this Section, a performance analysis of the proposed algorithm is presented. The BDAs used for comparison are:

- Move Average Algorithm (MA) [10] a loss-tolerant technique;

- Algorithm from Barreto and Aragão (BA) presented in [8], classified as loss intolerant technique;

- Dynamic Management of Dejitter Buffer (DMDB) presented in [13], considered a quality based technique.

For the tests, there were productions of two traces that simulates VoIP transmissions. Using as an example two traces of [3], in other words, the transmitter sent the same quantity of packets with the same time spacing observing in each traces model. The transmissions occurred between Londrina (PR) and São Carlos (SP), both in Brazil. The traces contain the sender and receiver timestamps of transmitted packets and 160 bytes of audio data. A new packet is sent at every 20 ms when there is speech activity [18]. Due to lack of synchronism between the transmitter and receiver clocks, the delay may present negative values, making these measurements useless for playback buffer setting. To avoid this problem the delays will be normalized, in other words, the delay of each packet will

319

be subtracted from the smaller delay value found in the same trace. The Fig. 12.4 presenting a normalized delay and an evolution of the average value on traces A and B.

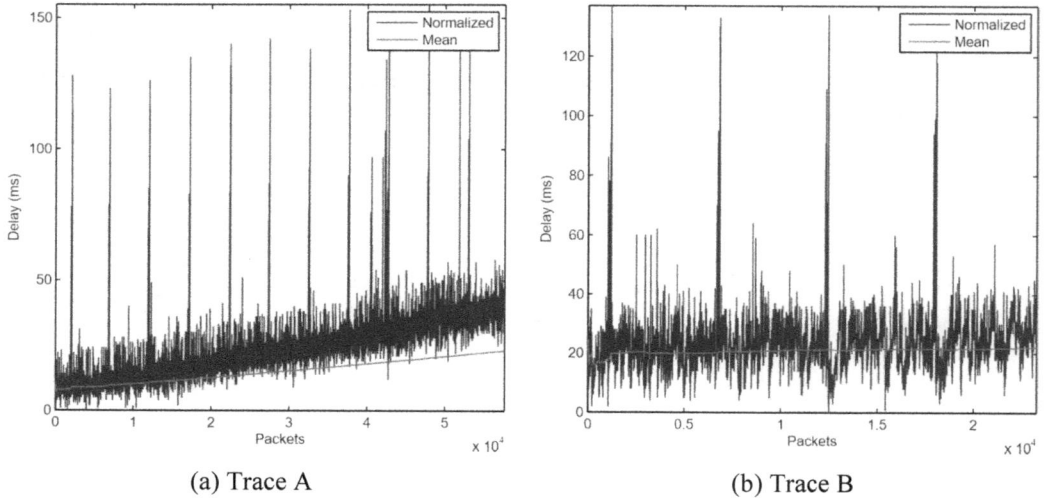

(a) Trace A   (b) Trace B

Fig. 12.4. Normalized Network Delay.

To assist in viewing the distribution of delay network, the Fig. 12.5 illustrate through its histograms the quantity occurred in each group of delay network on traces A and B. Through the pictures you can notice that the trace A brings up the biggest quantity of packets with network delay oscillating between 10 ms and 40 ms. On trace B, the biggest quantity of packets are found together in 20 ms.

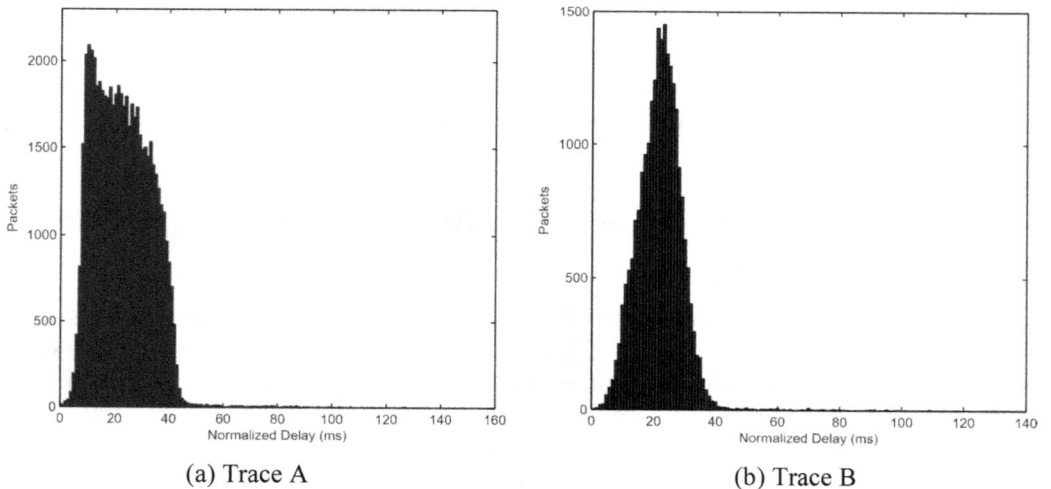

(a) Trace A   (b) Trace B

Fig. 12.5. Histogram of traces.

A description of the traces is depicted in Table 12.1.

Table 12.1. Traces Description.

trace	Talkspurts	Packets
A	832	57662
B	299	23257

In graphs of Fig. 12.6 we use the terms "With BDCA" to represent the original BDA running with BDCA. The used target percentage of packets loss (λ) is 1 %. These graphs are showing the evolution of packet loss rate in a voice call. For interactive audio, packet loss rate is considered suitable up to 1 % of call [19, 20].

The Figs. 12.7, 12.8 and 12.9 are demonstrating the Buffer Delay value used by BDA Barreto and Aragão (BA), Move Average (MA) and Dynamic Management of Dejitter Buffer (DMDB), respectively. Each one of them can be compared with the result obtained by Correction Algorithm (BDCA). The OBD illustrate the value expected for the BD in each one of the traces. The tests were made in Matlab [21].

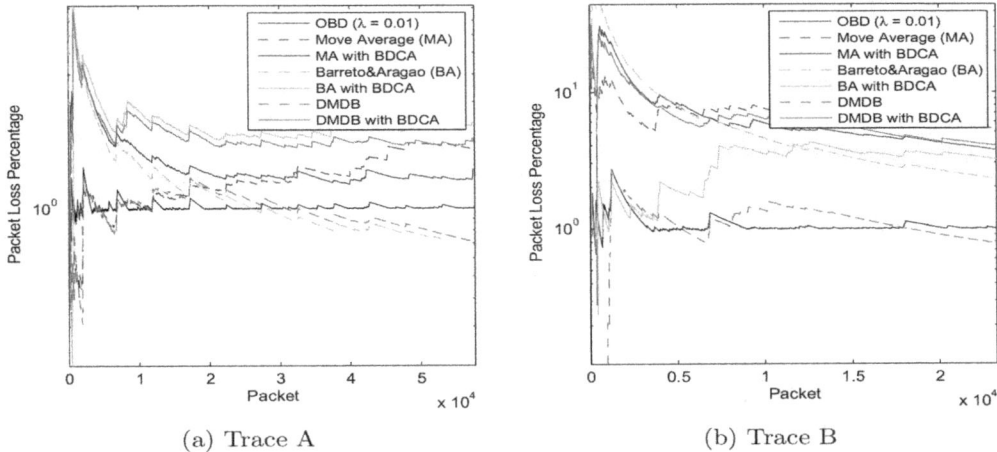

(a) Trace A   (b) Trace B

Fig. 12.6. Packet Loss Percentage.

12.5. Conclusion

This chapter presented the Buffer Delay Correction Algorithm (BDCA) to reduce the difference between packet loss rate of any BDA and the Optimum Buffer Delay (OBD). We have compared the BDA with and without BDCA using 1 % of packet loss rate.

Fig. 12.6 shows that packet loss percentage with each BDA and the value of Correction Algorithm (BDCA). These values are produced by $\lambda = 1$ %.

The Figs. 12.7, 12.8 and 12.9, shows that buffer delay defined by BDCA are closer to values from OBD than values produced only by BDA. This can show that a small extra buffer delay are used during a voice call, producing a better experience to end users.

We are currently expanding the definitions of Buffer Delay Adjustment to reach packet loss caused by latency, i.e., including the sum of packet discarded with playout time greater than the maximum threshold (L). To reach this new restriction, we are working in a new formulation of Adjust Factor.

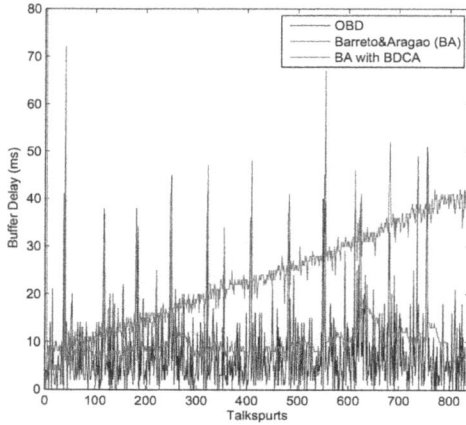
(a) Trace A

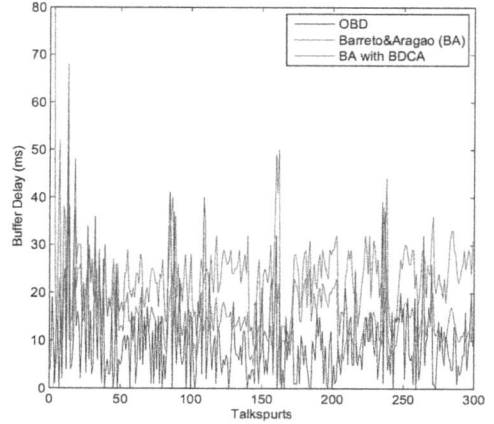
(b) Trace B

Fig. 12.7. Evolution of Buffer Delay of Barreto and Aragão Algorithm.

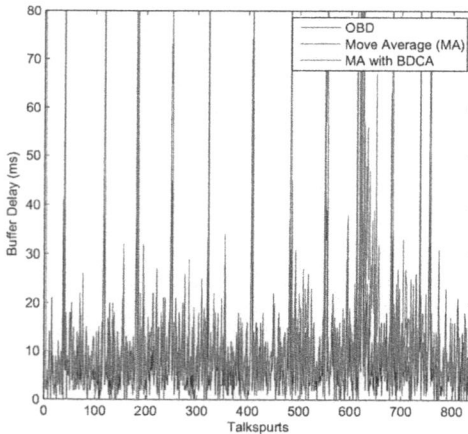
(a) Trace A

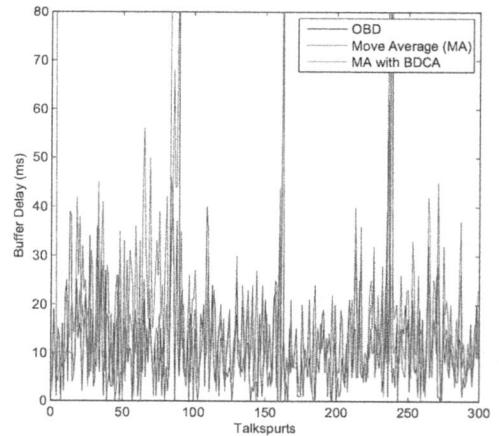
(b) Trace B

Fig. 12.8. Evolution of Buffer Delay of Move Average Algorithm.

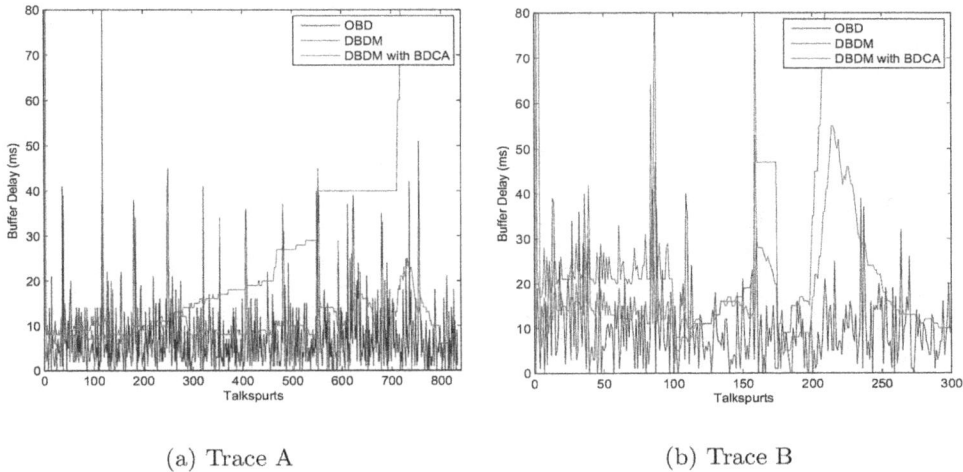

(a) Trace A (b) Trace B

Fig. 12.9. Evolution of Buffer Delay of DBDM Algorithm.

References

[1]. W. C. Hardy, VoIP Service Quality: Measuring and Evaluating Packet-Switched Voice, *McGraw-Hill*, 2003.

[2]. Z. Qiao, R. K. Venkatasubramanian, L. Sun, E. C. Ifeachor, A new buffer algorithm for speech quality improvement in VoIP systems, *Wirel. Pers. Commun.*, Vol. 45, No. 2, Apr. 2008, pp. 189-207.

[3]. S. B. Moon, J. Kurose, D. Towsley, Packet audio playout adjustment: performance bounds and algorithms, *ACM/Springer Multimedia Systems*, Vol. 6, No. 1, Jan. 1998, pp. 17-28.

[4]. Y. Zhang, D. Fay, L. Kilmartin, A. W. Moore, A garch-based adaptive playout delay algorithm for VoIP , *The International Journal of Computer and Telecommunications Networking*, Vol. 54, No. 17, Dec. 2010, pp. 3108-3122.

[5]. L. Atzori, M. L. Lobina, Playout buffering in IP telephony: a survey discussing problems and approaches, *Communications Surveys Tutorials*, IEEE, Vol. 8, No. 3, 2006, pp. 36-46.

[6]. C. Perkins, RTP: Audio and Video for the Internet, *Addison Wesley*, 2012.

[7]. R. Ramjee, J. Kurose, D. Towsley, H. Shulzrinne, Adaptive playout mechanisms for packetized audio applications in wide-area networks, in *Proceedings of the IEEE Conference on Computer Communications (INFOCOM'94)*, Vol. 2, Montreal, Canada, 1994, pp. 680-688.

[8]. J. B. Aragão Jr., G. A. Barreto, Novel approaches for online playout delay prediction in Voip applications using time series models, *Computers and Electrical Engineering*, Vol. 36, No. 3, 2010, pp. 536-544.

[9]. K. Fujimoto, S. Ata, M. Murata, Statistical analysis of packet delays in the internet and its application to playout control for streaming applications, *IEICE Transactions on Communications*, Vol. E84-B, No. 6, June 2001, pp. 1504-1512.

[10]. V. M. R. Ramos, C. Barakat, E. Altman, A moving average predictor for playout control in VoIP , in *Proceedings of the International Workshop on Quality of Service (IWQoS'03)*, Vol. 2707, June 2003, pp. 155-173.

[11]. K. Fujimoto, S. Ata, M. Murata, Adaptive playout buffer algorithm for enhancing perceived quality of streaming applications, *Telecommunication Systems*, Vol. 25, No. 3-4, April 2004, pp. 259-271.

[12]. Methods for Subjective Determination of Transmission Quality, ITU-T P.800, *Telecommunication Standardization Sector of International Telecommunication Union (ITU)*, August 1996.

[13]. R. F. Valle, L. S. G. de Carvalho, R. B. Aguiar, E. S. Mota, D. Freitas, Dynamical management of dejitter buffers based on speech quality, in *Proceedings of the IEEE Symposium on Computers and Communications (ISCC'10)*, Washington, DC, USA, 2010, pp. 56-61.

[14]. L. Sun, E. C. Ifeachor, Prediction of perceived conversational speech quality and effects of playout buffer algorithms, in *Proceedings of the IEEE International Conference on Communications (ICC'03)*, Vol. 1, 2003, pp. 1-6.

[15]. F. Sakuray, R. S. V. Hoto, L. S. Mendes, Analysis and estimation of playout delay in VoIP communications, *International Journal of Computer Science and Network Security*, Vol. 8, No. 3, pp. 98-105.

[16]. D. Florencio, L.-W. He, Enhanced adaptive playout scheduling and loss concealment techniques for voice over IP networks, in *Proceedings of the 2011 IEEE International Symposium on Circuits and Systems (ISCAS'11)*, 2011, pp. 129-132

[17]. One-Way Transmission Time, ITU-T G.114, *Telecommunication Standardization Sector of International Telecommunication Union (ITU)*, 2003

[18]. H. Schulzrinne, Voice communication across the internet: a network voice terminal, Tech. Rep., Dept. of Computer and Information Science, *University of Massachusetts at Amherst*, July 1992.

[19]. S. Nagireddi, VoIP Voice and Fax Signal Processing, 1st Ed., *Wiley Publishing*, 2008.

[20]. Telecommunications-IP Telephony Equipment - Voice Quality Recommendation for IP Telephony, TIA/EIA 116A, *Telecommunication Industry Association*, 2006.

[21]. MATLAB, MATLAB and Statistics Toolbox Release 2012b, Natick, Massachusetts, *The MathWorks Inc.*, 2012.

Chapter 13
Opportunistic Max²-Degree Network Coding for Wireless Data Broadcasting

Kui Xu, Jian Wang, Dongmei Zhang and Wei Xie

13.1. Introduction

Next generation broadband wireless access (BWA) networks such as WiMAX and Long Term Evolution (LTE) are expected to provide efficient, affordable and ubiquitous internet access for data multimedia broadcasting applications such as stock price, real-time traffic and weather information, high-definition (HD) digital television (TV) video streaming and video conferencing. To ensure users' satisfaction, the Quality of Service (QoS) provided by BWA networks should be comparable to that of wired data broadcasting networks. Although the QoS framework for the next generation BWA networks has been designed to support the requirements of multimedia applications, the inherent error-prone nature of the wireless channel often results in high packet error rate (PER) and hence degrades the system reliability. To transmit information reliably over wireless channel, many approaches are employed, such as Forward Error Correcting (FEC), Automatic Retransmission reQuest (ARQ) and Hybrid Automatic Retransmission request (HARQ) [1]. Among these approaches, FEC suffers from low transmission efficiency and ARQ faces large delay and low throughput caused by severe fading channel. HARQ, which combines both FEC and ARQ techniques, can contribute to an efficient utilization of the available resources.

13.1.1. Motivation and Related Work

HARQ has now become a fundamental tool of BWA networks, as it can significantly improve the reliability of wireless link. However, HARQ still encounters challenge in wireless broadcasting applications. As identical information is transmitted from one source to many receivers, it is impossible for each receiver to successfully receive all packets all the time. The traditional HARQ protocol retransmits one packet per slot.

Kui Xu
Army Engineering University of PLA, China

325

Therefore, HARQ may require large amount of retransmissions in broadcasting scenario to ensure correct reception for every packet in all the receivers, which leads to low bandwidth efficiency. In order to improve the transmission efficiency for wireless broadcasting (WBC) system, recently, network coding (NC) [3–7] has been applied to WBC system. By broadcasting, we refer to the scenario where a common sender serves multiple receivers with the same information. In NC based WBC scenario, lost packets or additional naive packets are NC combined before transmitted. According to the way NC combined packets are utilized, we could divide existing NC based WBC schemes into two types, i.e., wait-and-decode network coding (WDNC) [7–10] and instantly decodable network coding (IDNC) [14–20].

13.1.1.1. WDNC Based WBC

Typical WDNC applications include fountain codes and random linear. For WDNC based WBC, received packets that are not instantly decodable are stored for future decoding opportunity [7–10]. With WDNC, perfect throughput could be achieved, however, the receivers can decode all the lost packets only if they have received sufficient number of mixed packets, this kind of retransmission strategy usually has a greater computational complexity and incurs additional delay. In recent work, the trade-off between performance, delay and complexity is addressed to facilitate the application of network coding in multimedia services [11, 12].

13.1.1.2. IDNC Based WBC

IDNC refers to that the received packets are decoded only at their reception instant and cannot be stored for future decoding. For XORing NC based WBC, NC is performed on lost packets which are carefully selected to ensure the decodability at the receiver [14–19]. A dual-XOR HARQ retransmission scheme for wireless broadcasting is proposed in [21], which introduces an additional XOR operation between two lost packets from the individual receiver. In [22], a XORing network coding combining (XNCC) and distributed Turbo coding based type-III HARQ protocol is proposed for wireless broadcasting system. The broadcasting efficiency can be significantly improved by introducing the network and channel coding gain. The above literatures [8–10, 14–19, 21] address the problem of NC based retransmission for WBC system. Unfortunately, different lost packets are NC combined and retransmitted without considering the degree of each NC packet.

13.1.2. Contributions

In this chapter, an Opportunistic Max^2-Degree Network Coding (OM^{2}DNC) based WBC protocol is proposed to improve the overall system spectrum efficiency. At the access node (AN), lost packets of different user equipment (UEs) are combined by performing Max-Degree NC. Then, NC combined packets are broadcasted by utilizing the Max-Degree based opportunistic retransmission protocol. At each UE, the lost packet can be recovered by using the proposed joint network recursive systematic convolution (RSC) decoder (JNRD). Theoretical analyses and simulation results show that the average number of transmissions (ANTs) performance of the proposed OM²DNC based WBC protocol is superior to that of the maximum clique selection algorithm (MCSA) [16], traditional index NC (INC) [15] protocol and XNCC [22] protocol in severe fading channels.

13.2. System Model and Parameters

We consider a WBC system, where a wireless AN broadcasts a frame of λ packets to a set $\mathcal{R} = \{R_1, R_2, \cdots, R_M\}$ of M UEs within its coverage, and the UEs try to receive packets transmitted from BS. BS adopts TDMA protocol to broadcast these packets. During a TDMA slot, only one packet could be transmitted. In total, λ TDMA time slots are needed to broadcast λ data packets, and this λ time slots is referred to as Broadcast Phase. After decoding packets, UEs send feedback to report whether packets have been successfully decoded. After Broadcast Phase, BS starts Retransmission Phase, in which the BS retransmits packets and UEs send feedback to request retransmission of lost packets until all the packets are correctly received.

13.2.1. Mathematical Description

13.2.1.1. Signal Transmission

We assume that the original WBS frame $\mathbf{A}$ consisting of λ information vectors, i.e., $\mathbf{A} = \{\mathbf{a}_1, \mathbf{a}_2, \cdots, \mathbf{a}_\lambda\}$. A transmission information vector could be denoted by the XORing of a set of original. Let $\mathcal{C}$ denote the candidate set, which is defined as an index set of original packets that can be XORing NC combined. For the information vector set $\mathbf{A}_c = \{\mathbf{a}_i \mid i \in \mathcal{C}\}$ the AN combines all the information vectors by using NC, i.e., $\mathbf{a}_c = \mathbf{a}_i \oplus \mathbf{a}_j \oplus \cdots \oplus \mathbf{a}_k$, where $\oplus$ denotes the XOR operator. Specially, for broadcast phase, the candidate set is the corresponding original information vector itself. The information vector is encoded by the RSC encoder yielding the codeword $\mathbf{b}_c = \mathrm{RSC}(\mathbf{a}_c)$, where $\mathrm{RSC}(\bullet)$ denotes the RSC encoder. The code bits $b_c(i)$ are mapped to symbols $x_c(i) \in \mathbb{X}$ of the modulation alphabet $\mathbb{X}$ and broadcasted to the UEs. In case of block fading channel, the received signal vector at the lth UE can be expressed as

$$\mathbf{y}_{l,c} = h_l \sqrt{P_s} \mathbf{x}_c + \mathbf{n}_l,$$

$$(13.1)$$

where $\mathbf{x}_c$ denotes the transmitted signal packet and $\mathbf{n}_l$ is the noise vector whose elements are identically independent distributed (i.i.d) zero-mean Gaussian random variables with variance $\sigma_{n_l}^2$. P_s denotes the BS's transmission power. The coefficient h_l, $\forall_l$ denotes complex zero-mean circular symmetric Gaussian distributed variable with variance $\sigma_{h_l}^2$, i.e., $h_l \sim \mathcal{CN}(0, \sigma_{h_l}^2)$.

13.2.1.2. Signal Reception

After receiving packets, UEs try to decode these packets, and whether a packet is decoded successfully can be determined by calculating the CRC bits attached in the tail of the packet. We assume that the information vector $\mathbf{a}_i$ is lost at the lth UE. Based on the received NC combined signal $y_{l,c}(m)$, the LLRs for the participating coded bit $b_i(n)$ can be calculated as [23],

$$
L_{\text{Dem}}^l\left(b_i(n)\right) = sign\left(\hat{x}_{C/\{i\}}^m(n)\right)\ln\frac{\Pr\left\{b_i(n)=0\middle| y_{l,c}(m),\hat{h}_{l,c}\right\}}{\Pr\left\{b_i(n)=1\middle| y_{l,c}(m),\hat{h}_{l,c}\right\}}
$$

$$
= sign\left(\hat{x}_{C/\{i\}}^m(n)\right)\ln\frac{\sum_{x\in\mathbb{X},\; b_i(n)=0}\exp\left(-\frac{\left|y_{l,c}(m)-\hat{h}_{l,c}\sqrt{P_s}x\right|^2}{\sigma_{n_l}^2}\right)}{\sum_{x\in\mathbb{X},\; b_i(n)=1}\exp\left(-\frac{\left|y_{l,c}(m)+\hat{h}_{l,c}\sqrt{P_s}x\right|^2}{\sigma_{n_l}^2}\right)},\quad (13.2)
$$

where $\hat{h}_{l,c}$ is the estimate of the channel coefficient $h_{l,c}$ [25], sign(·) denotes the sign function, \ denotes the relative complement (also termed as set difference) operator, $\hat{x}_{C/\{i\}}^m(n) = \left[x_{C/\{i\}}(n)\right]_m$ denotes the modulated signal portion of $x_{C/\{i\}}(n)$ associated with the coded bit $b_{C/\{i\}}(m)$ and $sign\left(\hat{x}_{C/\{i\}}^m(n)\right)\in[1,-1]$. To calculate the probability of $b_i(n)=0$ (or $b_i(n)=1$), all possible symbols $x\in\mathbb{X}$ related to $b_i(n)=0$ (or $b_i(n)=1$) need to be considered. After getting the LLRs $L_{\text{Dem}}^l\left(b_i(n)\right)$, the RSC decoding can be done by using the maximum-likelihood decoder or the maximum a posteriori (MAP) decoder [24]. In this way, the soft value of the desired coded bit $b_i(m)$ can be obtained from the received combined signal.

13.2.2. Assumptions

In this chapter, we will consider the following assumptions:

- In both initial broadcast and retransmission phases, all the packets have the same fixed length and use the same modulation scheme.

- In both initial broadcast and retransmission phases, BS adopts TDMA protocol to broadcast these packets. During a TDMA slot, only one packet could be transmitted. We assume that the channel coefficient keeps constant during a TDMA slot, while independent with others in different slots, i.e. the channel is a block fading channel and $\mathbb{E}\left[h_{l,k}h_{m,n}^*\right]=\sigma_{h_{l,k}}^2\delta_{l,m}\delta_{k,n}$. $\mathbb{E}[\cdot]$ denotes the expectation, $(\bullet)^*$ denotes the complex conjugate and $\delta_{i,j}$ denotes the Kronecker delta function. Hence, the packet losses at different UEs are uncorrelated. Due to the randomness of UEs' location, we assume the distribution of channel between BS and different UEs are same, i.e., $\sigma_{h_{l,k}}^2 = \sigma_h^2, \forall\{l,k\}$.

- The BS knows the instantaneous states of all the packets at each UE. Each UE sends an acknowledgement signal (ACK) whenever the packet is received correctly, or else it sends a negative acknowledgement signal (NAK). For simplicity, we assume that all the ACK/NAKs are instantaneous and never lost, i.e. the feedback channel is error free and has no delay.

13.3. The Proposed OM²DNC Based WBC Protocol

The proposed OM²DNC based WBC protocol includes two phases to ensure reliable packets reception: *Broadcast Phase* and *Retransmission Phase*.

- *Broadcast Phase*: As illustrated in Fig. 13.1 (BS side), the BS first attaches FEC bits and CRC bits in the tail of the information vector. Then, the BS broadcasts the RSC encoded and modulated packets to the UEs. As illustrated in Fig. 13.1 (UE side), each UE receives packets and tries to decode these packets. Whether a packet is successfully decoded is determined by the CRC attached in the packet tail. Each UE sends a feedback to the BS after RSC decoding and CRC checking. Specifically, each UE sends an ACK signal whenever the packet is received correctly, or else it sends a NAK signal. If the CRC checking is correct, the received packet will be stored in the buffer. The decoding states of λ naive packets at UE m could be represented as a λ-dimensional decoding vector $\mathbf{d}_m$, that is

$$\mathbf{d}_m = \left[d_m^{(1)}, d_m^{(2)}, \cdots, d_m^{(\lambda)} \right]^T, \tag{13.3}$$

where $d_m^{(n)} \in \{0,1\}$ denotes the decoding state of the nth naive packet at UE m, i.e., $d_m^{(n)} = 0$ if the n th naive packet is correctly received; otherwise, $d_m^{(n)} = 1$.

- *Retransmission Phase*: As illustrated in Fig. 13.1 (BS side), the BS first forms an NC combined retransmission packet by using the proposed OM²DNC strategy. Information vectors of different lost packets are NC combined by several information vectors and could be denoted as a linear combination of all the information vectors over GF (13.2), i.e., the inner product of a coefficient vector and information vectors. The i th NC combined vector in the k th round of retransmission is given as

$$\begin{aligned} \mathbf{c}_{k,i} &= \mathbf{r}_{k,i} \cdot \mathbf{A} \\ &= \left[r_{k,i}^{(1)}, r_{k,i}^{(2)}, \cdots, r_{k,i}^{(\lambda)} \right] \left[\mathbf{a}_1, \mathbf{a}_2, \cdots, \mathbf{a}_\lambda \right]^T, \end{aligned} \tag{13.4}$$

where $1 \le k \le \mathcal{K}$, $1 \le i \le \mathcal{I}_k$. $r_{k,i}^{(n)} \in \{0,1\}$, $\mathbf{a}_n$, $\mathcal{K}$ and $\mathcal{I}_k$ denote the n th element of vector $\mathbf{r}_{k,i}$, the n th information vector, the total number of retransmission rounds and the number of network coded packets in k th round of retransmission, respectively. After passing the CRC encoder, RSC encoder in sequence, the network coded retransmission packets are broadcasted to the UEs by the BS.

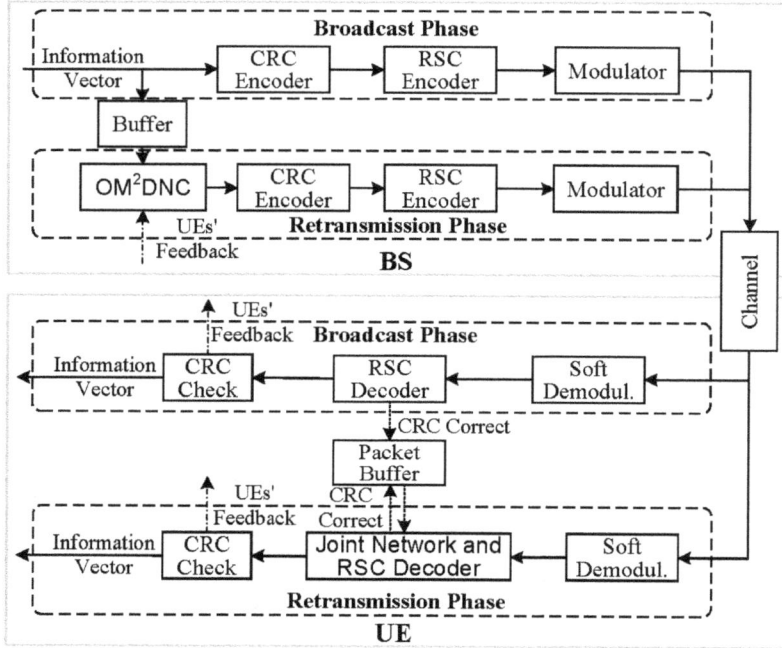

Fig. 13.1. The block diagram of the proposed OM²DNC based wireless broadcasting protocol.

After receiving a retransmission packet, UE m tries to obtain its own lost packet by using prior correctly received naive packets. Whether the received retransmission packet $\mathbf{c}_{k,i}$ could help recover a lost packet at UE m or not could be determined by checking the inner product of $\mathbf{r}_{k,i}$ and $\mathbf{d}_m$, denoted as $D_m^{\mathbf{r}_{k,i}}$, which represents the number of lost packets contained in $\mathbf{c}_{k,i}$ for UE m. $D_m^{\mathbf{r}_{k,i}}$ can be given as

$$D_m^{\mathbf{r}_{k,i}} = \mathbf{r}_{k,i} \cdot \mathbf{d}_m = \sum_{n=1}^{\lambda} r_{k,i}^{(n)} d_m^{(n)} . \tag{13.5}$$

If $D_m^{\mathbf{r}_{k,i}} = 0$ or $D_m^{\mathbf{r}_{k,i}} > 1$, the recovered information vector contains none lost packet or more than one lost packets corresponding to UE m. If $D_m^{\mathbf{r}_{k,i}} = 1$, UE m could obtain a lost packet after correctly decode the received retransmission packet. UE binary combine the correctly received information vectors that are contained in $\mathbf{c}_{k,i}$ (according to $\mathbf{r}_{k,i}$), and the recovered information vector of lost packet could relaxed as

$$
\begin{aligned}
t_{k,i} &= \left[\left(\tilde{\mathbf{r}}_{k,i} \odot \overline{\mathbf{d}}_m \right) \cdot \mathbf{A} \right] + \mathbf{c}_{k,i} \\
&= \left[\left(\mathbf{r}_{k,i} \odot \overline{\mathbf{d}}_m \right) \cdot \mathbf{A} \right] + \left[\left(\mathbf{r}_{k,i} \odot \overline{\mathbf{d}}_m \right) \cdot \mathbf{A} + \left(\mathbf{r}_{k,i} \odot \mathbf{d}_m \right) \cdot \mathbf{A} \right] \quad (13.6) \\
&= \left(\mathbf{r}_{k,i} \odot \mathbf{d}_m \right) \cdot \mathbf{A}
\end{aligned}
$$

where $\odot$ denotes Hadamard (Schur) product operation and $\overline{}$ denotes bitwise NOT (or complement) operation. As illustrated in Fig. 13.1 (UE side), each UE demodulates the received NC combined retransmission packet by using the proposed JNRD decoder and then feedbacks to the BS after CRC checking. If the CRC checking is correct, the received packet will be stored in the buffer.

We assume that the BS transmits a WBS frame $\mathcal{X}$, which consisting of λ packets $\mathcal{X}\{\mathbf{x}_1, \mathbf{x}_2, \cdots, \mathbf{x}_\lambda\}$, in the broadcast phase. In this chapter, an $M \times \lambda$ feedback matrix $\mathbf{T}$ is applied to store the state (correctly received or not) of all the packets at each UE. The (i, j)th entry of the matrix $\mathbf{T}$ ($1 \le i \le M$, $1 \le j \le N$) indicates whether the i th UE has received packet $\mathbf{x}_j$ or not, i.e., $\mathbf{T}(i, j) = 0$ if $\mathbf{x}_j$ is correctly received by the i th UE, otherwise $\mathbf{T}(i, j) = 1$. The ith column of matrix $\mathbf{T}$ represents the states whether the ith packet is received successfully or not by all the UEs. The ith row of matrix $\mathbf{T}$ represents the states whether all the packets are received successfully or not by the ith UE. The flowchart of the proposed OM²DNC based WBC protocol is given in Fig. 13.2.

13.3.1. OM²DNC Strategy

In this chapter, the OM²DNC strategy is proposed to guarantee the efficiency and decodability of the network coded packets. In the Retransmission Phase, BS forms a feedback matrix $\mathbf{T}$ based on UEs' HARQ feedback information, as shown in Fig. 13.3. When a packet is successfully received by one UE, the corresponding position of the feedback matrix $\mathbf{T}$ will be marked as "0", otherwise "1". Hence, matrix $\mathbf{T}$ can convey two meanings: (1) whether the packet is lost; (2) the index of UE that lost the packet. In order to guarantee the efficiency and decodability of the network coded retransmission packets, a set of lost packets to perform NC combining, named as candidate set (CS), is generated based on the following two principles: (1) at most one lost packet from each UE is contained in the CS, which is to ensure the decodability of the network coded retransmission packet in each UE's network decoder; (2) as many as lost packets of different UEs are contained in the CS. The proposed OM²DNC strategy consists of two phases: Phase I, Max-Degree NC; Phase II, Max-Degree opportunistic retransmission. In Phase I, Max-Degree XORing NC is applied to form retransmission packets. Each retransmission packet is obtained by performing XORing NC on a CS. The CS of lost packets is stored in the *CodingList* and the corresponding UEs' indexes of the lost packets are stored in the *UserList*. The *UserList_Temp* is a temporary list. To determine the CS of lost packets, a Max-Degree based two-step solution is designed:

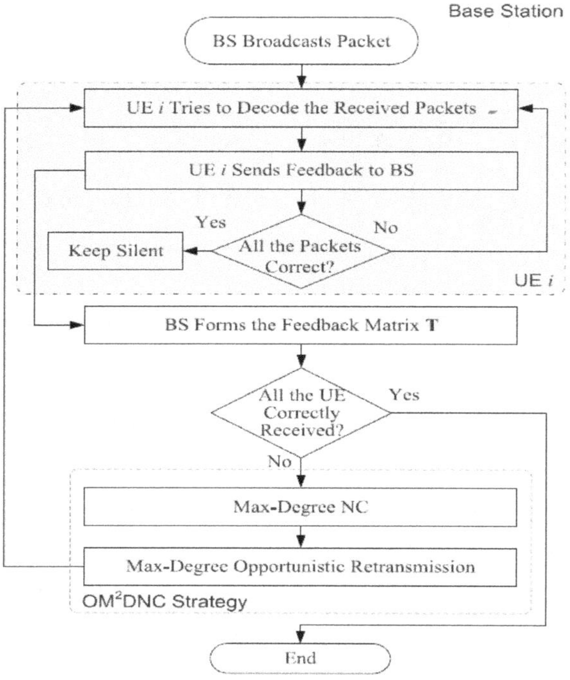

Base Station

BS Broadcasts Packet

UE *i* Tries to Decode the Received Packets

UE *i* Sends Feedback to BS

Yes   No

All the Packets Correct?

Keep Silent

UE *i*

BS Forms the Feedback Matrix **T**

All the UE Correctly Received?   Yes

No

Max-Degree NC

Max-Degree Opportunistic Retransmission

OM²DNC Strategy

End

Fig. 13.2. Flowchart of the proposed OM²DNC based WBC protocol.

- Step I: Selecting the first packet (Lost packet with maximum degree). Removing a column in the feedback matrix **T**, where the associated packet of this column is correctly received by all UEs, such as packet ID: 8 in Fig. 13.3. The packet ID with maximum degree (number of "1"s in its corresponding column, such as packet ID: 3 in Fig. 13.3) is stored to CodingList. The maximum degree is stored in MaxDegree and the indexes of UEs which lost the corresponding packet are stored in *UserList*. Removing the corresponding column in the feedback matrix **T**.

		Packet ID									
		P_1	P_2	P_3	P_4	P_5	P_6	P_7	P_8	P_9	P_{10}
UE ID	R_1	0	1	1	1	1	0	0	0	0	0
	R_2	0	0	0	0	0	1	1	0	0	1
	R_3	0	0	0	1	0	1	0	0	1	1
	R_4	1	1	1	0	0	0	1	0	0	0
	R_5	0	0	1	0	0	0	0	0	1	1

Fig. 13.3. The example of feedback matrix **T**.

- Step II: Updating the *CodingList* and *UserList*. For the *i*th, $i \in \{1, 2,...,\lambda\}$, column of matrix **T**, if the degree of the *i*th column is equal to *MaxDegree*, then store the corresponding UE IDs in *UserList_Temp*. Comparing *UserList_Temp* with *Userlist*, if there is no intersection, updating *UserList* through *UserList* = *UserList* ∪ *UserList_Temp*, updating *CodingList* through *CodingList* = *CodingList* ∪ *i* and removing the corresponding column in the feedback matrix **T**. Otherwise, *UserList* = *UserList* and *CodingList* = *CodingList*. If $(i + 1) \leq \lambda$, then empty *UserList_Temp* and repeat Step II for the $(i + 1)^{th}$ column of matrix **T**. Otherwise, if $i = \lambda$, then empty *UserList_Temp*, let *MaxDegree* = *MaxDegree* − 1, $i = 1$. If *MaxDegree* > 0, then repeat Step II for the *i*th column of matrix **T**, otherwise, repeat Step I to form another CS.

Example: Applying the Max-Degree XORing NC to the feedback matrix **T** given in Fig. 13.3, we can obtain the following four CSs: $C_1 = \{3, 6\}$, $C_2 = \{2, 10\}$, $C_3 = \{4, 7\}$, $C_4 = \{1, 5, 9\}$. Hence, the number of NC combined packets of the proposed scheme is four ($P_3 \oplus P_6$, $P_2 \oplus P_{10}$, $P_4 \oplus P_7$, $P_1 \oplus P_5 \oplus P_9$), whereas five NC combined packets ($P_1 \oplus P_4$, $P_2 \oplus P_6$, $P_3 \oplus P_7$, $P_5 \oplus P_9$, P_{10}) are formed for the scheme in [14-18], and nine retransmissions are required for traditional ARQ. In Phase II, NC combined packets are broadcasted by utilizing the Max-Degree based opportunistic retransmission protocol. We assume that all the lost packets form S CSs C_j, $j = 1, 2,..., S$. The CSs with maximum degree C_{max} can be obtained by

$$C_{max} = \arg \max_{C_j} \sum_{k \in C_j} \sum_{m=1}^{M} \left(\mathbf{T}(m,k) \sum_{u=1}^{\lambda} \mathbf{T}(m,u) \right), \tag{13.7}$$

where M denotes the number of UEs and λ is the number of transmitted packets. Then, the BS combines the lost packets in *CodingList* corresponding to CSs C_{max} by XORing NC and retransmit the NC combined packet. Upon receiving the feedback from UEs, the BS updates the feedback matrix **T** and starts the next round of OM²DNC strategy based on the updated matrix **T**. The OM²DNC strategy is shown in Algorithm 1, where M and λ denote the number of UEs and packets, respectively. Since each UE can only recover at most one lost packet by utilizing one NC combined retransmission packet, the number of total retransmissions is no less than the maximum number of lost packets among all the UEs. Hence, we note that the minimum number of NC combined packets by using Max-Degree NC equals to the maximum number of lost packets of all UEs, which is obvious for the case of two UEs.

13.3.2. Joint Network-RSC Decoding

After receiving a retransmission packet and the corresponding *CodingList*, the *l*th UE determines whether the received retransmission packet contains its own lost packet. If the desired lost packet of the *l*th UE is contained in the received NC combined packet, then the *l*th UE pushes out the corresponding prior successfully received information packets from their packet buffer and XORs them. The combined information packet passes CRC encoder and RSC encoder in sequence, which is the same as what has been done in the BS. The output of the RSC encoder $\mathbf{I}_c^l$ is the combination of the packets which are contained in *CodingList* and successfully received by the *l*th UE, namely local generated combination packet.

Algorithm 1. The proposed OM²DNC strategy.

Input: $M \times \lambda$ Feedback Matrix **T**

Export: NC Combined Packets

   for $i = 1$ To Limited Retransmission Number **do**

    · Phase I:

     while *sum* (**T** (:)) $\neq 0$ **do**

     · *sum_per_column* = *sum* (**T**,1)

     · *MaxDegree* = *max(sum_per_column)*

     · Step I: Select the first packet (lost packet with Maximum Degree)

     · Step II: Update the *CodingList* and *UserList*

     end while

    ·Phase II:

    · Select the CSs with Maximum Degree C_{max}

    · Perform XORing NC on lost packets in *CodingList* corresponding to C_{max}

    end for

We assume that the soft-demodulator generates the Log-likelihood ratio (LLR), namely soft information, for each bit of the combined information packet $C^l = \mathrm{RSC}\left(\mathbf{I}_L^l \oplus \mathbf{I}_C^l\right) = \mathrm{RSC}\left(\mathbf{I}_L^l\right) \oplus \mathrm{RSC}\left(\mathbf{I}_C^l\right)$, where $\mathbf{I}_L^l$ denotes the information of the lth UE's lost packet. The soft information of the lth UE's lost packet can be obtained by a soft-decision network decoder, which removes the local generated combination packet $\mathbf{I}_C^l$ from the soft information of the retransmission packet. Let $C^l(n)$ and $L\left[C^l(n)\right]$ denote the nth bit of the information C_1 and its LLR value, then the LLR value of $\mathrm{RSC}\left(\mathbf{I}_L^l\right)$ can be obtained from (13.2) by canceling the effect of $\mathrm{RSC}\left(\mathbf{I}_C^l\right)$ as:

$$L\left[\mathrm{RSC}\left(\mathbf{I}_L^l\right)\right](n) = \log \frac{\Pr\left\{\left[\mathrm{RSC}\left(\mathbf{I}_L^l\right)\right](n)=1\right\}}{\Pr\left\{\left[\mathrm{RSC}\left(\mathbf{I}_L^l\right)\right](n)=0\right\}}$$
$$= \begin{cases} L\left[C_L^l\right](n), & \left[\mathrm{RSC}\left(\mathbf{I}_L^l\right)\right](n)=0 \\ -L\left[C_L^l\right](n), & \left[\mathrm{RSC}\left(\mathbf{I}_L^l\right)\right](n)=1 \end{cases} \quad (13.8)$$

Hence, by using the soft-decision network decoder, the code-words which are prior known at the UE can be removed before RSC decoding. The soft information obtained by using the soft-decision decoder are applied to RSC decoding firstly and then passes the CRC decoder. If successful decoding, the received packet will be stored in the packet buffer.

13.4. Performance Analysis

In this section, the performance of the proposed OM²DNC based WBC protocol is theoretically analyzed. In [13], it was shown that the optimal solution of index network coding problem is an NP-hard problem. Hence, a lower bound of ANTs can be obtained through theoretical analysis. Suppose that there are M UEs with PERs $P_m, m \in \{1, \cdots, M\}$, in the Broadcast Phase and the Retransmission Phase. We have the following results:

Proposition 1. The ANTs lower bound of the NC based HARQ protocol with M UEs is

$$\Xi_{N1} = \max_{i=\{1,\cdots,M\}} \left\{ (1-P_i)^{-1} \right\}. \tag{13.9}$$

Proof. For the NC based WBC protocol, the number of transmissions needed to successfully deliver a packet to an UE with the PER P_i is a random variable Y. We have

$$P[Y \le k] = 1 - P[Y > k] = 1 - P_i^k, \tag{13.10}$$

and

$$P[Y = k] = 1 - P_i^k - \left(1 - P_i^{k-1}\right) \\ = P_i^{k-1}(1-P_i) \tag{13.11}$$

The ANTs per successful packet is

$$E[Y] = \sum_{k=1}^{\infty} kP[Y = k] \\ = 1 - P_i + \sum_{k=1}^{\infty} k\left(P_i^{k-1}(1-P_i)\right) \cdot \\ = (1-P_i)^{-1} \tag{13.12}$$

Thus the ANTs of NC based WBC protocol with M UEs is

$$\Xi_{N1} = \max_{i=\{1,\cdots,M\}} \left\{ (1-P_i)^{-1} \right\}. \tag{13.13}$$

13.5. Simulation Results

In this section, we carry out numerical simulations to evaluate the performance of the proposed OM²DNC based WBC protocol. We assume that the channels between BS and UEs subject to same complex Gaussian distribution $\mathcal{CN}(0,1)$ and BS's transmission power P_s is normalized. The channel code at BS is a rate $R = 1/2$ RSC code with constraint length $K = 5$. The feedforward generator is $F = 37$ and the feedback generator is $B = 21$ (both in octal). The length of information packet is L_I bits (containing CRC bits), and the length of coded packet is L_C bits.

Three existing schemes are adopted here for comparison, i.e., INC [15], MCSA [16], and XNCC [22]. In INC [15], original packets are divided into two subsets, i.e., ϕ_1 and ϕ_2

under two conditions: (1) any packet in ϕ_2 could be combined decodably with at least one packet in ϕ_1, and (2) all lost packets at the receivers with most lost packets is contained in ϕ_1. Packets in ϕ_1 are ordered according to the number of packets in ϕ_2 that could be separately decodably combined with them, from small to large. According to the order, each packet in ϕ_1 is decodably combined with packets in ϕ_2 by sequence, and packets in ϕ_2 failed to be combined are decodably combined among each other. Combination results are then scheduled to be retransmitted. In MCSA [16], each lost packet at each receiver is denoted as a vertex in a graph, where two vertexes are connected if the corresponding packets are lost in a same receiver or owned by different receivers. Each vertex is allocated with an initial weight proportional to the number of lost packets at the corresponding receiver. This weight is then updated as the sum of initial weight of connected vertexes. The vertex with maximum weight is selected out to add into the clique, and remain vertexes connected to this vertex form a new graph accordingly. This procedure continues until the newly formed graph is empty, and then the output clique is regarded as a candidate set to produce a coded packet.

Fig. 13.4 shows the simulation results of average targeted receivers (ATRs) versus SNR. Here, a receiver is regarded as a targeted receiver for a coded packet when coded packet could help the receiver recover a lost packet. From the figure, we could find that proposed OM^{2}DNC based WBC protocol achieves a significant ATRs performance gain when compared with the MCSA [16], INC [15] protocol and XNCC [22] protocol, respectively. That means in our proposed scheme, each coded packet could serve more receivers and therefore more NC gain could be achieved.

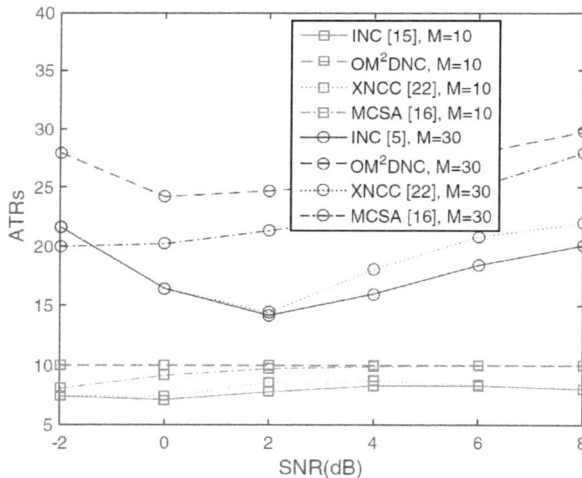

Fig. 13.4. ATRs versus SNR. $\lambda = 100$ and $L_c = 400$.

Fig. 13.5 shows the ATRs performance as a function of λ at SNR = −2 dB. We can see from the figure that the proposed OM^{2}DNC strategy outperforms MCSA [16], INC [15] protocol and XNCC [22] protocol on the ATRs performance, respectively. Moreover, the

ATRs performance can be observed as less influenced by the number of original packets as the curves of each scheme is almost flat.

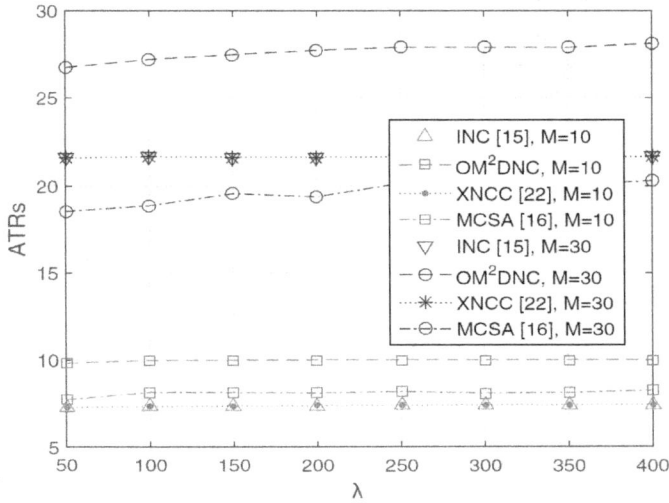

Fig. 13.5. ATRs versus λ. $L_c = 400$ and SNR $= -2$ dB.

Fig. 13.6 shows the ATRs versus the number of UEs M. As shown in the figure, the ATRs increases as the number of UEs M increases. The proposed OM^{2}DNC strategy outperforms MCSA [16], INC [15] protocol and XNCC [22] protocol on the ATRs performance, respectively.

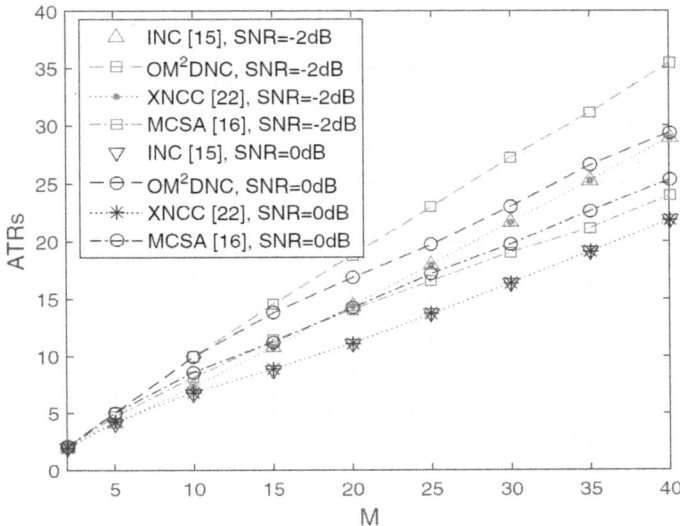

Fig. 13.6. ATRs versus *M*. $\lambda = 100$, $L_c = 400$.

337

Fig. 13.7 shows the simulation results of ANTs versus SNR. The numerical results of the theoretical analyses in (13.9) is also depicted for comparison. We can see from Fig. 13.7 that the exact ANTs performance of the proposed OM^2DNC strategy derived by Monte Carlo simulation for M = 2 matches well with the theoretical analyses in (13.9). Meanwhile, simulation results show that the proposed OM^2DNC based WBC protocol achieves a significant ANTs performance gain when compared with the MCSA [16], INC [15] protocol and XNCC [22] protocol, respectively.

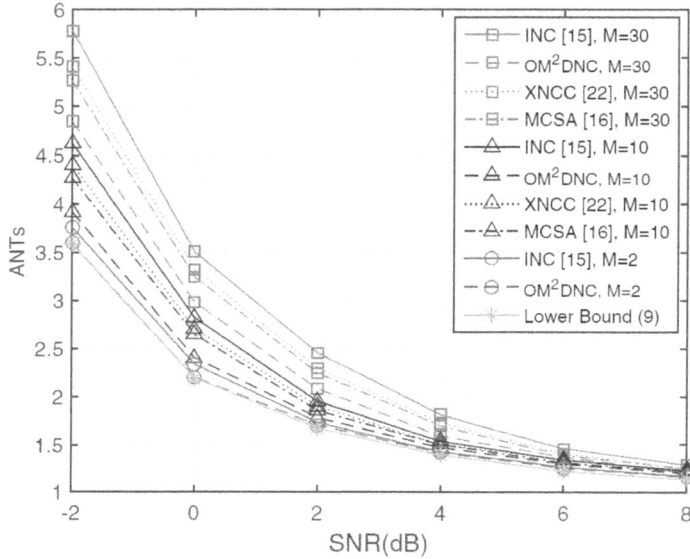

Fig. 13.7. ANTs versus SNR. λ = 100 and L_c = 400.

Fig. 13.8 shows the ANTs performance as a function of λ at SNR = −2 dB. We can see from Fig. 13.8 that the proposed OM^2DNC strategy outperforms MCSA [16], INC [15] protocol and XNCC [22] protocol on the ANTs performance, respectively. We can see from Fig. 13.8 that the ANTs performance of the proposed OM^2DNC strategy obtained by Monte Carlo simulation for M = 2 matches well with the theoretical analyses in (13.9). Moreover, ANTs performance floor can be observed in the large λ region when M = 10 and 30, which suggests the inherent performance gap between the proposed OM^2DNC strategy and the optimal index network coding solution.

Fig. 13.9 shows the ANTs versus the number of UEs M. As shown in the figure, the ANTs increases as the number of UEs M increases. The proposed OM^2DNC strategy outperforms MCSA [16], INC [15] protocol and XNCC [22] protocol on the ANTs performance, respectively. Meanwhile, simulation results show that the performance gap between the proposed OM^2DNC based WBC protocol and the theoretical lower bound in (13.9) increases as the increases of the number of UEs M.

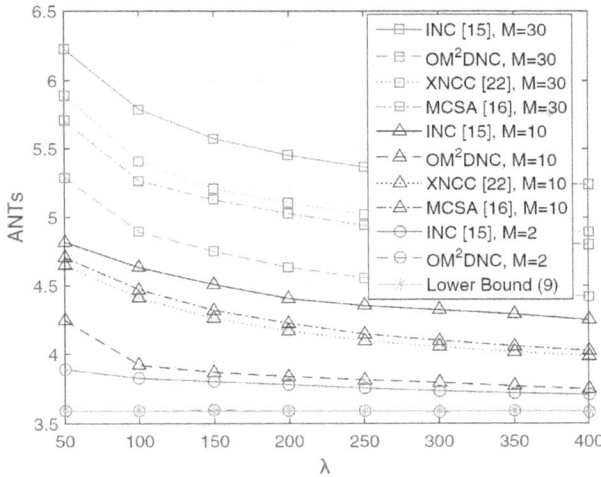

Fig. 13.8. ANTs versus λ. $L_c = 400$ and SNR $= -2$ dB.

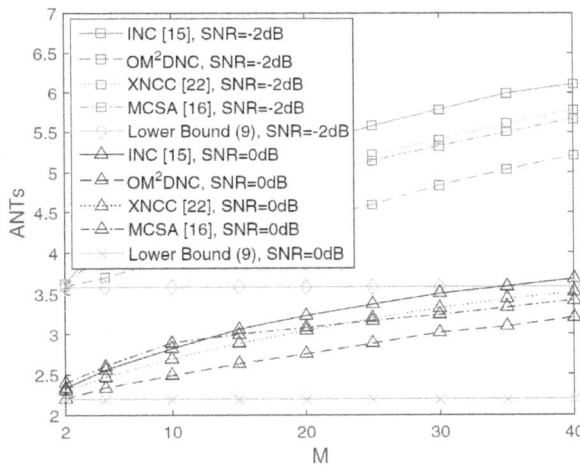

Fig. 13.9. ANTs versus M. $\lambda = 100$, $L_c = 400$.

From the above simulation comparison, it is clear that our proposed scheme achieve significant performance gain compared with existing schemes. It is nontrivial to give a brief explanation on why better performance could be achieved in our scheme and how to go further towards the lower bound. Priority plays an important role in coded packets determination. In general, two priorities are indeed important in order to achieve higher performance gain. One priority is to prioritize the receiver that demands most packet. Since retransmission packets suffer loss, NCB could roughly be regarded as a finite Markov decision procedure, with the numbers of lost packets at receivers regarded as state in the state space. Therefore to recover lost packets is to transform the current state to the all-zero state. Since only at most one step could be taken after receiving one coded packet

339

at each receiver, it is obvious to prioritize the receiver that need the most packets (for the purpose to transform to all-zero state faster). The other priority is to prioritize the packet that are needed most by receivers. In this way, there will be more chances for lost packets to be decodably combined, thus higher NC gain would be achieved. The performance advance of our scheme verifies the above priorities. Thus it is not clear yet that how to make a trade off between the two priorities. Further effort could be made to build a new rule that makes better balance between the two priorities, thus better performance would be expected to be achieved.

13.6. Conclusions

In this chapter, an OM^2DNC based WBC protocol is proposed to increase the bandwidth efficiency of wireless multimedia broadcasting systems. The advantages of the proposed scheme over the traditional NC based retransmission schemes are shown through theoretical analyses and simulations. Comparing with traditional NC based protocol, the proposed OM^2DNC based WBC protocol can significantly reduce the number of transmissions per successful packet, whose gains is attributed to the potential advantage of the OM^2DNC strategy.

References

[1]. J.-F. Cheng, Coding performance of hybrid ARQ schemes, *IEEE Trans. Commun.*, Vol. 54, No. 6, 2006, pp. 1017–1029.

[2]. J. Kim, H. Jin, D. K. Sung, et al., Optimization of wireless multicast systems employing hybrid ARQ with chase combining, *IEEE Trans. Vehic. Technol.*, Vol. 59, 2010, pp. 3342–3355.

[3]. X. C. Xia, Y. Y. Xu, K. Xu, et al., Outage performance of AF-based time division broadcasting protocol in the presence of co-channel interference, in *Proceedings of the IEEE Wireless Communications and Networking Conference (WCNC'13)*, 2013, pp. 3482–3487.

[4]. Y. Y. Xu, X. C. Xia, K. Xu, et al., Symbol error rate of two-way decode-and–forward relaying with co-channel interference, in *Proceedings of the IEEE International Symposium on Personal, Indoor and Mobile Radio Communications (PIMRC'13)*, 2013, pp. 138–143.

[5]. K. Xu, D. Zhang, Y. Xu, W. Ma, On the equivalence of two optimal power- allocation schemes for a-TWRC, *IEEE Trans. Vehic. Technol.*, Vol. 63, No. 4, 2014, pp. 1970–1976.

[6]. V. T. Muralidharan, B. S. Rajan, Performance analysis of adaptive physical layer network coding for wireless two-way relaying, *IEEE Trans. Wireless Commun.*, Vol. 12, No. 3, 2013, pp. 1328–1339.

[7]. F. Chiti, R. Fantacci, A. Tassi, An efficient network coding scheme with symbol combining: performance evaluation, optimization, and comparisons, *IEEE Trans. Vehic. Technol.*, Vol. 62, No. 3, 2013, pp. 1267–1275.

[8]. T. Ho, R. Koetter, M. Medard, et al., The benefits of coding over routing in a randomized setting, in *Proceedings of the IEEE International Symposium on Information Theory (ISIT'03)*, 2003, p. 442.

[9]. H. Xi, X. Wang, Y. Zhao, H. Zhang, et al., A reliable broadcast transmission approach based on random linear network coding, in *Proceedings of the IEEE Vehicular Technology Conference (VTC Spring'12)*, 2012, pp. 1–5.

[10]. A. Eryilmaz, A. Ozdaglar, M. Medard, et al., On the delay and throughput gains of coding in unreliable networks, *IEEE Trans. Inform. Theor.*, Vol. 54, No. 12, 2008, pp. 5511–5524.

[11]. E. Magli, M. Wang, P. Frossard, Network coding meets multimedia: A review, *IEEE Trans. Multimedia*, Vol. 15, No. 5, 2013, pp. 1195–1212.

[12]. A. Tassi, I. Chatzigeorgiou, D. Vukobratovic, Resource allocation frameworks for network-coded layered multimedia multicast services, *IEEE J. Sel. Areas Commun.*, Vol. 33, No. 2, 2014, pp. 141–155.

[13]. Z. Bar-Yossef, Y. Birk, T. Jayram, T. Kol, Index coding with side information, *IEEE Trans. Inform. Theor.*, Vol. 57, 2011, pp. 1479–1494.

[14]. S.K. Hong, J.-M. Chung, Network-coding-based hybrid ARQ scheme for mobile relay networks, *IET Electr. Lett.*, Vol. 46, No. 7, 2010, pp. 539–541.

[15]. L. Lu, M. Xiao, M. Skoglund, et al., Efficient network coding for wireless broad- casting, in *Proceedings of the IEEE Wireless Communications and Networking Conference (WCNC'10)*, IEEE, 2010, pp. 1–6.

[16]. S. Sorour, S. Valaee, Completion delay minimization for instantly decodable network codes, *IEEE/ACM Trans. Netw.*, Vol. 23, 2015, pp. 1553–1567.

[17]. F. Wu, C. Hua, H. Shan, A. Huang, Reliable network coding for minimizing de- coding delay and feedback overhead in wireless broadcasting, in *Proceedings of the Annual IEEE International Symposium on Personal, Indoor and Mobile Radio Communications (PIMRC'12)*, 2012, pp. 796–801.

[18]. D. Nguyen, T. Tran, T. Nguyen, B. Bose, Wireless broadcast using network coding, *IEEE Trans. Vehic. Technol.*, Vol. 58, No. 2, 2009, pp. 914–925.

[19]. J. Li, Z. Hu, Y. Wang, Compressed multicast retransmission in LTE-a eMBMS, in *Proceedings of the IEEE Vehicular Technology Conference (VTC Spring'10)*, 2010, pp. 1–4.

[20]. J. Wang, Y. Xu, C. Wang, K. Xu, A limited feedback based network coding retransmission scheme for machine-to-machine wireless broadcasting, *Trans. Emerg. Telecommun. Technol.*, 2015.

[21]. Z. Zhang, T. Lv, X. Su, H. Gao, Dual XOR in the air: A network coding based retransmission scheme for wireless broadcasting, in *Proceedings of the IEEE International Conference on Communications (ICC'11)*, 2011, pp. 1–6.

[22]. K. Xu, W. Ma, L. Zhu, Y. Xu, et al., NTC-HARQ: Network-turbo-coding based HARQ protocol for wireless broadcasting system, *IEEE Trans. Vehic. Technol.*, Vol. 64, No. 10, 2015, pp. 4633–4644.

[23]. J. Hagenauer, Communications, Coding and Cryprology, *Springer*, 1994, pp. 155–171.

[24]. B. Ezio, Coding for Wireless Channels, *Springer*, 2005.

[25]. H. Shin, S. Kim, J. H. Lee, Turbo decoding in a Rayleigh fading channel with estimated channel state information, in *Proceedings of the 52nd IEEE Vehicular Technology Conference (VTC'2000)*, 2000, pp. 1358–1363.

Chapter 14
Analytical Model for Vehicular Mobility: A Microscopic Approach

Mani Zarei and Amir Masoud Rahmani

14.1. Introduction

Vehicular ad hoc networks (VANETs) are one of the most interesting commercial applications of mobile ad hoc networks (MANETs) that would allow different types of vehicles on roads to form a self-organized wireless network. The growth rate in the variety of vehicular possibilities and increased applications of VANETs has certainly attracted much attention from researchers, vehicular experts and creative industries. Vehicular communication systems are described by a dynamic environment but relatively predictable movements [1].

The IEEE 802.11 working group continues to actively develop 802.11p [2] for supporting Intelligent Transportation System (ITS) applications. The 802.11p standard will provide wireless devices with the ability to perform the short-duration exchanges necessary to communicate between a high-speed vehicle and a fixed roadside unit (RSU) [3]. This mode of operation, called WAVE (wireless access in vehicle environments) will operate in a 5.9 GHz band and support the Dedicated Short Range Communications (DSRC) standard [4]. Utilizing DSRC [5, 6] enables a wide variety of driver-assisting applications such as vehicle-to-vehicle (V2V) communication and vehicle-to-roadside (V2R) messaging of traffic and accident information. The mobility pattern of each vehicle is affected by many factors such as highway traffic stream, movements of adjacent vehicles, traffic signs, and driver's reactions to these factors [7–10]. Since vehicular mobility is constrained by road or highway topology, modeling the movability in a VANET is quite involved. Using DSRC each vehicle as a mobile host can communicate with one another directly if and only if their Euclidean distance is no longer than the radio propagation range [11]. Vehicular interactions over DSRC allow timely and intelligent communication to improve road safety and traffic stream [12–15].

Mani Zarei
Department of Computer Engineering, Shahr-e-Qods branch, Islamic Azad University, Tehran, Iran

343

The broadcast nature of the wireless medium can be advantageous to support multi-path capabilities that information-centric networking (ICN) can enable for transport of information in VANETs [16]. The development of radio-frequency identification (RFID), wireless sensor networks (WSNs), and global positioning system (GPS) as tracking technologies has triggered the development of cloud based services that are rapidly growing and continuously changing human life. Using fixed/mobile wireless communications such as cellular communications, machine-to-machine (M2M) communications, vehicle-to-everything (V2X) communications play vital role in future internet and maturity of internet of things (IoT) technology that is quickly increasing and continuously changing human life [17-23].The taxonomy of cloud-based vehicular networks is addressed from the standpoint of the service relationship between cloud computing and VANETs [24].It is extremely necessary to efficiently provide comprehensive study for future vehicular networks especially cloud-based V2X communications.

14.2. Communication Patterns

Vehicles, due to their no uniformity and different types have dissimilar mobility patterns and adopt disparate velocities. This in turn results in variations in the density of a traffic stream, and the strong possibility of frequent splitting or merging intervehicle connectivity [25, 26]. Differences in movability patterns and vehicular velocity in such a network cause frequent variations in traffic flow rates in every lane of a multi-lane highway. Because of the dynamic nature of vehicles, transforming from a densely connected environment in a VANET to a sparsely disconnected situation may occur in a short time interval [27]. As a result, the topology of VANETs varies regularly and vehicles come into temporary interaction with other surrounding vehicles traveling in similar or dissimilar lane s/levels in the same directions as well as opposite directions. These opportunistic contacts can be employed to aid message dissemination via multihop forwarding using intermediate vehicles which overhear the message, regardless of being in a single lane /level or multiple lane s/levels, in the same or opposite directions in highways [28]. Networked environments that operate under such discontinuous and alternative connectivity are also referred to as irregularly connected, delay tolerant, or disruption tolerant networks (DTNs) [29]. Therefore, such DTN networks essentially follow a store–carry–forward scheme, where a piece of message is cached or buffered in a node's memory when the inter-vehicular communication is unavailable.

14.3. Information Dissemination

Information propagation In VANETs is straightforwardly affected by many aspects such as highway situation, mobility pattern of neighboring vehicles, traffic stream condition, etc. In V2V scenarios in a single direction, a cycle of information propagation process starts with physical movement of an informed leading vehicle which carrying buffered information as a *catch-up* process and ends with multihop transmission through uninformed faced vehicles as a *forwarding* process. According to these two common alternating processes information propagation cyclically renews in VANETs [30].

14.4. Vehicular Mobility Models

In the context of vehicular networks, some of previous studies use the assumption of mesoscopic mobility in which vehicular velocities are the same at all times [26, 31, 32]. In general, vehicular movability models in the literature can be classified into three methods (i.e. macroscopic, mesoscopic, and microscopic) in accordance with the detail level of the communications among vehicles that the model characterizes [33-36]. The macroscopic level is a simplified analysis wherein the intervehicle distance is invariant in a highway that is categorized based on the middling intervehicle distance. Mesoscopic mobility model is utilized for steady state traffic stream that is used to model the steady-state mobility in which the communication links and intervehicle distances are well established for a dense situation and vehicles follow the traffic, or move independently at a maximum velocity in a sparse situation in a road or highway [36]. It is evident that, vehicles, due to their non-uniform and different types contain different mobility patterns and adopt different velocities (i.e. the microscopic mobility pattern). The vehicular mobility in unsteady traffic flow is affected by many factors such as highway situation, movement of surrounding vehicles, information on the messaging signs along the highway, road signs, traffic lights and driver's reactions to these factors [8, 36]. The main contributions in this chapter use the assumption of microscopic level by which vehicles may possibly utilize dynamic and unstable mobility patterns in different traffic streams.

14.5. Network Architecture

In this section, the VANET construction is explained at microscopic level. This classification considers a set of common vehicular mobility such as slowing down, speeding up, changing lane s and overtaking. As shown in Fig. 14.1, following the convention in prior VANET literature, the highway is interpreted as eastbound and westbound [26]. Utilizing DSRC this chapter focuses on analyzing the V2V communication in one direction of a multilane highway (e.g. eastbound). However, the proposed microscopic models could be applied to V2R communications as well as V2V communications in opposite directions on both eastbound and westbound.

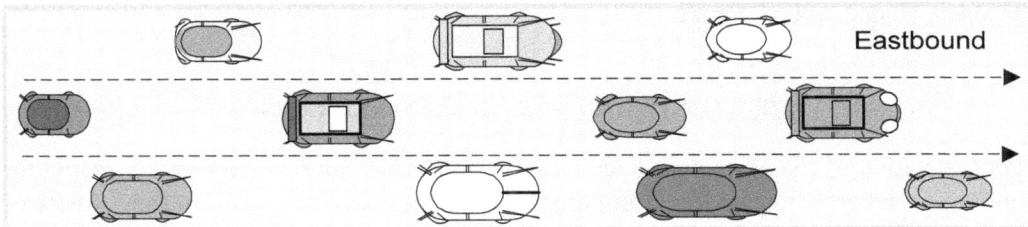

Fig. 14.1. VANET example. Vehicles following Poisson arrival travel in a single direction of a free-flow highway.

The vehicular distribution and arrival in each lane of a highway can be modelled as an independent Poisson process at different velocities. Considering a Poisson process, for $N(t) = n$ vehicles the following property holds:

$$p_r(N(t) = n) = \frac{e^{-\lambda t}(\lambda t)^n}{n!}.$$ (14.1)

In order to provide a dynamic vehicular mobility, a synchronized random walk mobility model is considered by which vehicles can change their speed at different times. It is considered that time is divided into equal intervals, and interval duration is referred to by τ, where the ith time slot is $t \in ((i-1)\tau, i\tau]$. The vehicle's velocity may differ at the beginning of each interval, independent of other vehicles as well as other intervals [25, 37]. Therefore, at any time interval, the spatial distribution of the vehicles on the road follows a homogeneous Poisson process with intensity $\rho = \lambda \int_{-\infty}^{\infty}(f_v(v)/v)dv$ in which $f_v(v)$ is a zero-mean normal speed distribution and λ is the Poisson intensity [veh/s]. The probability density function (pdf) of the vehicle speed, $f_v(v)$ [25, 37-41], utilizing the assumption of a zero-mean Normal speed distribution and standard deviation of σ is defined as

$$f_v(v) = \frac{1}{\sigma\sqrt{2\pi}}\exp\left(\frac{-v^2}{2\sigma^2}\right).$$ (14.2)

Let's consider that time is equally divide into i intervals of size τ such that $t = i\tau$. Inspired by [25], the velocity of each vehicle can vary at the start of each time slot independent of the other intervals as well as other vehicles. Let $p(x,\tau)$ be the probability that spatiotemporal displacement of a vehicle at time τ is equal to x. Because the speed is stable during a time interval, $p(x,\tau)$ can be easily obtained from $f_v(v)$. At the end of the first time slot, i.e. $t = \tau$, it is evident to show that [25]

$$p(x,\tau) = \frac{1}{\sigma\tau\sqrt{2\pi}}\exp\left(\frac{-x^2}{2(\sigma\tau)^2}\right).$$ (14.3)

In view of the fact that the vehicular velocities are modelled as independent random variables at different time intervals, their spatial movements are also independent random variables. Hence, at time $t = i\tau$:

$$p(x,t) = p(x,i\tau) = \overbrace{(p*p*\cdots*p*p)(x,\tau)}^{i-fold\ convolution}.$$ (14.4)

The calculation of the aforementioned i-fold convolution can be simplified by using the Fourier and inverse Fourier transformations. Note that convolution of i normal functions is a normal function [42],

$$p(x,i\tau) = \frac{1}{\sigma_i\sqrt{2\pi}}\exp\left(\frac{-x^2}{2\sigma_i^2}\right),$$ (14.5)

where $\sigma_i^2 = i(\sigma\tau)^2$.

14.6. Microscopic Analysis

This section shows the microscopic analysis of a dynamic VANET and completely explains the vehicular mobility aspects extracted from single direction of a multi-lane highway. In order to provide microscopic mobility perspective, this section considers a synchronized random walk model and mathematically investigates the mobility model of vehicles traveling in different lane s of a free-flow highway with a time-varying traffic flow assumption.

This section is organized as follows: First, the joint Poisson spatial distribution of vehicles in a multilane highway in a single direction is mathematically investigated. Next, the conditional probability of number of vehicles is shown for two-lane, three-lane and k-lane highways at microscopic level. Afterwards, the conditional expected number of vehicles is provided completely. Finally, the tail probability of number of vehicles is provided for different scenarios in a single direction.

14.6.1. Joint Poisson Spatial Distribution

This subsection provides a widespread spatiotemporal vehicular mobility analysis in which the time is equally slotted and the spatial distribution of the vehicles on a multilane highway follows a homogeneous joint Poisson process with intensity ρ [veh/m].

14.6.1.1. Two-Lane Highway

In this subsection microscopic mobility analysis for investigating the joint Poisson distribution in one direction of a VANET is provided. For the sake of simplicity, let's focus on vehicles passing eastbound of a highway including two lane s in which the spatial distribution of vehicles traveling in both lane s follows a homogeneous Poisson process with intensity ρ [veh/m]. Without loss of generality ρ_i is defined as vehicle density of lane i, in which $\sum_{i=1}^{2}\rho_i = \rho$. The probability that a vehicle travels in lane 1 or 2 is $P_1 = \rho_1/\rho_1 + \rho_2$ or $P_2 = 1 - P_1 = \rho_2/\rho_1 + \rho_2$, respectively, where $\sum_{i=1}^{2}P_i = 1$. Moreover, let $N_1(t) = n_1$ and $N_2(t) = n_2$ be the number of vehicles travelling in lane 1 and lane 2 at time t, respectively. Therefore, the number of vehicles passing eastbound is $N(t) = N_1(t) + N_2(t)$. Furthermore, $N_1(t)$ and $N_2(t)$ are independent Poisson variables with means $\rho_1\zeta(t)$ and $\rho_2\zeta(t)$, respectively. In accordance with the assumptions, the joint probability that exactly n_1 vehicles from lane 1 and n_2 vehicles from lane 2, at time t, travel in a spatial distribution with a length of $\zeta(t)$ and mean $\rho\zeta(t)$ is:

$$\Pr\{N_1(t) = n_1, N_2(t) = n_2\} = \binom{n_1 + n_2}{n_1} p_1^{n_1} p_2^{n_2} e^{-\rho\zeta(t)} \frac{(\rho\zeta(t))^{n_1+n_2}}{(n_1 + n_2)!}$$

$$= e^{-\rho_1 \zeta(t)} \frac{(\rho_1 \zeta(t))^{n_1}}{n_1!} \times e^{-\rho_2 \zeta(t)} \frac{(\rho_2 \zeta(t))^{n_2}}{n_2!}.$$

(14.6)

Eq. (14.6) can be studied completely using [43]. Eq. (14.6) is separated into two products; one of which depends only on n_1 and other only on n_2. It is clear that when each of a Poisson number of vehicles is independently classified as either belonging to lane 1 with probability $P_1 = \rho_1/\rho_1 + \rho_2$ or belonging to lane 2 with probability $P_2 = 1 - P_1 = \rho_2/\rho_1 + \rho_2$, the numbers of vehicles from lane 1 and lane 2 are independent Poisson random variables. Hence, the joint Poisson distribution of vehicles for a single direction of a two-lane highway is mathematically investigated by (14.6).

To check the validity and accuracy of the analytical results, using MATLAB (similar to previous research done in [1, 38, 43-46]), an interrupted 1-D highway is simulated. It should be noted that the execution of vehicular mobility in a free-flow highway through the simulation, can be isolated from complicated V2V interactions. Therefore, the simulation generates a free-flow vehicular traffic and assumes that, at any time slot, the spatial distribution of the vehicles on the road travelling in the same direction remains a homogeneous Poisson process with spatial intensity ρ veh/m. The simulation closely tracks the unidirectional highway in which n vehicles following Poisson arrival traveling in a multilane free-flow highway with temporal intensity λ veh/s. In the rest of this chapter similar to the previous publications, in order to configure the simulations the following assumptions are made: The average vehicle speed is $E[v] = 25$ m/s, the standard deviation is $\sigma = 7.5$ m/s [25], the time slot is $\tau = 5$ s [25], and the simulation time is $t = 2000$ s up to $t = 8000$ s. Simulation results depicted in the figures are obtained from averaging across 100 iterations.

To investigate the joint Poisson distribution let's simulate a two-lane highway in which 100 vehicles following Poisson process pass eastbound with density of $\rho = \rho_1 + \rho_2 = 0.004$ veh/m. Fig. 14.2 shows the simulation and analytical results of joint Poisson distribution in a spatial distribution with a length of $\zeta(t)$ (i.e. joint Poisson spatial distribution). As mentioned earlier, time is equally slotted and at the beginning of each time interval velocity variation is allowed. As a result, vehicles move faster/slower independent of other vehicles as well as other intervals. As depicted in Fig. 14.2, whenever the Poisson intensity of eastbound direction is equal to $\rho = \rho_1 + \rho_2 = 0.004$ veh/m, the expected intervehicle distance is $1/\rho = 250$ m. Furthermore, for $N = 100$ veh, results verify that the maximum probability belongs to the spatial distribution with a length of $\zeta(t) = 25000$ m. The results in Fig. 14.2 are depicted based on three choices of Poisson intensity for ρ_1 and ρ_2 consisting of $(\rho_1, \rho_2) = (0.002, 0.002)$, $(0.001, 0.003)$ and $(0.0005, 0.0035)$ in which the spatial distribution of the vehicles on the road for these three scenarios is equivalent and remains a homogeneous Poisson process with density $\rho = \rho_1 + \rho_2 = 0.004$ veh/m. The minimum joint Poisson distribution belongs to the scenario that both lane s have the same Poisson intensities. As the intensity difference of the traffic flows in two lane s increases, the difference in the number of vehicles in two lane s also increases; moreover, probability increases accordingly.

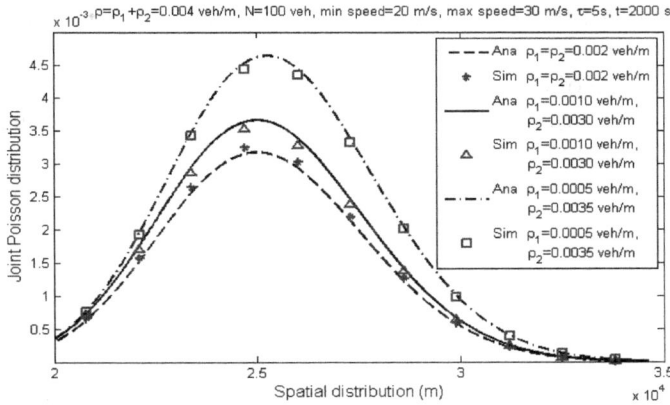

Fig. 14.2. Variations of joint Poisson distribution based on three different traffic flows where $\rho = \rho_1 + \rho_2 = 0.004$ veh/m, the analytical (Ana) and simulation (Sim) results are shown.

Fig. 14.3 shows the variations of spatial distribution, $\zeta(t)$ given that $N = 200$ vehicles following Poisson process travel in two dissimilar lane s in a single direction with Poisson intensity of $\rho_1 = \rho_2 = 0.001$, 0.0015, and 0.002 veh/m in three different scenarios. As illustrated in Fig. 14.3, if the vehicle intensity (i.e. $\rho = \rho_1 + \rho_2$) increases, the intervehicle distance and therefore the spatial distribution decreases. Considering the simulation time of $t = 8000$ s, and time varying vehicular speed assumptions of $v_{min} = 20$ m/s and $v_{max} = 30$ m/s the analytical results perfectly similar the simulation results.

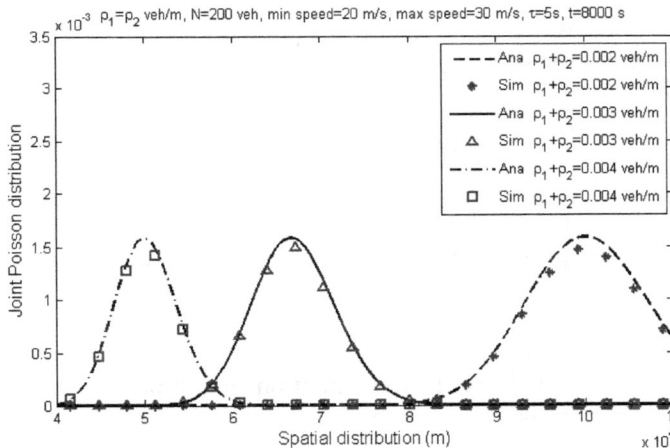

Fig. 14.3. Variations of vehicular spatial distribution based on variations in traffic flow.

In Fig. 14.4, in order to investigate the joint Poisson distribution of a free-flow highway, it is assumed that 100 vehicles following the Poisson process travel in two different lane s on equal direction with intensity of $(\rho_1, \rho_2) = (0.001, 0.002)$, $(0.001, 0.003)$ and

(0.001, 0.004) in three different test scenarios. As depicted in Fig. 14.4, when the vehicular density is $\rho = \rho_1 + \rho_2 = 0.003$ veh/m, the maximum probability belongs to the spatial distribution with length of $\zeta(t) = 33000$ m. But whenever the vehicular intensity increases (e.g. $\rho = 0.004$ veh/m), the maximum joint Poisson distribution also increases (i.e. depends to the spatial distribution with length of $\zeta(t) = 25000$ m). As expected, if the density increases, the vehicular spatial distribution clearly decreases.

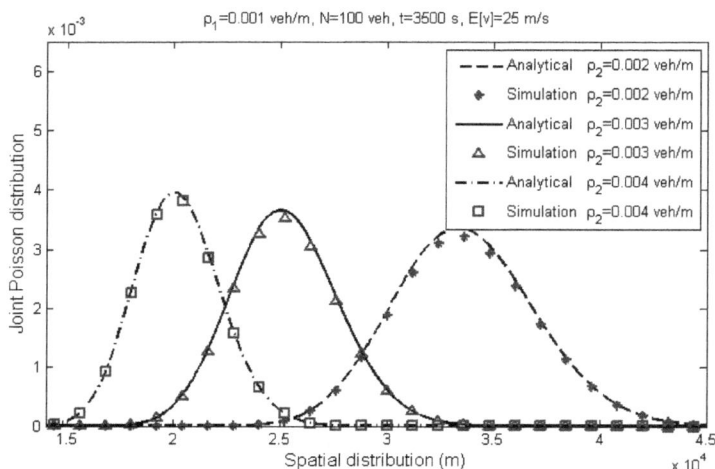

Fig. 14.4. Variations of joint Poisson distribution and spatial distribution based on three choices of ρ_2. Traffic flow in lane 1 is constant (i.e. $\rho_1 = 0.001$ veh/m) and vehicle size is $N = 100$ veh.

Fig. 14.4 confirmed that if the vehicle density of at least one lane varies, the joint Poisson distribution in a single direction also varies. Fig. 14.5, investigates that the impact of the number of vehicles on joint Poisson distribution in a highway wherein $N = 200$ vehicles following Poisson process travel in different lane s with Poisson intensity of $(\rho_1, \rho_2) = (0.0025, 0.002)$, $(0.0026, 0.002)$ and $(0.0027, 0.002)$ in three different replications. Results show that whenever the vehicle intensity of lane 1 increases, the number of vehicles traveling in this lane among $N = 200$ veh increases too. However, it is interesting to note that joint Poisson distribution decreases slightly. Fig. 14.5 illustrates that the predicted behavior from the analytical model closely tracks the simulation results.

In Fig. 14.6, unlike the assumption made in Fig. 14.5, the Poisson intensity of lane 1 is invariable (i.e. $\rho_1 = 0.001$ veh/m). In this situation, variations in the probability of the joint Poisson distribution is determined by changes in vehicle intensity of lane 2 (i.e. $\rho_2 = 0.00230$, 0.00235, and 0.00240 veh/m). It is straightforward to show that by increasing the traffic flow of lane 2 the expected number of vehicles from lane 1 among $N = 200$ veh, decreases. Moreover, the probability of joint Poisson distribution slightly decreases. This is because of the increasing dissimilarity between the traffic flow of lane 1 and lane 2. In this experiment, simulation time is set to $t = 2500$ s, minimum speed is set to 20 m/s, maximum speed is set to 30 m/s, time is equally slotted and speed variation is allowed at the beginning of each interval.

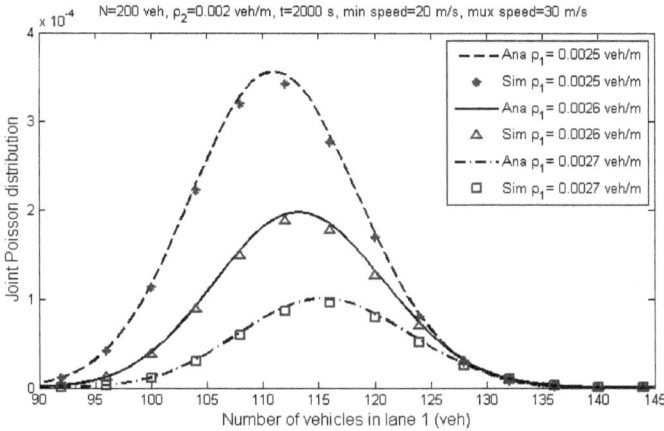

Fig. 14.5. Variations of joint Poisson distribution and number of vehicles inside lane 1 based on traffic flow of lane 1.

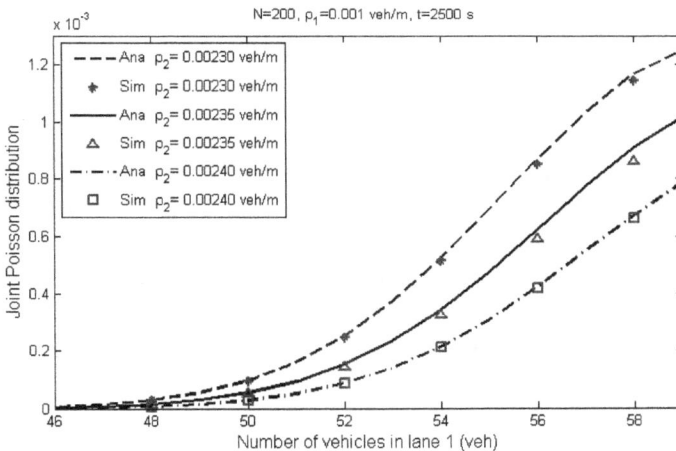

Fig. 14.6. Variations of joint Poisson distribution and number of vehicles in lane 1 based on traffic flow of lane 2.

Fig. 14.7 shows the impact of variation in number of vehicles on joint Poisson distribution in three different scenarios corresponding to the choices of $N = 200$, 210, and 220 veh. Vehicles following the Poisson process travel in the same direction inside two different lane s with equivalent density of $\rho_1 = \rho_2 = 0.002$ veh/m. It is obvious that by increasing the total number of vehicles from $N = 200$ veh to 220 veh, the expected number of vehicles travelling in these two lane s increases uniformly. As shown in Fig. 14.7, when the vehicle quantity of lane 1 is $N_1 = 100$ veh, the maximum probability is 0.0014, i.e. belongs to scenario that the total number of vehicles is $N = 200$ veh. In other words, whenever the vehicle intensity in both of lane 1 and 2 remains equal, the joint Poisson distribution confirms that the number of vehicles traveling in these two lane s is similar.

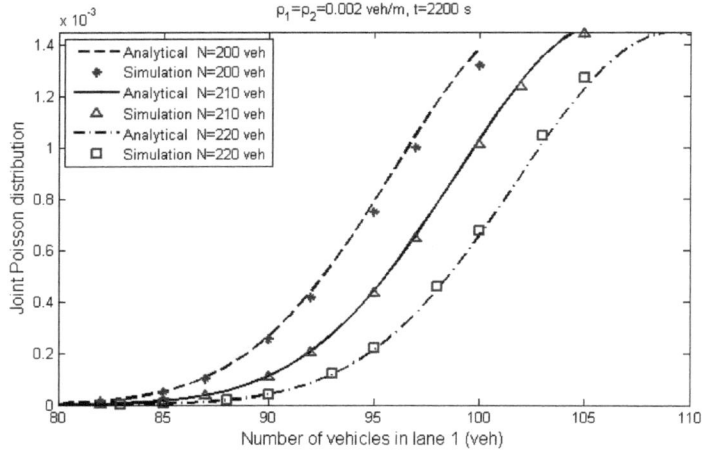

Fig. 14.7. Variations of joint Poisson distribution and number of vehicles inside lane 1 based on total number of vehicles.

14.6.1.2. K-Lane Highway

Let's consider a highway model including k lane s in each direction in which the spatial distribution of vehicles inside lane i ($i = 1, 2, ..., k$), follows a homogeneous Poisson process with intensity ρ_i [veh/m]. The result of (14.6) could be further generalized to include a highway consisting of k lane s. Inside lane i vehicles are Poisson distributed with mean $\rho_i \zeta(t)$, and the probability $P_i i = 1, ..., k$, $\sum_{i=1}^{k} P_i = 1$. If $N_i(t)$ is the number of vehicles of lane i, then $N_1(t), ..., N_k(t)$ are independent Poisson random variables with respective means $\rho_1 \zeta(t), ..., \rho_k \zeta(t)$. Therefore, the joint probability distribution of $n = \sum_{i=1}^{k} n_i$ vehicles in k lane s which are Poisson distributed in a Euclidean distance of $\zeta(t)$ and mean $\rho \zeta(t)$, is

$$\Pr\{N_1(t) = n_1, ..., N_k(t) = n_k\} = \frac{n!}{n_1! ... n_k!} p_1^{n_1} ... p_k^{n_k} e^{-\rho \zeta(t)} \frac{(\rho \zeta(t))^n}{n!}, \quad (14.7)$$

$$= \prod_{i=1}^{k} e^{-\rho_i \zeta(t)} \frac{(\rho_i \zeta(t))^{n_i}}{n_i!} \quad (14.8)$$

Fig. 14.8 captures the results for the conditional probability of number of vehicles inside lane 1. The spatial distribution of the vehicles inside lane 1 follows a heterogeneous Poisson process with density $\rho_1 = \lambda \int_{-\infty}^{\infty} (f_v(v)/v)dv = 0.007$ veh/m that increases to $\rho_1 = 0.013$ veh/m. However, the vehicle density of the two other lane s remain stable and equal (i.e. $\rho_2 = \rho_3 = 0.01$ veh/m). In this simulation 200 vehicles following Poisson arrival, travel in a single direction of a three-lane highway. The results are shown for three different number of vehicles in three different scenarios (i.e. $N_1 = 70$, $N_1 = 75$ and $N_1 = 80$ vehicles).

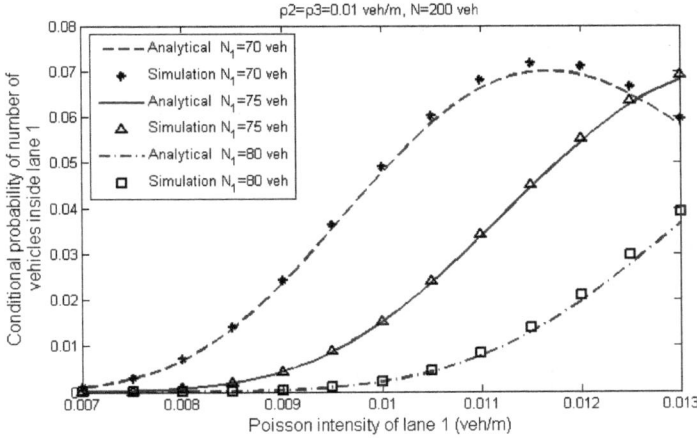

Fig. 14.8. Conditional probability of number of vehicles inside lane 1 based on variations in traffic intensity of this lane.

Fig. 14.9 shows joint probability of number of vehicles based on three choices of Poisson intensity for lane 1, lane 2 and lane 3 consisting of (ρ_1, ρ_2, ρ_3) =(0.002, 0.002, 0.002), (0.001, 0.002, 0.003) and (0.0005, 0.0020, 0.0035) in which the spatial distribution of the vehicles on the road for these three scenarios remains a homogeneous Poisson process with density $\rho = \rho_1 + \rho_2 + \rho_3 = 0.004$ veh/m. In this simulation 200 vehicles travel in a single direction of a free-flow highway. As depicted in Fig. 14.9, the analytical and simulation results of joint Poisson distribution are shown for variations in spatial distribution from a length of $\zeta(t) = 28000$ m up to $\zeta(t) = 33500$ m. It is clear that for a given $\zeta(t)$, if the difference between vehicle intensity of these three lane s increases the joint Poisson distribution also increases.

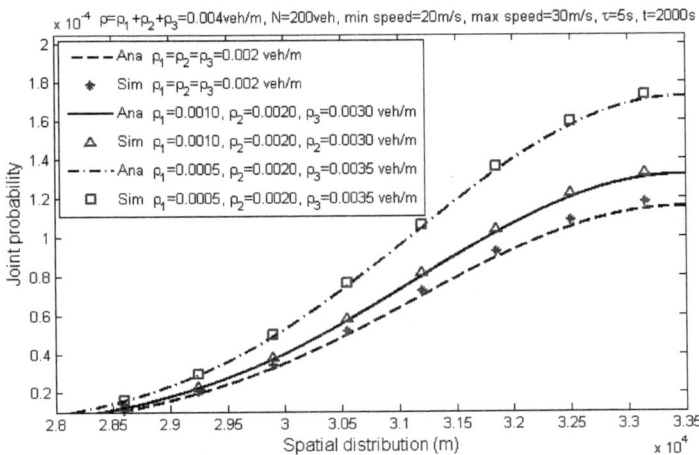

Fig. 14.9. Variations of joint probability based on three choices of traffic flows.

The simulation and analytical results conclude that the mobility pattern of the vehicles in different lane s follows a heterogeneous Poisson process with different densities. These differences in vehicular movability is mainly because of the differences in the speeds of vehicles (e.g. cars move fast but trucks and buses move slow) and/or difference in type of vehicles (e.g. bus and truck movements not allowed in lane 1). Using the determined Poisson intensity for a 1-D three-lane highway through at least 100 simulations and duration time of $t = 2000$ s, a minimum speed of 20 m/s and a maximum speed of 30 m/s, simulation results confirm the accuracy of the analytical model.

14.6.2. Conditional Probability of Number of Vehicles

14.6.2.1. Two-Lane Highway

Conditional probability of number of vehicles inside a lane in a two-lane highway is mathematically investigated at this subsection. Let's consider that the spatial distribution of vehicles traveling in part of a free-flow highway in a single direction with a length of $\zeta(t)$ follows a homogeneous Poisson process with an intensity of $\rho = \rho_1 + \rho_2$ [veh/m]. It is interesting to note that:

$$\Pr\{N_1(t) = l \mid N_1(t) + N_2(t) = n\} = \frac{\Pr\{N_1(t) = l\} \Pr\{N_2(t) = n - l\}}{\Pr\{N_1(t) + N_2(t) = n\}}, \qquad (14.9)$$

where (14.9) is the result of assuming $N_1(t)$ and $N_2(t)$ are independent. Recall that $N_1(t) + N_2(t)$ has a Poisson distribution with a mean $\rho_1 \zeta(t) + \rho_2 \zeta(t)$. Hence, (14.9) could be expressed as:

$$\Pr\{N_1(t) = l \mid N_1(t) + N_2(t) = n\} = \frac{n!}{(n-l)! l!} \frac{(\rho_1 \zeta(t))^l (\rho_2 \zeta(t))^{n-l}}{(\rho_1 \zeta(t) + \rho_2 \zeta(t))^n}$$

$$= \binom{n}{l} \left(\frac{\rho_1}{\rho_1 + \rho_2} \right)^l \left(\frac{\rho_2}{\rho_1 + \rho_2} \right)^{n-l}. \qquad (14.10)$$

14.6.2.2. Three-Lane Highway

In this section the conditional probability of number of vehicles inside a lane in a three-lane highway in a single direction is mathematically investigated. Let's consider that the spatial distribution of $N(t) = n$ vehicles traveling in three lane s follows a homogeneous Poisson process with intensity ρ [veh/m]. Without loss of generality let's define ρ_i as vehicle density of lane i, in which $\sum_{i=1}^{3} \rho_i = \rho$. The probability that a vehicle travels in lane 1, 2 and 3 are $P_1 = \rho_1/\rho$, $P_2 = \rho_2/\rho$ and $P_3 = 1 - P_1 - P_2 = \rho_3/\rho$, respectively, where $\sum_{i=1}^{3} P_i = 1$. Let $N_1(t)$, $N_2(t)$ and $N_3(t)$ be the number of vehicles travelling in lane 1, 2 and 3 at time t, respectively, in which $N_1(t) + N_2(t) + N_3(t) = n$.Therefore, the conditional probability of $N_1(t)$ given that $N_2(t) = m$ is,

$$\Pr\{N_1(t) = l \mid N_2(t) = m\} = \binom{n-m}{l}\left(\frac{P_1}{1-P_2}\right)^l\left(1-\frac{P_1}{1-P_2}\right)^{n-m-l} .\tag{14.11}$$

Proof, for $l \le n - m$,

$$\Pr\{N_1(t) = l \mid N_2(t) = m\} = \frac{\Pr\{N_1(t) = l, N_2(t) = m\}}{\Pr\{N_2(t) = m\}} .\tag{14.12}$$

If $N_1(t) = l$ and $N_2(t) = m$, then it follows that$N_3(t) = n - l - m$. However,

$$\Pr\{N_1(t) = l, N_2(t) = m, N_3(t) = n-l-m\} = \frac{n!}{l!m!(n-l-m)!}P_1^l P_2^m P_3^{(n-l-m)} .\tag{14.13}$$

This follows since any particular sequence of the $N(t) = n$ vehicles having outcome ρ_1 appearing l vehicles, outcome ρ_2 appearing m vehicles, and finally outcome ρ_3 appearing $(n - l - m)$ vehicles has probability $P_1^l P_2^m P_3^{(n-l-m)}$ of occurring. Since there are $n!/[l!\, m!\,(n-l-m)!]$ such sequences, Eq. (14.13) follows. Therefore,

$$\Pr\{N_1(t) = l \mid N_2(t) = m\} = \frac{n!}{l!m!(n-l-m)!}P_1^l P_2^m P_3^{(n-l-m)} \times \left(\frac{n!}{m!(n-m)!}P_2^m(1-P_2)^{n-m}\right)^{-1} .\tag{14.14}$$

We used the fact that $N_2(t) \ll \infty$ has a binomial distribution with parameters n and P_2. Therefore,

$$\Pr\{N_1(t) = l \mid N_2(t) = m\} = \frac{(n-m)!}{l!(n-m-l)!}\left(\frac{P_1}{1-P_2}\right)^l\left(\frac{P_3}{1-P_2}\right)^{n-m-l} .\tag{14.15}$$

Then, using$P_3 = 1 - P_1 - P_2$ in (14.15), Eq. (14.11) is readily proved.

14.6.3. Conditional Expected Number of Vehicles

14.6.3.1. Two-Lane Highway

The conditional distribution for expected number of vehicles traveling in lane 1, given that the $N_1(t) + N_2(t) = n$ vehicles travel in two lane s in a single direction, is a binomial distribution with parameters n and $\rho_1/\rho_1 + \rho_2$expressed as

$$E\{N_1(t) \mid N_1(t) + N_2(t) = n\} = n\frac{\rho_1}{\rho_1 + \rho_2} .\tag{14.16}$$

Fig. 14.10 shows conditional expected number of vehicles inside a lane based on Poisson intensity variations in a VANET in which 200 vehicles travel in a two-lane highway in a single direction. Let's consider the spatial distribution of the vehicles inside lane 1 follows a heterogeneous Poisson process with density $\rho_1 = 0.002$ veh/m that increases to $\rho_1 = 0.06$ veh/m. As depicted in Fig. 14.10, when the traffic flow of lane 1 increases, the intervehicle distance decreases and the expected number of vehicles in lane 1 among $N = 200$ veh increases accordingly. The analytical and simulation results are shown for two different traffic flows of lane 2 (i.e. $\rho_2 = 0.002$ veh/m and 0.004 veh/m). It is clear that if the vehicle intensity of lane 2 increases, the conditional expected number of vehicles in lane 1 among 200 vehicles decreases.

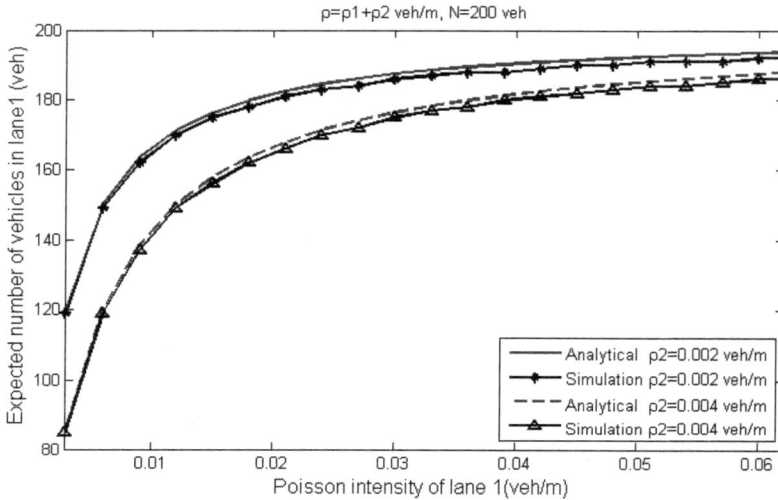

Fig. 14.10. Expected number of vehicles in lane 1 based on Poisson intensity variations.

14.6.3.2. Three-Lane Highway

The conditional distribution of $N_1(t)$, given that $N_2(t) = m$, is binomial with parameters $n - m$ and $P_1/(1 - P_1)$.

$$E\{N_1(t) \mid N_2(t) = m\} = (n - m)\frac{P_1}{1 - P_2}. \tag{14.17}$$

Fig. 14.11 illustrates the conditional expected number of vehicles inside lane 1. In this simulation, let's consider a time-varying vehicular intensity assumption for lane 2 in a single direction of a three-lane highway. In other words, the spatial distribution of the vehicles inside lane 2 follows a heterogeneous Poisson process with density $\rho_2 = 0.0$ veh/m that increases to $\rho_2 = 0.07$ veh/m. However, the Poisson intensity of the two other lane s remains invariant (i.e. $\rho_1 = 0.005$ veh/m and $\rho_3 = 0.015$ veh/m). The results in Fig. 14.11 are depicted based on three choices of number of vehicles consisting of $N = 200$ veh, $N = 250$ veh and $N = 300$ veh. Since the traffic flow of lane 1 remains invariant, it is evident that the conditional expected number of vehicles inside this lane for the three mentioned values decreases as the traffic flow of lane 2 increases.

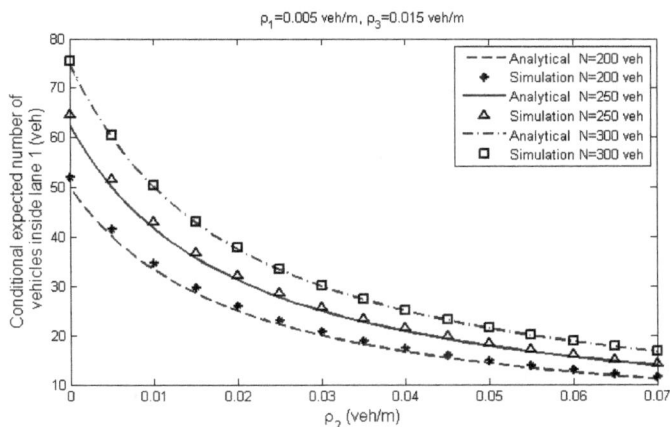

Fig. 14.11. Conditional expected number of vehicles in lane 1 based
on Poisson intensity of lane 2.

Fig. 14.12 shows the conditional expected number of vehicles inside lane 1 in a VANET based on Poisson intensity variations of this lane. In practice different lane s of a highway follow a heterogeneous mobility pattern. Accordingly, let's consider time-varying vehicular traffic intensity for lane 1 in a single direction of a three-lane highway. In order to investigate the conditional expected number of vehicles of lane 1, as shown in Fig. 14.12, let's consider three choices of number of vehicles (i.e. $N = 200$ veh, $N = 250$ veh and $N = 300$ veh) in three different scenarios. In this simulation, vehicle intensity of lane 2 and lane 3 remain invariant (i.e. $\rho_2 = 0.005$ veh/m and $\rho_2 = 0.015$ veh/m). It is evident that the conditional probability of number of vehicles traveling in lane 1 increases as the Poisson intensity of this lane increases. Moreover, for a given vehicle intensity, increasing the total number of vehicle increases the related conditional expected number of vehicles traveling inside lane 1.

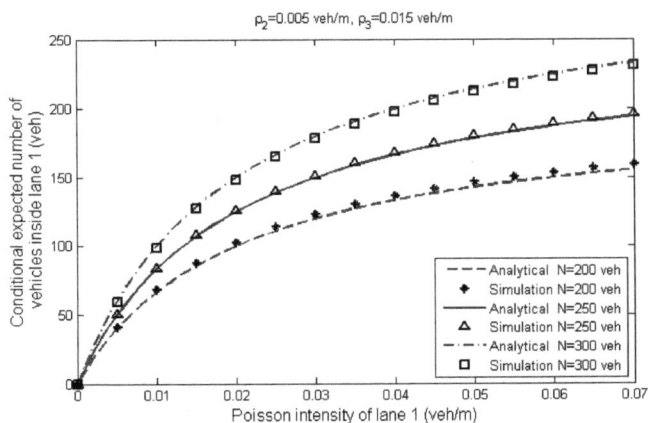

Fig. 14.12. Conditional expected number of vehicles inside lane 1 versus variations
in traffic intensity of this lane.

357

14.6.3.3. K-Lane Highway

Consider that vehicles traveling in a k lane highway with a size of $N_1(t) + \cdots + N_k(t) = n$. Using (14.16), the conditional expected number of vehicles in each lane has a multinomial distribution with parameters $(n, p_1, \ldots, p_k)$. Consequently, the conditional expected size of lane 1 in a k lane highway could be expressed as:

$$E\{N_1(t) \mid N_1(t) + \ldots + N_k(t) = n\} = np_1 = n \frac{\rho_1}{\rho_1 + \ldots + \rho_k}. \tag{14.18}$$

14.6.4. Tail Probability of Number of Vehicles

14.6.4.1. Two-Lane Highway

The tail probability of the number of vehicles inside lane 1, $N_1(t)$ (i.e., the probability that at least l vehicles belong to lane 1), in the condition that $N_1(t) + N_2(t) = n$ vehicles travel in two lane s in a single direction, is given by:

$$\Pr\left(N_1(t) > l \mid N_1(t) + N_2(t) = n\right) = 1 - \sum_{j=0}^{l} \binom{n}{j} \left(\frac{\rho_1}{\rho_1 + \rho_2}\right)^j \left(\frac{\rho_2}{\rho_1 + \rho_2}\right)^{n-j}. \tag{14.19}$$

Using an invariable traffic flow of $\rho = \rho_1 + \rho_2 = 0.06$ veh/m, the simulation and analytical results for the tail probability of the expected number of vehicles of lane 1 for a two-lane highway is shown in Fig. 14.13. In this scenario the vehicle intensity of lane 1 is $\rho_1 = 0.004$ veh/m, 0.006 veh/m and 0.008 veh/m. It is evident that the tail probability of three mentioned values decreases as the number of vehicles in lane 1 increases. Moreover, for a given vehicle count, increasing the Poisson intensity increases the related tail probability.

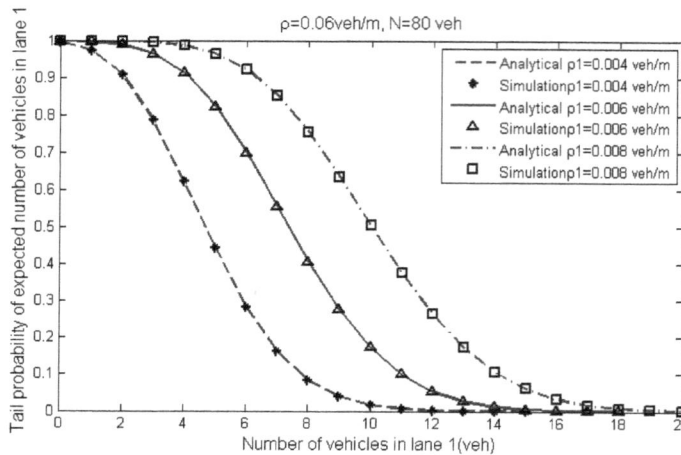

Fig. 14.13. Tail probability of the expected number of vehicles in lane 1.

14.6.4.2. Three-Lane Highway

The tail probability of number of vehicles inside lane 1, $N_1(t)$, in the condition that $N_2(t) = m$, $(N_1(t) + N_2(t) + N_3(t) = n)$ is given by:

$$\Pr\{N_1(t) > l \mid N_2(t) = m\} = 1 - \sum_{j=0}^{l} \binom{n-m}{j} \left(\frac{P_1}{1-P_2}\right)^{j} \left(1 - \frac{P_1}{1-P_2}\right)^{n-m-j} . \quad (14.20)$$

In Fig. 14.14, in order to investigate the tail probability of number of vehicles inside lane 1 in a single direction of a free-flow highway, let's considered three different scenarios in which the total number of vehicles is defined as $N = 135$ veh, $N = 180$ veh, and $N = 200$ veh. In each scenario, vehicles travel in a three-lane highway in a single direction in which the traffic flow of each lane remains invariant (i.e. $\rho_1 = 0.00025$ veh/m, $\rho_2 = 0.00125$ veh/m and $\rho_3 = 0.0013$ veh/m). In this simulation the number of vehicles in lane 2 varies from $N_2 = 0$ veh to $N_2 = 90$ veh. As shown in Fig. 14.14, it is evident that for a constant number of vehicles inside lane 2, if the total number of vehicles increases (e.g. from $N = 135$ veh, to $N = 180$ or $N = 200$), the tail probability of number vehicles inside lane 1, increases accordingly. Fig. 14.14 illustrates that the predicted behavior from the analytical model closely tracks the simulation results.

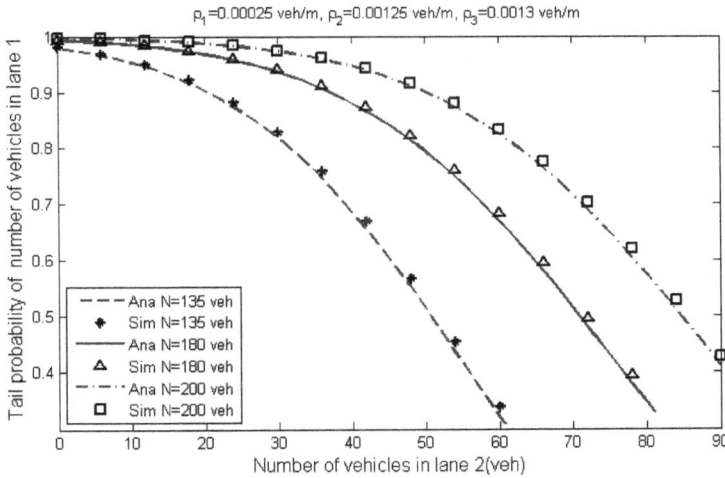

Fig. 14.14. Tail probability of number of vehicles inside lane 1 versus number of vehicles in lane 2.

14.7. Conclusions and Future Work

This chapter using the time-varying vehicular speed assumption proposed a comprehensive microscopic mobility model in VANETs. In suggested formulations, a single direction of a free-flow multilane highway was considered in which n vehicles were Poisson distributed and vehicular speeds followed a generic pdf $f_v(v)$. The proposed model used a time-varying velocity assumption and considered that time was equally divided into i intervals of size τ such that $t = i\tau$, by which allowed vehicles to change their speed at the start of a time slot independent of other time intervals as well as other vehicles. According to this the joint Poisson spatial distribution of vehicles in a multilane highway, the conditional probability of number of vehicles, the conditional expected number of vehicles and also the tail probability of number of vehicles inside a lane in a multilane highway were mathematically investigated. The accuracy of the proposed analytical models was validated through extensive simulation results. The proposed microscopic model developed in this study can be easily extended to determine the mobility model and communication aspects between vehicles traveling in opposite directions of a free-flow highway. This fundamental approach can provide a novel insight for further development of VANETs. Much more helpful would be a comparison with actual traffic data in order to validate the proposed model. Therefore, the future research includes extending simulation of the proposed model using a hybrid simulation framework Veins (Vehicles in Network Simulation) [47], i.e. composed of the network simulator OMNet++ [48] and the road traffic simulator SUMO [49].

References

[1]. M. Zarei, A. M. Rahmani, Analysis of vehicular mobility in a dynamic free-flow highway, *Vehicular Communications*, Vol. 7, 2016, pp. 51–57.

[2]. L. Armstrong, W. Fisher, Status of ProjectIEEE 802.11, *Task Group p: Wireless Access in Vehicular Environments (WAVE)*,May 2010.

[3]. K. Lim, D.Manivannan, An efficient protocol for authenticated and secure message delivery in vehicular ad hoc networks, *Vehicular Communications*, Vol. 4, 2016, pp. 30–37.

[4]. NHTS Administration, Vehicle Safety Communications Project – Final Report, Tech. Rep., *U.S. Department of Transportation*, Apr. 2006.

[5]. R. Florin, S. Olariu, A survey of vehicular communications for traffic signal optimization, *Vehicular Communications,* Vol. 2, 2015, pp. 70–79.

[6]. Y. Allouche, M. Segal, Cluster-based beaconing process for VANET, *Vehicular Communications*, Vol. 2, 2015, pp. 80–94.

[7]. J. Jakubiak, Y. Koucheryavy, State of the art and research challenges for VANETs, in *Proceedings of the 5th IEEE Consumer Communications and Networking Conference (CCNC'08)*, 2008, pp. 912–916.

[8]. J. Blum, A. Eskandarian, L. Hoffman, Challenges of intervehicle ad hoc networks, *IEEE Transactions on Intelligent Transportation Systems*, Vol.5, No. 4, 2004, pp. 347–351.

[9]. F. Li, Y. Wang, Routing in vehicular ad hoc networks: A survey, *IEEE Vehicular Technology Magazine*, Vol. 2, No. 2, 2007, pp. 12–22.

[10]. Y. Toor, P. Muhlethaler, A. Laouiti, Vehicle ad hoc networks: Applications and related technical issues, *IEEE Communications Surveys & Tutorials*, Vol. 10, No. 3, 2008, pp. 74–88.

[11]. Y.S. Yena, H.C. Chaob, R.S. Changd, A.V. Vasilakos, Flooding-limited and multi-constrained QoS multicast routing based on the genetic algorithm for MANETs , *Mathematical and Computer Modeling*, Vol. 53, No. 11-12, 2011, pp.2238–2250.

[12]. S. Zeadally, R. Hunt, Y.S. Chen, A. Irwin, A. Hassan, Vehicular ad hoc networks (VANETS): status, resultsandchallenges, *Telecommunication Systems*, Vol. 50, No. 4, 2012, pp. 217–241.

[13]. Z. Ahmed, H. Jamal, S. Khan, R. Mehboob, A. Ashraf, Cognitive communication device for vehicular networking, *IEEE Transactions on Consumer Electronics*, Vol. 55, No. 2, 2009, pp. 371–375.

[14]. H. Vahdat-Nejada, A. Ramazania, T. Mohammadib, W. Mansoor, A survey on context-aware vehicular network applications, *Vehicular Communications*, Vol. 3, 2016. pp. 43–57.

[15]. N. Jabera, W.G. Cassidy K. E. Tepeb, E.Abdel-Raheemb, Passivecooperative collision warning (PCCW) MAC designs for reliable vehicular safety messaging, *Vehicular Communications*, Vol. 2, No. 2015, pp. 95–109.

[16]. P. TalebiFard, V.C.M. Leung, M. Amadeo, C. Campolo, A. Molinaro, Chapter Vehicular ad hoc Networks, in Information-Centric Networking for VANETs, *Springer*, Switzerland, 2015, pp. 503–524.

[17]. M. Yang, Y. Li, D. Jin, L. Zeng, X. Wu, A.V. Vasilakos, Software-defined and virtualized future mobile and wireless networks: A survey, *ACM/Springer Mobile Networks and Applications*, Vol. 20, No. 1, 2015, pp. 4–18.

[18]. M. Zarei, A.M. Rahmani, R. Farazkish, CCTF: congestion control protocol based on trustworthiness of nodes in wireless sensor networks using fuzzy logic, *International Journal of Ad Hoc and Ubiquitous Computing*, Vol. 8, No. 1-2, 2011, pp. 54–63.

[19]. M. Zarei, A.M. Rahmani, R. Farazkish, S. Zahirnia, FCCTF: fairness congestion control for a distrustful wireless sensor network using fuzzy logic, in *Proceedings of the IEEE 10th International Conference on Hybrid Intelligent Systems (HIS'10)*, 2010, pp. 1–6.

[20]. M. Zarei, A.M. Rahmani, A. Sasan, M. Teshnehlab, Fuzzy based trust estimation for congestion control in wireless sensor networks, in *Proceedings of the IEEE International Conference on Intelligent Networking and Collaborative Systems (INCoS'09)*, 2009, pp. 233–236.

[21]. H. Zhou, The Internet of Things in the Cloud: A Middleware Perspective, 1st Ed., *CRC Press*, 2012.

[22]. C. Perera, A. B. Zaslavasky, P. Christen, D. Georgakopoulos, Contextaware computing for the Internet of Things: A survey, *IEEE Communications Surveys & Tutorials,* Vol. 16, No. 1, May 2013, pp. 414–454.

[23]. M. A. Razzaque, M. Milojevic-Jevric, A. Palade, S. Clarke, Middleware for Internet of Things: A survey, *IEEE Internet of Things Journal*, Vol. 3, No. 1, Feb. 2016, pp. 70–95.

[24]. M.K. Jiau, S.C. Huang, J.N. Hwang, A.V. Vasilakos, Multimedia services in cloud-based vehicular networks, *IEEE Intelligent Transportation Systems Magazine*, Vol. 7, No. 3, 2015, pp. 62–79.

[25]. Z. Zhang, G. Mao, B.D.O. Anderson, On the information propagation process in mobile vehicular ad hoc networks, *IEEE Transactions on Vehicular Technology*, Vol. 60, No. 5, 2011, pp. 2314–2325.

[26]. A. Agarwal, D. Starobinski, T.D.C. Little, Phase transition of message propagation speed in delay-tolerant vehicular networks, *IEEE Transactions on Intelligent Transportation Systems*, Vol. 13, No. 1, 2012, pp. 249–263.

[27]. N.H. Khiadani, S.M. Safavi Hemami, F. Hendessi, Analysis of acceleration effect in data dissemination in vehicular networks using rateless codes, *Wireless Personal Communications*, Vol. 77, No. 2, 2014, pp. 991–1006.

[28]. T. Meng, F. Wu, Z. Yang, G. Chen, A.V. Vasilakos, Spatial reusability-aware routing in multi-hop wireless networks, *IEEE Transactions on Computers*, Vol. 65, No. 1, 2015, pp. 244–255.

[29]. T. Spyropoulos, R.N.B. Rais, T. Turletti, K. Obraczka, A.V. Vasilakos, Routing for disruption tolerant networks: taxonomy and design, *Wireless Networks,* Vol. 16, No. 8, 2010, pp. 2349–2370.

[30]. M. Zarei, A. M. Rahmani, Renewal process of information propagation in delay tolerant VANETs ,*Wireless Personal Communications*, Vol. 89, No. 4, 2016, pp. 1045–1063.

[31]. E. Baccelli, P. Jacquet, B. Mans, G. Rodolakis, Highway vehicular delay tolerant networks: information propagation speed properties, *IEEE Transactions on Information Theory*, Vol. 58, No. 3, 2012, pp. 1743–1756.

[32]. R. Nagel, The effect of vehicular distance distributions andmobility on VANET communications,in *Proceedings of the IEEE Intelligent Vehicles Symposium (IV'10)*, San Diego, CA,USA, 2010, pp.1190–1194.

[33]. A. May, Traffic Flow Fundamentals, *Prentice-Hall*, NJ, USA, 1990.

[34]. M. Krbálek, K. Kittanova, Theoretical predictions for vehicular headways and their clusters, *Journal of Physics A: Mathematical and Theoretical*, Vol. 46, No. 44, 2013.

[35]. L. Li, W. Fa, J. Rui, H. Jian-Ming, J. Yan, A new car-following model yielding log-normal type headways distributions, *Chinese Physics B*, Vol. 19, No. 2, 2010, 020513.

[36]. K. Abboud, W. Zhuang, Stochastic analysis of a single-hop communication link in vehicular ad hoc networks, *IEEE Transactions on Intelligent Transportation Systems*, Vol. 15, No. 5, Oct. 2014, pp. 2297–2307.

[37]. Z. Zhang, G. Mao, B.D.O. Anderson, Stochastic characterization of information propagation process in vehicular ad hoc networks, *IEEE Transactions on Intelligent Transportation Systems*, Vol. 15, No. 1, 2014, pp. 122–135.

[38]. S. Yousefi, E. Altman, R. El-Azouzi, M. Fathy, Analytical model for connectivity in vehicular ad hoc networks, *IEEE Transactions on Vehicular Technology*, Vol. 57, No. 6, 2008, pp. 3341–3356.

[39]. W. Leutzbach, Introduction to the Theory of Traffic Flow, *Springer-Verlag*, NewYork, 1998.

[40]. M. Rudack, M. Meincke, M. Lott, On the dynamics of ad hoc networks for inter vehicle communications (IVC), in *Proceedings of the International Conference on Wireless Networks (ICWN'02)*, Las Vegas, NV, 2002.

[41]. P. P. Roess, E. S. Prassas, W. R. McShane, Traffic Engineering, *Pearson/Prentice Hall*, New Jersey, 2004.

[42]. S. M. Ross, Introduction to Probability Models, 10th Ed., *Elsevier*, New York, 2011.

[43]. M. Zarei, A. M. Rahmani, H. Samimi, Connectivity analysis for dynamic movement of vehicular ad hoc networks, *Wireless Networks*, Vol. 23, No. 3, 2016, pp. 843–858.

[44]. Y. Liu, N. Xiong, Y. Zhao, A. V. Vasilakos, J. Gao, Y. Jia, Multi-layer clustering routing algorithm for wireless vehicular sensor networks, *IET Communications*, Vol. 4, No. 7, 2010, pp. 810–816.

[45]. K. Abboud, W. Zhuang, Impact of microscopic vehicle mobility on cluster-based routing overhead in VANETs , *IEEE Transactions on Vehicular Technology*, Vol. 64, No. 12, 2015, pp. 5493–5502.

[46]. K. Abboud, W. Zhuang, Stochastic modeling of single-hop cluster stability in vehicular ad hoc networks, *IEEE Transactions on Vehicular Technology*, Vol. 65, No. 1, 2016, pp. 226–240.

[47]. C. Sommer, R. German F. Dressler, Bidirectionally coupled network and road traffic simulation for improved IVC analysis, *IEEE Transactions on Mobile Computing*, Vol. 10, No. 1, Jan. 2011, pp. 3–15.

[48]. A. Varga, OMNet++ discrete event simulation system user manual, 2011, http://citeseerx.ist.psu.edu/viewdoc/download?doi=10.1.1.111.6503&rep=rep1&type=pdf

[49]. M. Behrisch, L. Bieker, J. Erdmann, D. Krajzewicz, SUMO -simulation of urban mobility: An overview, in *Proceedings of the 3rd International Conference on Advances in System Simulation(SIMUL'11)*, 2011, pp. 63–68.

Chapter 15
USRP-based Implementations of Various Scenarios for Spectrum Sensing

Zdravka Tchobanova, Galia Marinova and Amor Nafkha

15.1. Introduction

Radio Frequency Spectrum is a resource that is used irrationally. Cognitive radio is a technology built on a software-defined radio (SDR) that allows a more efficient use of the spectrum through dynamic access. It uses the methodology of environmental observation and study and autonomous adaptation to real-time parameter changes. The network or the wireless node modify their transmission or reception parameters to communicate effectively anywhere and at any time, avoiding the interference of unlicensed users in licensed users' broadcasts for efficient use of the radio spectrum. Cognitive radio has two main features that distinguish it from a traditional radio: cognition and reconfiguration. Fig. 15.1 illustrates how the cognitive radio characteristics interact with the surrounding radio environment.

The concept was first presented by Dr. Joseph Mitola [1] and the result of this concept is the IEEE 802.22 standard, which regulates the Cognitive Radio using for Wireless Regional Area Network (WRAN), using white spaces in the frequency television spectrum, ensures that no interference is caused to existing licensed devices.

Users with access, called primary users (PU), do not use the spectrum permanently. Users without access privileges called secondary users (SU), could use the spectrum when it's not used by PU, without impeding PU access. Therefore, it is very important for every user to be able to recognize licensed users and can find spectral holes, so called white spaces, to be able realizes its own transmission.

This chapter discusses various methods for spectrum sensing in the context of cognitive radio (CR). An overview of the theoretical foundations of the presented detectors has been made. Simulation results in GNU Radio are introduced. Various types of USRP implementations are developed and presented.

Zdravka Tchobanova
Faculty of Telecommunications, Technical University Sofia, Bulgaria

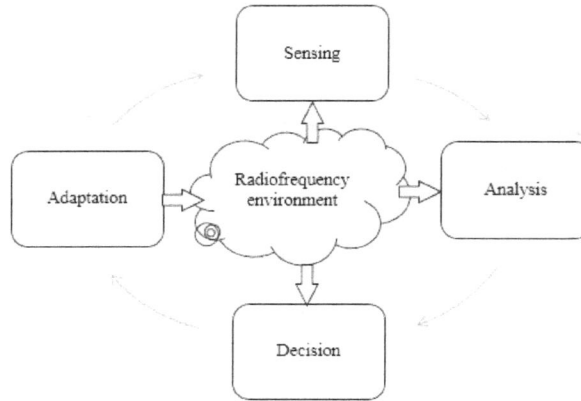

Fig. 15.1. Cognitive cycle.

15.2. Theoretical Aspects

One of the important issues related to the use of the spectrum is spectrum detection, i.e. the spectrum sensing: is there a signal of the channel or not. There are different detection techniques, some of them require prior knowledge of the state of the environment, i.e. signal to noise ratio (SNR). These are energy detection, cyclostationary feature detection, matched filter detection, compressed sensing based detectors, eigenvalue based detectors and others. The most popular spectrum sensing technique is detection of the energy level in the tested channel.

15.2.1. Energy Detection

Energy detection technique is easy to implement and does not lead to complex mathematical calculations [2]. The receiver does not need a prior information about the primary user signal parameters, but the value of the noise power in the channel must be known. This makes it possible to detect very weak signals, even with negative SNR regimes, since the detector can detect a very small power increase due to the sum of power noise and the signal. The energy detection includes applying a threshold of collected energy from the channel. The threshold is used to decide whether the primary user is present or not.

15.2.1.1. System Model

Depending on the busy or idle state of the primary user, with the presence of the noise, the signal detection problem at the secondary user can be modelled as a binary hypothesis testing problem:

H_0: - signal is absent, H_1: - signal is present

The PU signal, notated s, is a complex signal and it has a real and imaginary component: $s = s_r + js_i$. If the received signal y is discrete $n = 1, 2,.., y(n)$, it is given as:

$$y(n) = \begin{cases} w(n) : H_0 \\ x(n) + w(n) : H_1 \end{cases}, \tag{15.1}$$

where $x(n) = h. s(n)$, and h is the channel gain is the complex value.

$w(n) = w_r(n) + jw_i(n)$ are noise samples. The noise is assumed as Gaussian random variable with a mean value $E[w(n)] = 0$ and variance $2\sigma_w^2$. The channel gain $h = hr + jh_i$ is a constant value for all detection time.

The transmitted signal can be described as:

$$y(n) = \theta. x(n) + w(n), \tag{15.2}$$

where $\theta = 0$ for H_0 and $\theta = 1$ for H_1, when allowing perfect synchronization between the transmitter and the receiver. When there is heavy traffic in a multi-user network in which the PU signals arrive at SU with incoming misses of duration n0, then the signal is modelled as:

$$H_1: y(n) = \begin{cases} w(n): 1 \leq n \leq n_0 - 1 \\ x(n) + w(n): n_0 \leq n \leq N, \end{cases} \tag{15.3}$$

where N is the length of the observed interval corresponding to the total number of samples.

15.2.1.2. Conventional Energy Detector

The conventional energy detector is based on the idea that with the presence of a signal in the channel, there will be significantly more energy than if there is no signal [3]. This concept is applicable without knowledge on the PU signal characteristics. The detection of energy involves the application of a suitably selected threshold of collected energy from the channel. The threshold is used to decide whether the primary user is present or not. Therefore, the ED block diagram shown in Fig. 15.2 shows the simple detection mechanism.

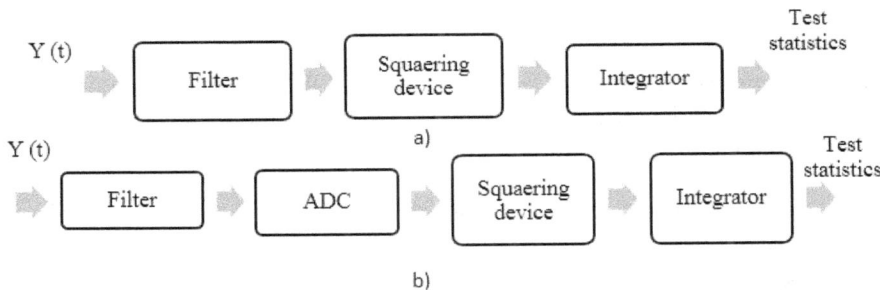

Fig. 15.2. Energy Detector Block Diagram: a) Analog, and b) Digital.

For the theoretical analysis, two models of the conventional energy detector were considered as a realization in the time domain:

- Analogue energy detector. It also contains a pre-filtration block that limits noise and normalizes the noise variance. The output of the integrator is proportional to the power of the transmitted signal.

- Digital energy detector. It consists of a low-pass filter that limits noise and signals from adjacent frequency bands; an analogue-to-digital converter (ADC) that converts continuous signals into discrete ones; A block that squaring the input signal value and finally an integrator.

15.2.1.2.1. Test Statistics

Integrator output is called Test statistics, and it is compared with a threshold, to take final decision: is there in the channel PU signal, or not.

In the digital realization of an energy detector, the test statistics is given by:

$$\Lambda = \sum_{n=1}^{N} |y(n)|^2 = \sum_{n=1}^{N} (e_r(n)^2 + e_i(n)^2, \tag{15.4}$$

where:

$$e_r(n) = \theta h_r s_r(n) - \theta h_i s_i(n) + w_r(n), \tag{15.5}$$

$$e_i(n) = \theta h_r s_i(n) - \theta h_i s_r(n) + w_i(n). \tag{15.6}$$

The test statistic is equivalent to

$$\Lambda = \sum_{k=1}^{N} |Y(k)|^2, \tag{15.7}$$

where Y (k) is the representation of the y (n) signal in the frequency domain. Signal representation in frequency domain is important for spectrum observation through an energy detector on an OFDM system.

The analogue detector test statistics is given by:

$$\Lambda = \frac{1}{T} \int_{t-T}^{t} y(t)^2 dt, \tag{15.8}$$

where T is the time. The bandwidth sampling function W and the duration of time T can be described by approximately a set of samples $N \approx 2TW$, where TW is a product of time and bandwidth. Therefore, the analog test statistics can be described like the digital detector test statistics.

The exact form of test statistics may vary for different applications. For example, in heavy network traffic with many users, test statistics can be determined using hypothesis H_1 as:

$$\Lambda = \sum_{n=1}^{n_0-1} |y(n)|^2 + \sum_{n=n_0}^{N} |y(n)|^2, \tag{15.9}$$

where there is only a noise signal in the range [1, n0 − 1].

In addition, for optimization parameter analysis or noise evaluation error, L is normally normalized with respect to the number of discrete N and the noise variance $2\sigma_w^2$

$$\Lambda = \frac{1}{2\sigma_w^2 N} \sum_{n=1}^{N} |y(n)|^2. \tag{15.10}$$

The energy detector is characterized by metrics based on the test statistics in the binary hypothesis:

Detection probability (Pd): This is the probability that there will be a signal in the channel when the hypothesis H1 is true: Pd = Pr [$\Lambda > \lambda | H_1$]. Pr is probability for event, λ is the threshold for comparison, Λ is the tested statistics.

False alarm probability (Pfa): The probability that the signal is present, when the hypothesis H0 is true: Pfa = Pr [$\Lambda > \lambda | H0$]. From the point of view of detection, these are the undiscovered holes in the spectrum - missed opportunity.

Probability of missed-detection Pmd: The probability that a signal will not be available when H1 is true Pmd = Pr [$\Lambda < \lambda | H1$] = 1 - Pd. If there is a signal in the channel, but detector indicates that it is free. These are messages for free channel, when it is busy.

In terms of reliability and efficiency, Spectrum sensing techniques are expected to have higher Pd (lower Pmd) and lower Pfa.

Based on Λ 's available knowledge, the receiver can adopt a suitable model that helps analyze the distribution of test statistics below H_1.

15.2.1.2.2. Energy Detector Parameters

The main energy detector parameters are the number of samples N and the threshold for comparison.

Choosing the number of samples, it is important parameter to meet the requirements of the probabilities of detection and false alarm. For given Pf and Pd, the minimum number of samples is given as a function of the signal-to-noise ratio SNR

$$N = [Q^{-1}(Pfa) - Q^{-1}(Pd)\sqrt{2\gamma + 1}]^2 \gamma^{-2}, \tag{15.11}$$

where their number is not a function of the threshold. Since the Q-1 function is monotonically decreasing, the signal can be detected even in the very low SNR range by increasing the N when the power noise is fully known. The approximate number of samples required to achieve the probability of false alarm and probability detection is in the order of $\theta(\gamma^{-2})$ i.e. the energy detector requires more discretion at very low levels of SNR. Since $N \approx \tau f_s$, where τ is the observation time and fs is the sampling rate, the observation time increases when N increases. This is a major disadvantage in monitoring the low-frequency SNR spectrum due to the limitation of the maximum allowable observation time (since IEEE 802.22 specifies that the observation time should be less than 2 seconds). Hence, the choice of N represents an optimization task.

Although the energy detector operation depends on the SNR and the noise variance σ_w^2, there is very limited control over them because these parameters depend on the behavior of the wireless channel.

Choosing a suitable threshold is particularly important for the good performance of the energy detector. If the threshold value is high, SU may find free space in the spectrum when it does not exist and may interfere with PU transmission. If the value is low, the detector can react and detect the presence of a signal in the channel without it, and omit the possibility of using the spectrum.

The threshold depends on Pd, Pmd and Pfa. If the threshold value increases (decreases), then Pd and Pfa decrease (increase). When the number of samples and noise variance is known, the threshold is calculated at a constant value of Pfa. IEEE limits Pfa to 10 %

$$\lambda = (Q^{-1}(P_f) + \sqrt{N} 2\sigma_w^2, \tag{15.12}$$

where σ_w^2 is the Noise variance; N is the Number of samples;

$$Q^{-1}(Pfa) = \sqrt{2} erf^{-1}(1 - 2Pfa), \tag{15.13}$$

Q-function is the queue of the probability of the standard normal distribution, erf^{-1} is the inverse error function.

This threshold cannot ensure that the energy detector achieves the target detection probability. Thus, the choice of threshold can be considered as an optimization task between the two conflicting goals (i.e. maximum Pd and at the same time minimal Pf).

15.2.1.3. Energy Detection with Uncertainty

The exact value of the noise variance is difficult to determine, even if the system is calibrated. This lack of prior information is known as a noise uncertainty. The effect of noise uncertainty does not allow a single energy detector to meet observation requirements, even for a very long observation time. The threshold λ used in the energy detector depends on the noise variance. Therefore, any noise variance errors can lead to significant spectrum sensing failures that can limit the efficiency of the energy detector for cognitive radio. There is a SNRwall value of the signal-to-noise ratio beyond which detection is theoretically impossible [4, 5].

Error in the assessment of noise spectral density N_0 is supposed to be limited [6]. In practice N_0 will have to be evaluated by the detector. There are several factors that increase noise uncertainty such as temperature change due to temperature and / or calibration errors. The noise uncertainty can be much greater than 1 dB and cannot be reduced by increasing the sensing time [7]. It is important to say that in hardware equipment the noise uncertainty is in general between 2 dB and 3 dB.

When uncertainty is included in the noise model [6, [8] the noise distribution function can be summarized in a limited interval $[(1 + \rho) - 1\sigma_w^2, (1+\rho)\,\sigma_w^2]$, where σ_w^2 is the nominal noise power and ρ is the parameter in dB, $\rho \geq 0$, that determines the level of uncertainty. Then Pd and Pfa are calculated using the formulas:

$$P_{fa} \approx Q\left(\frac{(\lambda - (1+\rho)\sigma_w^2)}{\sqrt{\frac{2}{N}}(1+\rho)\sigma_w^2}\right), \qquad (15.14)$$

$$P_d \approx Q\left(\frac{\lambda - ((1+\rho)^{-1}\sigma_w^2 + \sigma_s^2)}{\sqrt{\frac{2}{N}((1+\rho)^{-1}\sigma_w^2 + \sigma_s^2)}}\right), \qquad (15.15)$$

where $Q(x) = \frac{1}{2}erfc(\frac{x}{\sqrt{2}})$ is the queue of the probability of the standard normal distribution function,

σ_s^2 is the signal variance, σ_w^2 is the noise variance, N is the number samples, ρ is the parameter that determines the level of uncertainty.

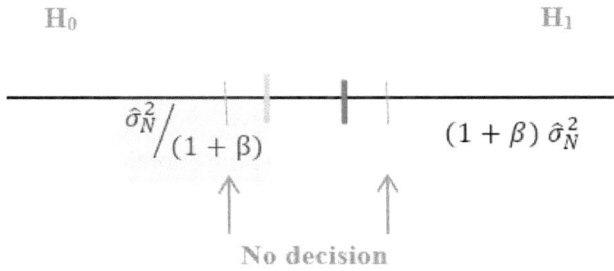

σ_N^2 is the exact value of the noise variance;

$\hat{\sigma}_N^2$ is the estimated value of the noise variance.

The test statistics is determined by the noise variance that can be estimated when there is no signal:

$$\Lambda = \frac{1}{N}\sum_{i=1}^{n} X_i^2, \Lambda \to \sigma_N^2, \qquad (15.16)$$

when N$\to \infty$.

In general case, the ED takes the decision, as follows:

$\lambda > (1 + \beta)\,\hat{\sigma}_N^2,$ then H$_1$ – ED generates 1;

$\lambda < \frac{\hat{\sigma}_N^2}{(1+\beta)},$ then H$_0$ – ED generates 0;

$\lambda \in \left[\frac{\hat{\sigma}_N^2}{(1+\beta)}, (1 + \beta)\,\hat{\sigma}_N^2\right],$ then no decision – ED generates -1;

369

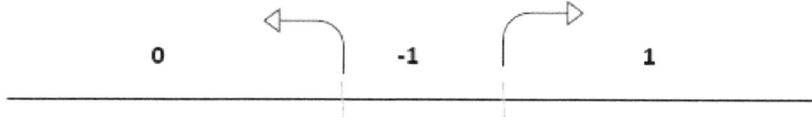

$N \neq \infty$, so an estimate of the noise variance is determined - $\hat{\sigma}_N^2$.

The exact value of the variance is in the interval with a very high probability: $\left[\frac{\hat{\sigma}_N^2}{(1+\beta)}, (1+\beta) \hat{\sigma}_N^2 \right]$. In cooperative sensing, each secondary user SUi generates a different estimate of the noise variance: $\hat{\sigma}_{NSU_i}^2$. The corresponding noise uncertainty interval can be determined for each SUi: $\left[\frac{\hat{\sigma}_{NSU_i}^2}{(1+\beta)}, (1+\beta) \hat{\sigma}_{NSU_i}^2 \right]$. So, in the FC a new reduced noise uncertainty interval can be defined, as follows: $\left[Max \frac{\hat{\sigma}_{NSU_i}^2}{(1+\beta)}, Min(1+\beta) \hat{\sigma}_{NSU_i}^2 \right]$, where i is the indices of the SUs.

The corresponding value of β might be transmitted as feedback to SUi, thus reducing the noise uncertainty i interval by a kind of learning process. In time, the estimate is expected to converge to the exact value of σ_N^2.

15.2.2. Eigenvalue Based Detection

In [9], two spectrum sensing algorithms based on the distribution of eigenvalues in large dimensional random matrix theory. The discrete-time domain received signal r[m] under H_1 can be given as

$$r[m] = \sum_{k=0}^{N_h} h[k]s[m-k] + w[m], \qquad (15.17)$$

where N_h is the channel filter length. At the cognitive radio's receiver, the received samples are split into M vectors each of length N_s. Let us consider the following $M \times N_s$ matrix consisting of the stacking of the M vectors.

$$Y = \begin{bmatrix} r_1[1] & r_1[2] & \cdots & r_1[N_s] \\ r_2[1] & r_2[2] & \cdots & r_2[N_s] \\ \vdots & \vdots & \ddots & \vdots \\ r_M[1] & r_M[2] & \cdots & r_M[N_s] \end{bmatrix}. \qquad (15.18)$$

In the absence of primary user signal (H_0 hypothesis) all the received samples are uncorrelated whatever fading channel model. Moreover, the non-diagonal element of the received covariance matrix is theoretically zero, whereas the diagonal elements contain the noise variance. Hence, for a fixed M and $N_s \to \infty$, the sample covariance matrix $\frac{1}{N_s} YY^*$ converges to the true covariance matrix $\sigma^2 I_M$.

In this chapter, we assume that the noise is additive white Gaussian noise and, furthermore, the noise and the transmitted signal are uncorrelated. Then if the number of received samples Ns are large enough, it can be shown that

$$R_r(N_s) \approx R_r = E\left[\frac{1}{N_s}YY^*\right] = HR_sH^* + \sigma^2 I_M, \tag{15.19}$$

where R_r and R_s matrices represent the covariance matrices of the received and transmitted signals, respectively. I_M is the identity matrix of size M. Let λ_{max} and λ_{min} represent the maximum and minimum eigenvalues of R_r, respectively. Now suppose that ρ_{max} and ρ_{min} are the maximum and minimum eigenvalues of the matrix HR_sH^* then

$$\lambda_{max} = \rho_{max} + \sigma^2, \tag{15.20}$$

$$\lambda_{min} = \rho_{min} + \sigma^2. \tag{15.21}$$

If $Z = HR_sH^* = \delta I_M$, then $\rho_{min} = \rho_{max}$ where δ is a positive integer. In practice, it is highly unlikely that the matrix Z will be equal to δI_M. Hence, if there is no signal present then $\lambda_{max}/\lambda_{min} = 1$ otherwise $\lambda_{max}/\lambda_{min} > 1$. Therefore, this ratio can be used to detect the presence or absence of the signal. Based on the eigenvalues of the received covariance matrix, the following two methods are proposed in literature.

15.2.2.1. Maximum-Minimum Eigenvalue (MME) Detection Method

The steps of the MME algorithm are stated below:

- Compute the received sample covariance matrix

$$R_r(N_s) = \frac{1}{N_s}YY^*. \tag{15.22}$$

- Compute the Maximum and Minimum Eigenvalues of the matrix $R_r(N_s)$ (i.e. λ_{max} and λ_{min}).

- Sensing decision is given by:

$$D = \begin{cases} H_1, & \frac{\lambda_{max}}{\lambda_{min}} \geq \gamma MME \\ H_0, & otherwise \end{cases}. \tag{15.23}$$

Note that γMME represents the threshold for the MME method. It is shown that the covariance matrix of the received signal in the absence of any signal at the receiver approximates to one class of random matrices called Wishart random matrix. The probability density function of Wishart random matrix has no marginal defined expression, finite dimensional case, and present complex mathematics. Some studies on the spectral distribution of eigenvalues found the limiting values for maximum and minimum eigenvalues. The distribution of the (properly rescaled) largest eigenvalue of the complex (real) Wishart matrix converges to the Tracy-Widom law as M, N_s tend to $+\infty$ in some ratio $M/N_s > 0$. As a result, the threshold γMME given by:

$$\gamma MME = \frac{\chi^2}{\delta^2}\left(1 + \frac{\chi^{-2/3}}{(MN_s^2)^{1/6}}F_\beta^{-1}(1 - P_{fa})\right), \tag{15.24}$$

where:

371

$$\chi = \sqrt{N_S} + \sqrt{MN_S}, \qquad (15.25)$$

$$\delta = \sqrt{N_S} - \sqrt{MN_S}, \qquad (15.26)$$

and F_β is the cumulative distribution function (CDF) of the Tracy - Widom distribution of order β (i.e. $\beta = 1$ for real signal, $\beta = 2$ for complex signal).

15.2.2.2. Energy with Minimum Eigenvalue (EME) Detection Method

The steps of the EME algorithm are stated below:

- Compute the covariance matrix R_r as given in (15.16).

- Compute the average power of the received signal

$$\Gamma(N_s) = \frac{1}{MN_s} \sum_{i=1}^{M} \sum_{n=0}^{N_s-1} |r_i[n]|^2. \qquad (15.27)$$

- Then the eigenvalues of R_r are obtained.

- Decision: if

$$\frac{\Gamma(N_s)}{\lambda_{min}} > \gamma EME, \qquad (15.28)$$

where γEME represents the threshold for EME method defined by:

$$\gamma EME = \left(\sqrt{\frac{2}{MN_s}} Q^{-1}(P_{fa}) + 1 \right) \frac{N_s}{\delta^2}, \qquad (15.29)$$

where Q(.) is the Gaussian Q-function and Pfa is the probability of false alarm.

In the above discussion, we can notice that the threshold is not depending on the noise property but rather on the probability of false alarm, the number of segments M and the number of samples N_s per segment.

15.2.2.3. Cooperative Spectrum Sensing

Cognitive radio allows opportunistic access to unused licensed frequency bands. For example, unlicensed users (SU or cognitive user CU) first detect the radio spectrum to check for PU activity and then access the spectrum holes (white intervals) if no primary activity is detected. The accuracy of spectrum sensing is important to avoid interference in the transmissions of the PUs. But reliable spectrum sensing is not always guaranteed due to multipath, shadow fading, and the hidden terminal problem. To avoid these problems, cooperative spectrum sensing has been introduced, resulting in a rapid and reliable detection of the PU signal. It has been shown that when multiple SUs are employed using spatial diversity, the cooperative spectrum sensing exceeds the single spectrum sensing [10].

The main idea of cooperative spectrum sensing is to increase the detection efficiency by using spatial diversity in the observations of spatially located SU. By collaborating SUs can share their information of detection making a combined solution that is more accurate than individual decisions.

To confirm the benefits of collaborative monitoring in individual reporting in [11], a routing algorithm is implemented to hardware the USRPs with 2 SUs to implement the longest test. A common energy and frequency band for spectrum sensing with the GNU radio and USRP is proposed in [12]. The [13] proposed Kullback-Leibler optimization based on the distance to determine the cooperative spectrum sensing decision thresholds and its scalability according to the number of SUs. These conversions are quite limited and more in-depth learning is needed.

In [14], a classification of the cooperative spectrum sensing in three categories was made based on how collaborative SU have shared the network detection data: centralized, distributed and relay-assisted.

Centralized cooperative sensing has a single Fusion Center FC, which controls and manages the three-step cooperative sensing process. First, FC chooses a channel or bandwidth of interest for detection and instructs all cooperating SU to make an individual local detection. Second, all cooperating SUs report their results on the control channel. Thirdly, FC combines received information from local observation, selects SUs, determines the presence of PU, and distributes the decision back to CR users in co-operation. [14, 15]. For local detection, all SUs are tuned to the selected licensed channel or bandwidth. This connection between the PU transmitter and each SU in cooperation to observe the primary signal is called the sensing channel. For data reporting, all SU are set to a control channel. The physical point-to-point link between each SU in cooperation and FC is called a reporting channel, and it sends the sensor result. Centralized cooperative spectrum sensing can also take place in a centralized and decentralized CR network. In centralized CR networks, the CR base station (BS) is FC. In CR Ad Hoc networks (CRAHNs), where no CR BS, each SU can act as FC, to coordinate cooperative spectrum sensing and combine information from monitoring the cooperating neighbours. SUs report to FC the results of the spectrum sensing by one of the following methods.

15.2.2.4. Soft Data Fusion

Each SU boosts the received signal from the PU and transmits it to FC. Although the SU does not require a complicated detection process, the reporting channel bandwidth should be the same as the sensing channel bandwidth. In the FC center, different consolidation techniques can be applied, such as:

- Maximum ratio combining MRC. When MRC is used, channel state information CSI is required from PU to SU and from each SU to FC [15, 16].

- Square-law combining SLC. When SLC with fixed gain factor of each SU is used, only CSI from SU to FC is needed. But if a variable gain factor is applied CSI from PU to SU and from SU to FC is needed [16].

- Quantized Soft Combining [16].

- Optimal Soft Combination Scheme [17].

Soft combination provides better performance than hard combination, but requires greater bandwidth control channel.

15.2.2.5. Hard Decision Fusion

Each SU takes a binary decision for the PU activity and the individual decisions are reported in FC through the reporting channel. The main advantage of this method is that the reporting channel may have a narrow bandwidth. Complex signal processing of each SU is required. The merging rules in FC may be OR, AND, or the MAJORITY rule, which can be summarized as a "k-out of-n rule." Two basic assumptions have been admitted of simplification in the literature:

- The reporting channel is error-free;

- The SU has information about the SNR statistics of the received PU signals.

However, in the cognitive network these assumptions are practically not real.

A system is considered that consists of one PU and four SUs. SUs detect the spectrum for the presence of the PU signal and a FC takes the global decision after a preliminary selection of the SUs to be involved in the decision process. Hard decision combining rules are applied in the FC, for making decision about the presence of PU signal. Each SUs makes a local-level spectrum detection, takes its decision and sends it in 1 bit of information to FC - 0 or 1. 1 means that the signal is available and 0 means there is no transmission in the channel

$$\Delta_k = \begin{cases} 1, E_k > \lambda_k \\ 0, E_k \leq \lambda_k \end{cases}. \tag{15.30}$$

FC combines the results of all SUs and takes a general decision through the AND, OR and the majority rule. The OR function determines that the PU signal is present when at least one SU reports "1"

$$\begin{vmatrix} H_1 & \sum_{k=1}^{K} \Delta_k \geq 1 \\ H_0 & \sum_{k=1}^{K} \Delta_k < 1 \end{vmatrix}. \tag{15.31}$$

The "AND" function decided that the PU signal is present when all SUs have reported a "1" decision

$$\begin{vmatrix} H_1 & \sum_{k=1}^{K} \Delta_k = K \\ H_0 & \sum_{k=1}^{K} \Delta_k \neq K \end{vmatrix}. \tag{15.32}$$

In the voting rule, if at least M of K users have reported a local decision "1" of $1 \leq M \leq K$, FC decides that PU signal is present. The test is formulated as:

$$\begin{vmatrix} H_1 & \sum_{k=1}^{K} \Delta_k \geq M \\ H_0 & \sum_{k=1}^{K} \Delta_k < 1 \end{vmatrix}. \tag{15.33}$$

The rule of majority is a special case of the voting rule for M = K / 2, i.e. If half or more SUs are reported as local solutions "1", FC decides that a PU signal is available. Similarly, rules "AND" and "OR" are a private case for M = K and M = 1.

The probability of cooperative detection Qd and the common probability of a false alarm Qfa are defined as:

$$Q_d = \Pr\{\Delta = 1 | H_1\} = \Pr\left\{\sum_{i=1}^{K} \Delta_k \geq M | H_1\right\}, \tag{15.34}$$

$$Q_{fa} = \Pr\{\Delta = 1 | H_0\} = \Pr\left\{\sum_{i=1}^{K} \Delta_k \geq M | H_0\right\}, \tag{15.35}$$

where Δ is the final solution. Since "OR" corresponds to M = 1, hence:

$$Q_{d,or} = 1 - \prod_{k=1}^{K}(1 - P_{d,k}), \tag{15.36}$$

$$Q_{fa,or} = 1 - \prod_{k=1}^{K}(1 - P_{fa,k}). \tag{15.37}$$

The probability of detection and false alarm of the OR rule corresponding to M = K is calculated by the expressions:

$$Q_{d,and} = \prod_{k=1}^{K} P_{d,k}, \tag{15.38}$$

$$Q_{fa,and} = \prod_{k=1}^{K} P_{fa,k}, \tag{15.39}$$

15.3. Experimental Setup

15.3.1. GNU Radio

GNU Radio is open source software and free access for signal processing applied to various communications developments [18]. It is a platform using a graph of data streams. This graph is built from blocks that are available when installing the GNU Radio Companion graphical environment.

15.3.2. GNU Radio Blocks

The block description is done in the Python or C++ programming languages. One block typically describes one type of signal processing - for example, filter blocks or demodulators are available for filtering or demodulation. If there is no block available in libraries that responds to a given user signal processing, it can be created. Several blocks have been created that perform different functions. The blocks have one input and one or two outputs that support float data, put the samples at each input M (in our case one), and allow this number to be set during the creation of graph. The number of samples will be included as a parameter. Fig. 15.3 shows the created blocks.

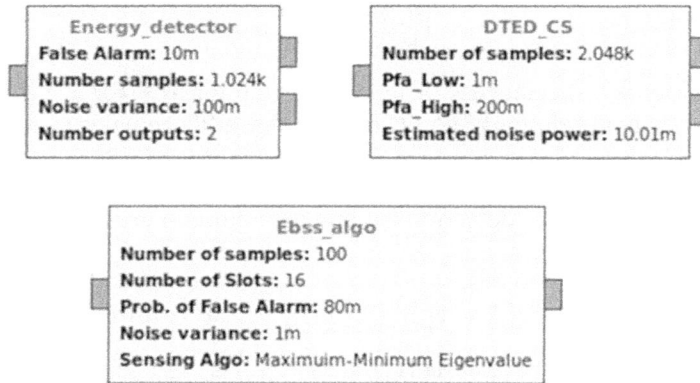

Energy_detector		DTED_CS
False Alarm: 10m		**Number of samples:** 2.048k
Number samples: 1.024k		**Pfa_Low:** 1m
Noise variance: 100m		**Pfa_High:** 200m
Number outputs: 2		**Estimated noise power:** 10.01m

Ebss_algo
Number of samples: 100
Number of Slots: 16
Prob. of False Alarm: 80m
Noise variance: 1m
Sensing Algo: Maximuim-Minimum Eigenvalue

Fig. 15.3. GNU Radio blocks.

The block Energy detector calculates the energy of the channel Eq. (15.7) and a single threshold Eq. (15.12). Input parameters are Probability of false alarm, Number of samples, Noise variance. Block DTED has input parameters Number of samples, Estimated noise power and two Probability of false alarm values Pfa_Low and Pfa_High, that specified two thresholds. GNU Radio block EBSS input parameters are Number of samples, Number of slots, Probability of false alarm, Noise variance and sensing algorithm.

The block is set as follows: Number of samples – 128, Number of slots – 16, False alarm probability - 0.08, Noise variance - 0,001, but any value can be put, because the MME algorithm does not require prior information about the channel noise and Sensing algo - Maximum-Minimum Eigenvalue

15.3.3. Results

15.3.3.1. Simulation Results

The simulation is done in GNU Radio Companion. Energy detector block diagram in GNU Radio is shown in Fig. 15.4. To investigate the energy detector's performance, characteristic of the probability of detection Pd as a function of SNR [3] is used. For this purpose, the Energy detector is fed random signal from block Signal Source and noise signal from the block Noise Source. The noise signal is set to Amplitude 1 and Gaussian distribution. The Energy detector block is tuned: Probability of false alarm Pfa = 0.1; Noise variance = 0.9. The probability of detection was deducted from a different number of samples: N = 100; N = 500; N = 1000 and N = 2000. The signal amplitude from the Signal Source block changes from 0 to 2 V in step 0.05 V. This changes the SNR.

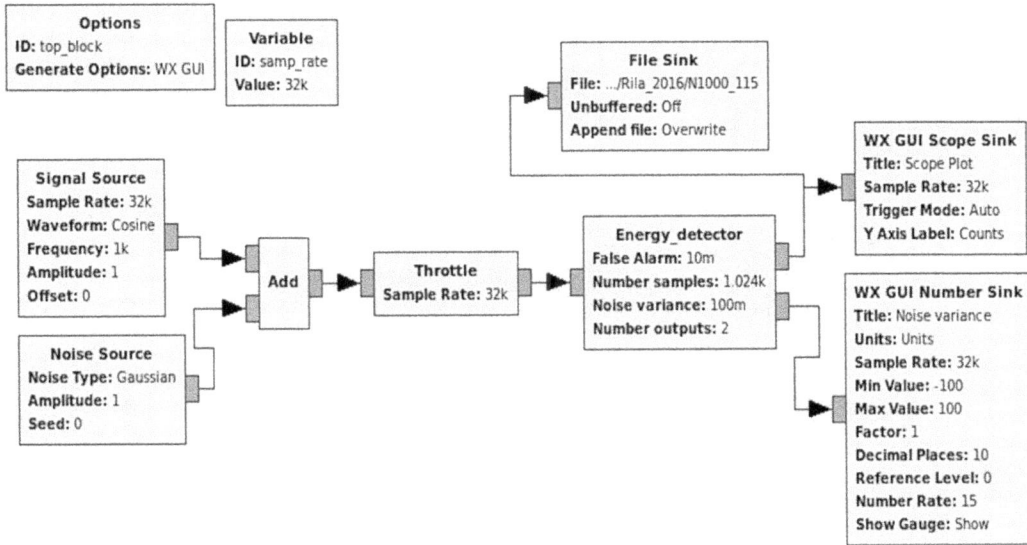

Fig. 15.4. Energy detector simulation flowgraph in GNU Radio Companion.

At the energy detector output appears the decision: Is there a signal in the channel (1) if the energy in the channel is greater than the calculated threshold or not (0) if the energy is less than the threshold and the SU could use the channel.

A series of 0 and 1, leaving the output of the detector are collected from the block File Sink and recorded in files for different values of SNR. The amplitude of the noise signal is constant - 1 V. The amplitude of the input signal varies from 1 to 2 V with a step of 0.05 V. The flowgraph is started. During the time, it works (20s), the mean value of the signal noise ratio SNR remains constant.

Files are binary type and can be opened via Octave (a mathematical and engineering computational program that can also open.m files) or MATLAB. An Octave program has been developed that opens, reads data from files, and draws the graph of Pd depending on SNR.

Fig. 15.5 shows a graph of the probability of detection, depending on the signal to noise SNR, at different values of the number of samples, N: N = 100; N = 500; N = 1000 and N = 2000.

The same simulation is done by varying the SNR ratio from 0 dB to 5 dB with step 0, 25 dB with samples 1000, 2000, 4000, 8000 and noise variance 1.029. The results of the measurements are stored in files and the graph built on the new values is shown in Fig. 15.6. When comparing the two graphs it is seen that with increasing number of samples the steepness of the curves increases, which leads to a clear solution whether there is any signal in the channel or not.

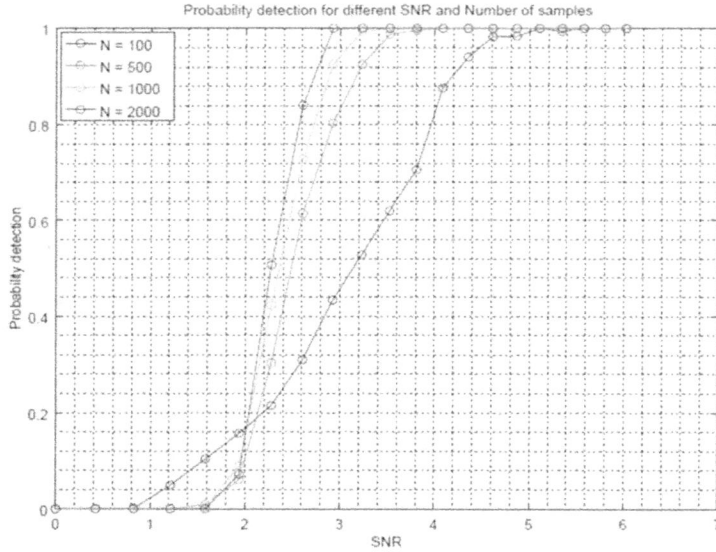

Fig. 15.5. Graph of the probability of detection depending on the SNR at a small number of samples – simulation.

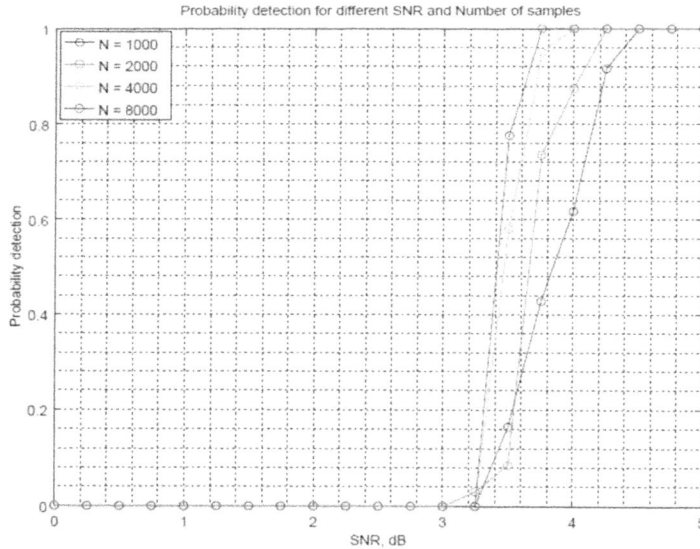

Fig. 15.6. Graph of the probability detection depending on the SNR of a larger number of samples – simulation.

15.3.3.2. Experimental Results for Single Energy Detector

A block diagram of the experimental setup is shown in Fig. 15.7. As a transmitter, a signal generator SMY01 9 kHz - 1.040 GHz Rohde & Schwarz is used, connected to an antenna through a coaxial cable with BNC connectors. The generator is set at 433 MHz, with FM modulation with 2 MHz deviation. The generator sets different transmit signal values in the range -21.9 dBm / 18 mV to -17.4 dBm / 30.2 mV with a step 0.5 dBm.

The receiving side was implemented with the USRP N210 of Ettus Research, connected to the PC via Gigabit Ethernet with GNU Radio installed platform [4, 5]. The energy detector is implemented as a GNU Radio block from the GNU Radio Companion libraries.

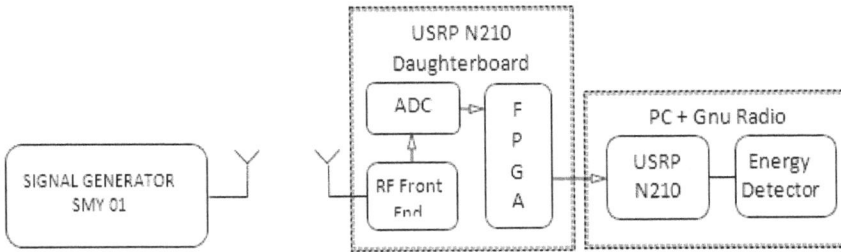

Fig. 15.7. Block diagram of experimental set-up for testing the energy detector.

Channel detection, whether it is free or not, is made in the receiving side of the Energy detector block in the following sequence:

- The signal from the generator is switched off. This is how the channel noise is evaluated. The second output of the detector goes exactly the noise variance in the channel - channel energy squared.

- This value is reported - in this case is 192.10-6.

- The block is calibrated with the new value of the noise variance, because at any moment σ_n^2 depends on the state of the channel. In case the noise spectrum is not uniform, as expected, an offset correction parameter is inserted, which adds extra noise to obtain a uniform noise distribution. In this case noise from the Noise source block with 400 µV amplitude was added. (Then, when calculating the signal to noise ratio, this value must be subtracted).

- From the generator, the signal level increases and the measurements are carried out.

The results of the measurements are recorded in files. The ED block settings are: Number of samples – 1000, 2000, 4000 and 8000; Noise variance - 192.10-6; Probability of False Alarm – 0.01.

Fig. 15.8 shows that the results of the measurements confirm the results of the simulation. At low levels of the SNR signal, the probability of detection is very small. With the increase in SNR, this probability increases rapidly and at levels above 3 dB, it is equal to 1.

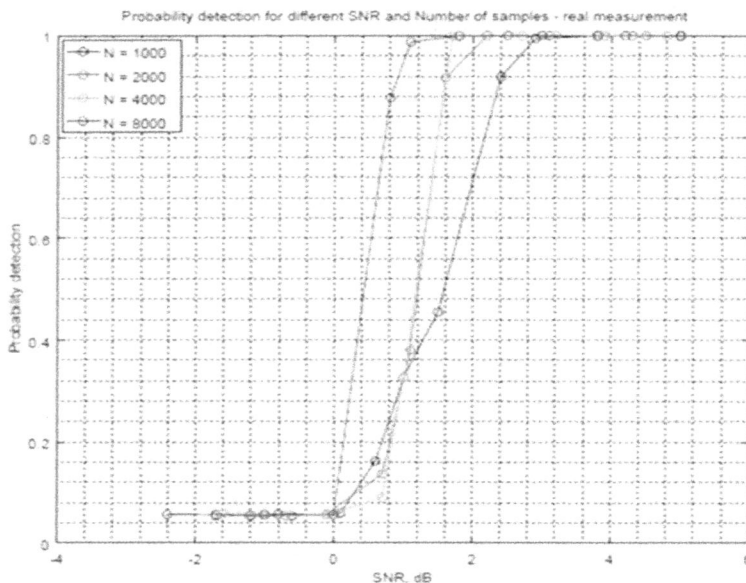

Fig. 15.8. Probability of detection as a function of the SNR – measurement.

15.3.3.3. Eigenvalue Based Spectrum Sensing

The block diagram of the measurements with eigenvalue based detector is shown in Fig 15.9. The generator is set at a frequency of 433 MHz, with FM modulation with a 75 kHz deviation. A comparison between an eigenvalue based detector, an energy detector with uncertainty of 1 dB and 0.4 dB and the ideal energy detector is made.

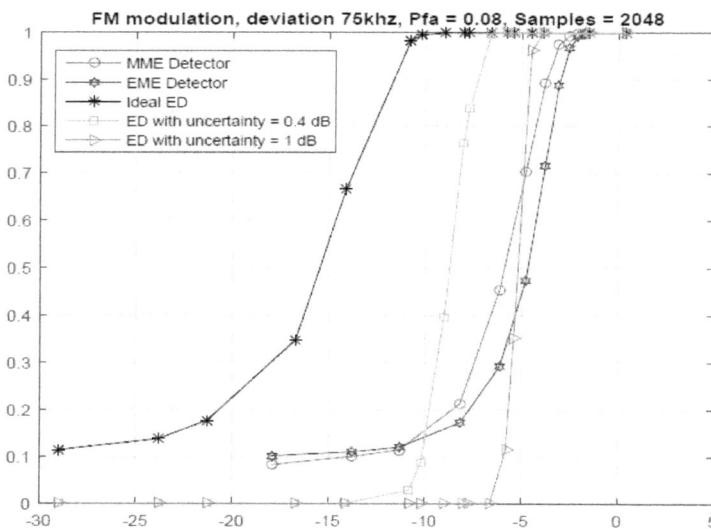

Fig. 15.9. Comparison between various detectors.

15.3.3.4. Cooperative Spectrum Sensing Results

15.3.3.4.1. One-Bit SU Decision Model Cooperative Spectrum Sensing

The block diagram of the experimental setup is shown in Fig. 15.7. For the transmitter, a signal generator SMY01 9 kHz - 1,040 GHz Rohde & Schwarz is used. It is connected to a transmitting antenna via a BNC coaxial cable. The generator is set at a frequency of 433 MHz, with FM modulation with a 1 MHz deviation. The generator gives different values of the transmitted signal in the range - from -30 dBm to 10 dBm with a step of 2 dBm / 5 dBm.

The receivers are implemented by 4 USRP N210 from Ettus Research, connected to computers with GNU Radio installed, via Gigabit Ethernet. The data are recorded 20 seconds for each value of the transmitted signal in binary files, the first record is when the generator is off and there is only noise in the channel. The energy detector processes the received signal and calculates the noise variance σ_w^2. The estimated value of the noise variance in the channel is used as an input parameter of the Energy detector block. For the four channels, these values are shown in Table 15.1.

Table 15.1. Estimated values of the noise variance.

SU	SU1	SU2	SU3	SU4
σ_w^2	14.7 nW	46.5 nW	43.6 nW	46.1 nW

A block diagram of cooperative spectrum sensing system model is shown in Fig. 15.10. Each SU detects the spectrum and decides about PU signal in the channel and sends it to FC. The SU3 solution is visualized in the GNU Radio and is shown in Fig. 15.11.

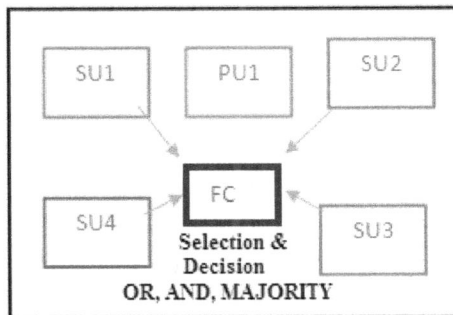

Fig. 15.10. Block diagram of cooperative spectrum sensing system model.

Fig. 15.12 and Fig. 15.13 show the flowgrapf of FC decision-making under the various rules implemented by blocks in GNU Radio. The characteristic probability of detection as a function of the SNR of the energy detector is captured, based on data from the first 700,000 bits of recorded binary files. Mathematical calculations are done in MATLAB.

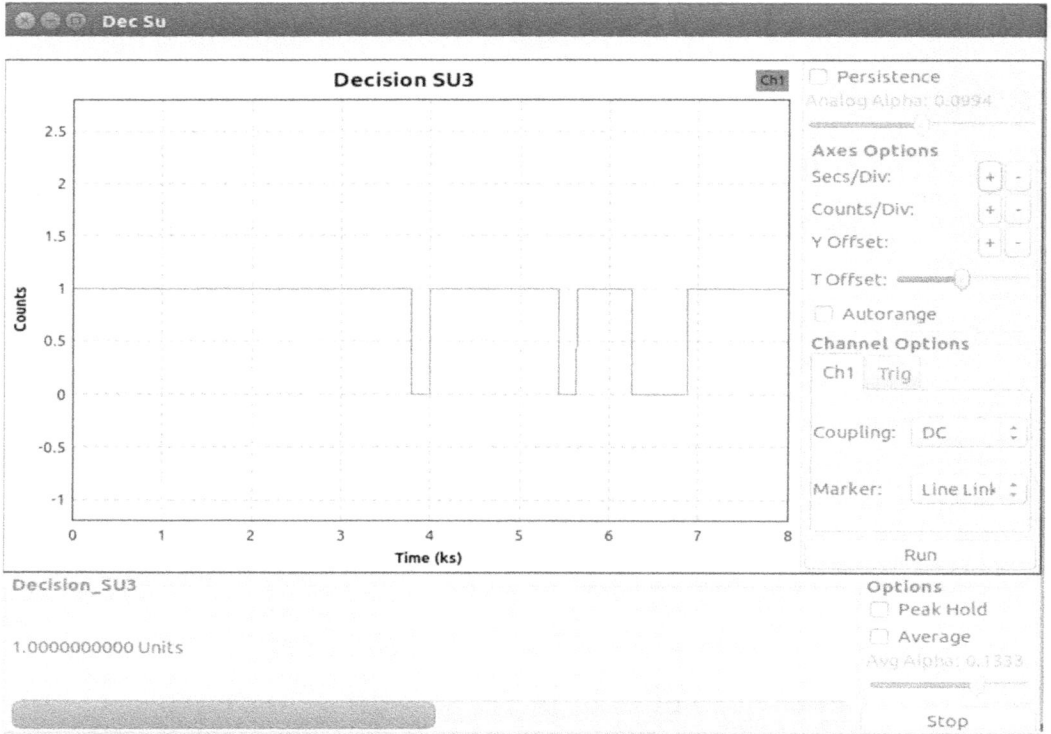

Fig. 15.11. SU3 makes local decision.

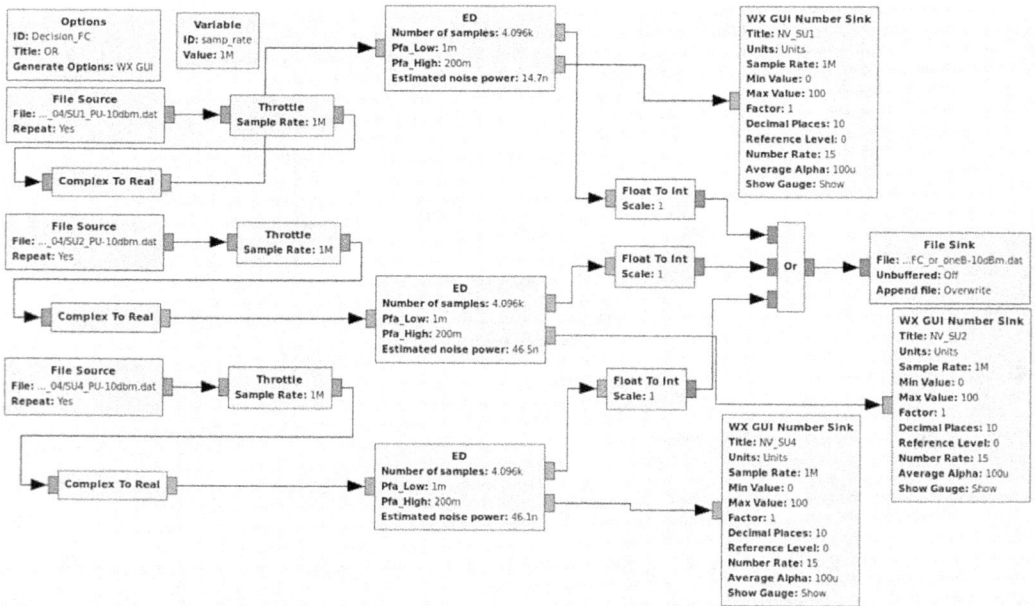

Fig. 15.12. Taking decision flowgraph in the FC by rule "OR".

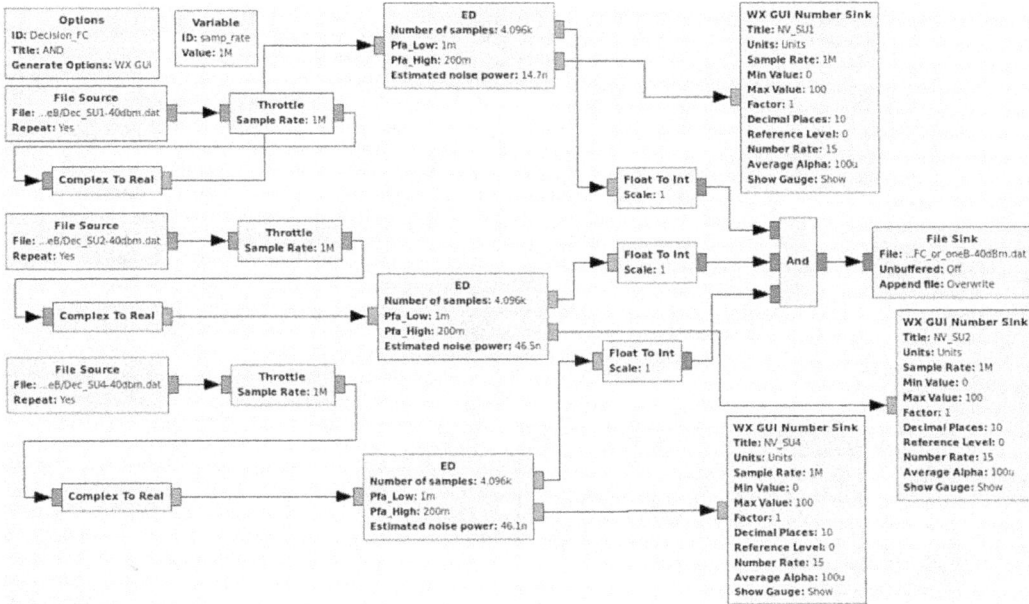

Fig. 15.13. Taking decision flowgraph in the FC by rule "AND ".

Fig. 15.14 shows the Pd (SNR) curves for all 4 SUs. The result analysis shows that 3 of the curves are typical of the observed characteristic, and the SU 3 curve is almost unchanged. This is an illustration of the necessity of a SUs selection procedure in the base station for security and reliability considerations. Sharply distinct results should be ignored to avoid results distortion of the general decision-making. In this case, the decision taken by FC is based on the results of SU 1, SU 2 and SU 4.

The results show that the OR and MAJORITY rule exceeds the AND rule and gives better results: FC may decide to have a PU signal at a lower SNR.

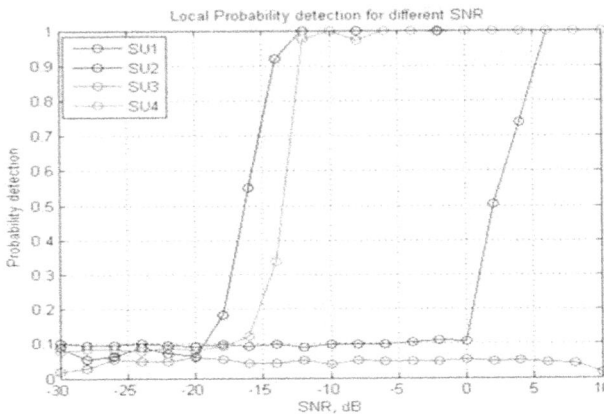

Fig. 15.14. Pd (SNR) curves for all 4 SUs.

383

15.3.3.4.2. Two-Bits SU Decision Model Cooperative Spectrum Sensing

A block DTED takes a two-bit decision based on a double-threshold energy comparison.

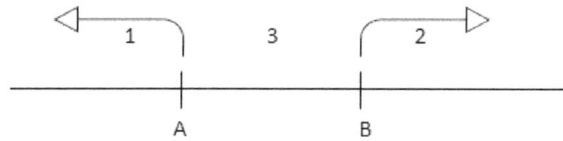

This block has two outputs: Out and Out1.

1. If $E < A$ Out = 0   Out1 = 0,
2. If $E > B$ Out =1   Out1 = 0,
3. If $E < B$ and $E > A$ Out = 0   Out1 = 1.

In third case, DTED_CS can't make decision.

Block arguments are:

Pfa_Low = 0.001;
Pfa_High = 0.2 and 0.5;

Estimated_noise_power:

SU 1 (USRP 104)	Noise Power:	2.0437e-6,
SU 2 (USRP 106)	Noise Power:	2.4960e-6,
SU 3 (USRP 107)	Noise Power:	2.0828e-6.

Fig. 15.15 shows a flowgraph, describing the operation of the two-threshold cooperative spectrum model. The flowgraph is applied to every SU and results are stored in files for further processing.

Fig. 15.16 shows the SUs probability of detection – local decision.

Fig. 15.17 shows the probability that the detector does not to make a decision on the availability of a channel signal.

Let ρ denotes the signal to noise ratio and $P(\rho)$ denotes the theoretical probability of no decision after DTED processing. We can theoretically define this probability as

$$P(\rho) = \text{Prob}(\Lambda > \lambda_{low}) - \text{Prop}(\text{statistic} > \lambda_{high}). \qquad (15.40)$$

Fig. 15.17 shows the theoretical and the simulated probability that after the DTED processing no sensing decision is made by the spectrum sensing detector. The tiny difference between the theoretical result and experimental result might be caused by the Gaussian approximation of the collected power (statistic Λ), and the real experimental environment which differs from the considered theoretical Rayleigh fading channel. Moreover, it can be observed that the probability of no decision increases dramatically in mid-SNR range. In this case, the utility of cooperative sensing (diversity) and the fusion center will have its great importance to make a global decision for all secondary users.

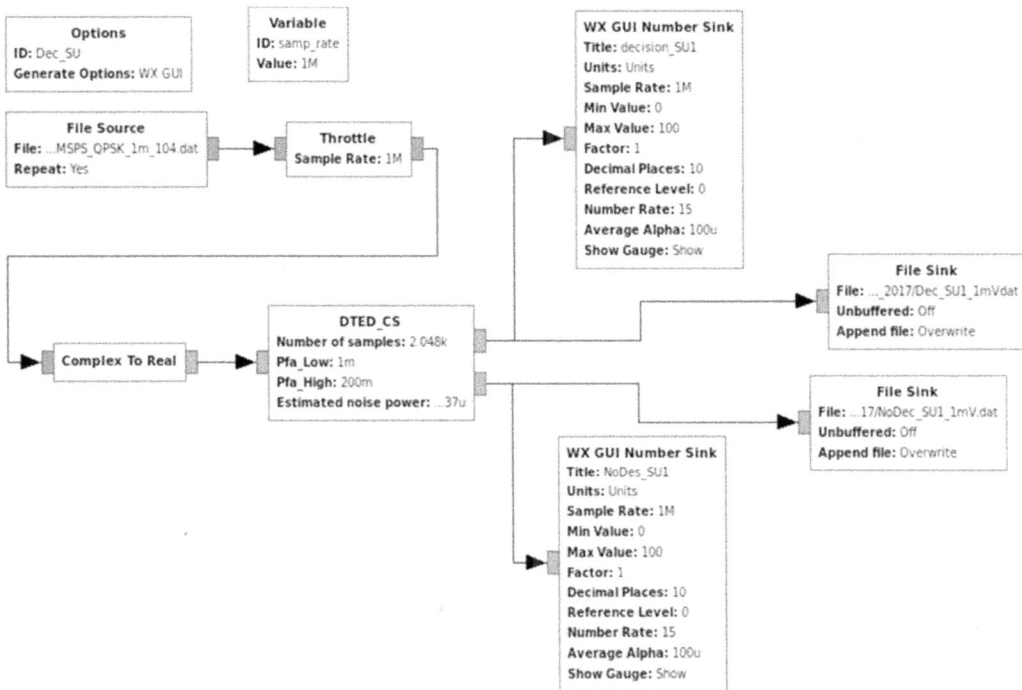

Fig. 15.15. Flougraph, applied to every SU. The results are stored in files.

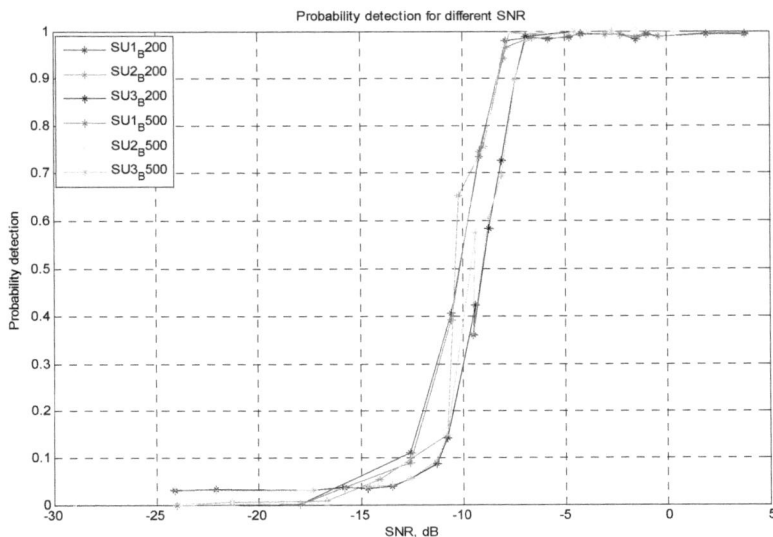

Fig. 15.16. Pd (SNR) curves for 3 SUs. Pfa_High = 0.2 and 0.5.

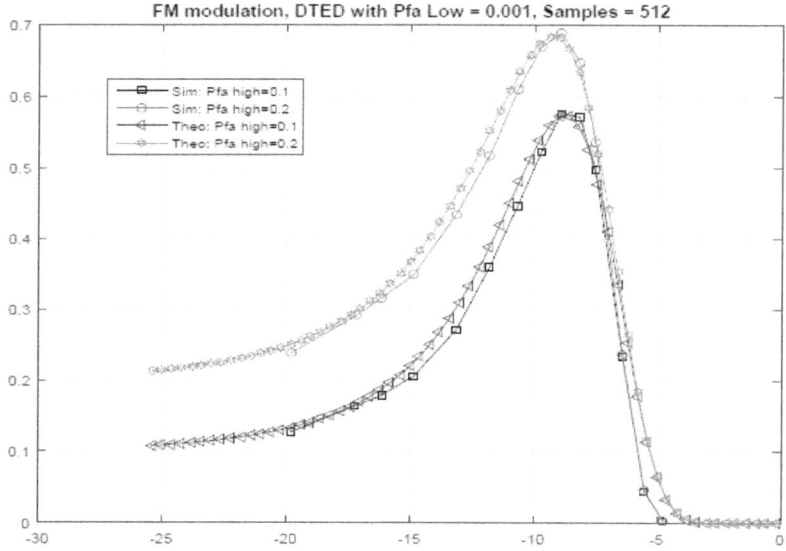

Fig. 15.17. Probability of no decision (SNR) curves. Pfa_High = 0.1 and 0.2.

In Cooperative spectrum sensing, we have the same spatial dependence between SU1, SU2 and SU3. Considering this fact, we can assume that these users belong to the same cluster. In FC, we can calculate an average SNR to create the FC curves. Fig. 15.18 shows the FC curves for SNR_average and compares it whit the curves of the three SUs.

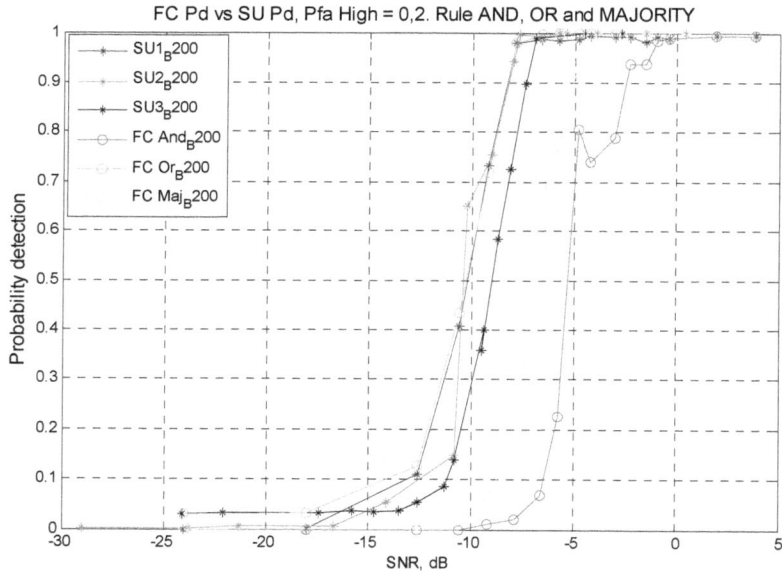

Fig. 15.18. Pd (SNR) curves for 3 SUs and FC Pd(SNR_average). Pfa_High = 0.2.

15.4. Conclusions

This chapter discusses various methods for spectrum sensing. It introduces the results of a research on the implementation on USRP and GNU Radio of single spectrum sensing system and cooperative sensing system with energy detectors and eigenvalue based detectors. The experimental setup realized permits to perform experiments using different scenarios as OR, AND, Majority rules for decision making in a fusion center based on local decisions of primary users. Here is presented a study of the noise uncertainty patterns used in the detection of the spectrum based on energy detection. These models are compared to spectrum sensing algorithms based on the distribution of eigenvalues in large dimensional random matrix theory. The efficiency of these spectrum sensing methods is analyzed by measuring detection probabilities as a function of the SNR for a given probability of false alarm. Further work is foreseen on a deeper study of the uncertainty influence and uncertainty overcoming through cooperative spectrum sensing. Additional research is planned on Fusion center impact with non-lossless feedback reporting and with feedback reporting (Gaussian Feedback channel and Rayleigh Feedback channel).

Acknowledgements

This research is partly supported by the project Bg: DRILA01/10/17.04.2015 from Fund "Nauchni izsledvania", Fr: Project №34281ZM. Additional information on the project can be found on http://www.phcrila15.free.bg/

References

[1]. J. Mitola III, Cognitive Radio an Integrated Agent Architecture for Software Defined Radio, http://www.diva-portal.org/smash/get/diva2:8730/FULLTEXT01.pdf

[2]. S. Hossain, I. Abdullah, M. A. Hossain, Energy detection performance of spectrum sensing in cognitive radio, *I. Journal Information Technology and Computer Science*, Vol. 11, 2012, pp. 11-17.

[3]. S. Atapattu, C. Tellambura, H. Jiang, Conventional Energy Detector, Chapter 2, in Energy Detection for Spectrum Sensing in Cognitive Radio, *Springer*, 2014, pp. 11-26.

[4]. R. Tandra, A. Sahai., SNR walls for signal detection, *EEE Journal of Selected Topics in Signal Processing*, Vol. 2, Issue 1, February 2008, pp. 4-17.

[5]. R. Tandra, A. Sahai, Fundamental limits on detection in low snr under noise uncertainty , in *Proceedings of the International Conference on Wireless Networks, Communications and Mobile Computing (WIRELESSCOM'05)*, Vol. 1, June 2005, pp. 464-469.

[6]. S. Bahamou, A. Nafkha et al., Noise uncertainty analysis of energy detector: Bounded and unbounded approximation relationship, in *Proceedings of the 21st European Signal Processing Conference (EUSIPCO'13)*, 2013, pp. 1-4.

[7]. mentor.ieee.org/802.22/dcn/06/22-06-0134-00-0000-performance-of-the-power-detector-with-noise-uncertainty.ppt

[8]. W. Lin, Q. Zhang, A design of energy detector in cognitive radio under noise uncertainty , in *Proceedings of the 11th IEEE Singapore International Conference on Communication Systems (ICCS'08)*, 19-21 November 2008, Guangzhou, China, pp. 213-217.

[9]. A. Nafkha, B. Aziz, M. Naoues, A. Kliks, Cyclostationarity-based versus eigenvalues-based algorithms for spectrum sensing in cognitive radio systems: experimental evaluation using

GNU radio and USRP, in *Proceedings of the IEEE 11ᵗʰ International Conference on Wireless and Mobile Computing, Networking and Communications (WiMob'15)*, Abu Dhabi, United Arab Emirates, 19-21 October 2015, pp. 310-315.

[10]. S. M. Mishra, A. Sahai, R. W. Brodersen, Cooperative sensing among cognitive radios, in *Proceedings of the IEEE International Conference on Communications (ICC'06)*, Istanbul, Turkey, 11-15 June 2006, pp. 1658-1663.

[11]. Y. Zhao, J. Pradhan, Y. Luo, L. Pu, Joint energy-and-bandwidth spectrum sensing with GNU radio and USRP, *ACM SIGAPP Applied Computing Review*, Vol. 14, Issue 4, December 2014, pp. 40-49.

[12]. R. Yoshimura et al., A USRP based scheme for cooperative sensing networks, in *Proceedings of the IV Workshop de Redes de Acesso em Banda Larga (WRA'14)*, May 2014, pp. 67-76.

[13]. D. Bielefeld, G. Fabeck, M. Zivkovi, Optimization of cooperative spectrum sensing and implementation on software defined radios, in *Proceedings of the 3ʳᵈ International Symposium on Applied Sciences in Biomedical and Communication Technologies (ISABEL'10)*, November 2010, Rome, Italy, pp. 1-5.

[14]. I. F. Akyildiz, B. F. Lo, R. Balakrishnan, Cooperative spectrum sensing in cognitive radio networks: A survey, *Physical Communication*, Vol. 4, 2011, pp. 40–62.

[15]. H. Rifà-Pous, M. J. Blasco, C. Garrigues, Review of Robust Cooperative Spectrum Sensing Techniques for Cognitive Radio Networks, *Springer Science+Business Media, LLC*, 2011.

[16]. D. Teguig, B. Scheers, V. Le Nir, Data fusion schemes for cooperative spectrum sensing in cognitive radio networks, in *Proceedings of the Military Communications and Information Systems Conference (MCC'12)*, Gdansk, Poland, Oct. 2012, pp. 1-7.

[17]. J. Ma, G. Zhao, Y. Li, Soft combination and detection for cooperative spectrum sensing in cognitive radio networks, *IEEE Transactions on Wireless Communications*, Vol. 7, No. 11, Nov. 2008, pp. 4502-4507.

[18]. GNU Radio is a Free & Open Software Radio Ecosystem, http://www.gnuradio.org

Index

2-Connected component, 92, 97, 100
3GPP, 247
4G/5G Networks, 247
802.11p, 343

A

Access Network Layer, 260
achtophorous node, 41, 43, 45-51, 57-59, 63, 67, 70, 81, 82
Action Archetypes, 141
Ad hoc wireless network, 115
adaptive modulation, 218
additive white Gaussian noise, 220
ADSL, 218
Agent Archetypes, 139
Algorithm from Barreto and Aragão, 319
ANTs, 327
AODV (Ad hoc On-demand Distance Vector), 116
articulation, 89, 91
assessment
    evalueted criteria, 207
attacks, 34, 35, 72, 73
Authentication, 164
    and Integrity Checking protocol (AICP), 164
average
    path lengths, 41, 52, 56, 58, 72
    symbol error rate, 218
AvgLQI, 45, 46, 56

B

BA, 319, 321
bandwidth efficiency, 218
Base Station, 40, 41, 43, 47, 54, 58, 61, 67, 68, 77, 80, 81
BD, 314, 321
BDA, 312, 318, 319, 321, 322
BDCA, 317-319, 321, 322
BER (Binary Error Rate), 115
Biconnected component, 92, 95, 99, 101
Big Data analysis, 155
binary phase-shift keying, 218
blind equalization, 218
bridge, 89, 91, 100, 101
buffer, 312, 314, 315, 319, 322
    algorithm, 314
    delay, 311, 317, 318, 321-323

Bus Network Layer, 260

C

CallGen323, 314
CARCC. *See* Connectivity Auto-Recovering via Cooperative Communication
catch-up process, 344
category of the events, 210
CC. *See* Cooperative Communication
CC2420, 40, 45, 56
CC-link. *See* Cooperative Communication link
cellular network, 247
Certificate Authority, 160
certificates, 160
channel state information, 217
cloud computing, 171-177, 181, 184
co-channel interference, 258, 288
code
    confidentiality, 158
    mobility, 155
    obfuscation, 158
Cognitive radio, 363, 372
cold chain monitoring, 36-38, 40, 48, 55, 68, 83
Collisions, 33, 73, 81
conditional probability, 347, 352, 354, 357, 359
Connected component, 91
Connectivity
    Augmentation, 110
    Auto-Recovering via Cooperative Communication, 105
    Recovery, 104
Convergence Layer, 260
convolution, 346
Cooperative Communication, 90, 102
    link, 103
    communication systems, 217
    sensing, 370, 373, 385, 387
Co-Tier Interference, 288
crisis, 187, 189, 198
criteria of the capacity, 209
critical
    infrastructure, 187-189, 197, 198
    links, 91
    nodes, 91
Crossbow, 28, 29, 30
Cross-layer approach, 116
crowded places, 201
cryptography, 34, 72, 73, 83

cut-edges, 89
cut-vertices, 89
cycle of information propagation, 344

D

DAS, 97, 99-101, *See also* Distributed
    Articulation Search
data rate, 217, 247
DBS, 101, *See also* Distributed Bridge Search
DDBC. *See also* Distributed Search for
    Biconnected Components
Decision
    support, 145
    tree method, 175
Dedicated Short Range Communications
    (DSRC), 343
degree of connectivity, 41, 45, 55, 56, 61-63, 65,
    67, 81
delay, 312-315, 318-320, 322
Deming circle, 203
direct link, 102
disruption tolerant networks (DTNs), 344
Distributed
    Articulation Search, 97
    Bridge Search, 100
    Search for Biconnected Components, 101
DMDB, 314, 319, 321
Dynamic
    assessment, 209
    Management of Dejitter, 314, 319, 321
    modelling of the road transport, 187

E

Effective protection, 211
eigenvalue based detectors, 364, 387
E-MOS, 313
Empirical Cumulative Distribution Function, 274
energy
    detection, 364, 387
    detector, 365-368, 376, 377, 379-381
    with Minimum Eigenvalue, 372
Environment Knowledge, 130, 135-139, 143,
    148
ETX, 118
EVENT, 202

F

Femtocells, 249
    technology, 247
Fire
    Protection, 201
    Safety, 203
FMEA

G

Failure Mode and Effect Analysis, 206
forwarding process, 344
FR4 epoxy, 233

general
    coefficient, 209
    secure coefficient, 207
GNU Radio, 363, 375-377, 379, 381, 387
GPS, 344

H

HaaS, 172
Hard Decision Fusion, 374
HARQ, 325
helper
    nodes, 103
    set, 102
    set selection routine, 106
Heterogeneous Network, 247, 256
High-Speed Packet Access, 247
HSS. *See* helper set selection routine
hybrid
    cloud computing, 171-175, 184
    networks, 256

I

IaaS, 172
IDNC, 326
IEEE 802.15.4, 26-28, 30, 40, 45
IEEE 802.16j, 217
IEEE 802.16m, 217
incidents, 204
Inference Engine, 136, 137, 138
Information propagation, 344
information-centric networking (ICN), 344
Informed Virtual Geographic Environments
    (IVGE), 130
input's parameters, 207
inputs parameters
    "locality", "category" and "the number of the
        visitors", 207
intelligent
    Base Station, 247
    Transportation System (ITS), 343
Inter-Cell Interference Coordination, 256
intrusion detection, 155
ISMS, 177
ISO International Organization for
    Standardization, 202

J

jitter, 311, 313, 314-318
JNRD, 327
joint Poisson, 347-353, 359

K

Knowledge about
Actors and Behaviours Archetypes, 143
Application Domain, 145
Environment, 143

L

L2RP, 23, 41, 43, 50, 53-56, 62, 67, 71, 73, 74, 81, 82
latency, 314, 315, 322
law's requirements, 203
LIF, 41, 52, 59-61, 70, 72
lifetime, 23, 25, 29, 33, 34, 40-42, 52, 65-70, 72, 73, 81, 82
LIKELIHOOD, 202
linear autoregressive, 312
link quality indicator, 23, 40, 41, 51, 55
listening, 33, 34, 73, 81
load
balancing, 23, 40-43, 47, 48, 50, 51, 56, 58, 59, 61, 63, 65, 67, 70-72, 81, 82
imbalance factor, 52, 72
locating articulation, 92, 93, 97, 104
Location of
Articulation and Bridges, 92
Biconnected component, 101
Long Term Evolution, 247
LQI, 23, 40-43, 45, 46, 50, 51, 54-56, 58, 81, 82
link quality indicator. *Voir*

M

MA, 313, 319, 321
MAC, 33, 45
machine-to-machine (M2M), 344
MANETs, 343
maximal ratio combining, 218
Maximum and Minimum Eigenvalues, 371
MaxLQI, 45, 46, 56, 58-61, 63, 65, 67, 68, 81, 82
MCSA, 327
mesoscopic, 345
MICAz, 28, 30
Microcells, 249
microstrip antenna, 233, 243
array, 233, 234, 239
high-gain, 234, 239
low-cost, 234, 239

MinLQI, 45, 46, 54, 56, 58-61, 63, 65-68, 70, 71, 82, 83
mobile ad hoc networks (MANETs), 343
Mobile agent
Protecting mobile agents, 156
Mobile agent computing paradigm, 155
Mobile agents, 155
open multi-agent system, 157
securing mobile agents, 157
mobile
broadband communication, 217
code. *See* Code Mobility
communication networks, 248
cryptography, 158
Mobile-Femto, 251
mobility pattern, 343-345, 354, 357
MOS, 314
Move Average Algorithm, 313, 319, 322
MSA. *See* microstrip antenna
Multi Point Relays Mechanism, 116
MultiHopLQI, 42, 46, 58, 81, 82
multilevel amplitude-shift keying, 218
multimedia wireless applications, 218

N

Nakagami-m fading channels, 219
NC, 326
noise uncertainty, 368, 370, 387
Normal speed distribution, 346

O

OBD, 319, 321, 322
Occupational Health and Safety, 201
requirements, 203
OLSR (Optimized Link State Routing), 116
OM^2DNC, 326
OMNet++, 359
OneShotBehaviour, 167
OpenH323, 314
orthogonal binary frequency-shift keying, 218
outside effects
"number of the visitors" and "population" and "undesired situation in the narrow area", 209
overhearing, 33, 61, 63, 65, 68, 80-82
Overmitting, 34, 73, 81

P

PaaS, 172
packet, 311-322
Delivery Ratio, 116, 119
Pareto, 313
pattern symmetry, 234

Pico-cells, 249
Poisson process, 346-354, 356
preventive
  actions, 207
  solutions
    "immediate and permanent corrective actions,
    207
primary users, 363, 387
private cloud, 171, 172, 174, 178
probability density function, 219
process management, 204
project management, 175
public cloud, 171, 172, 178, 182

Q

quadrature amplitude modulation, 217, 218

R

Radio Access Network, 247
radome, 235
random
  variables, 346, 348, 352
  walk mobility, 346
rate-adaptive transmission strategies, 218
Rayleigh fading channels, 217, 219, 229
RBS, 175
reference monitor, 159
relay node, 217, 219, 249
relaying scheme, 217
Request To Send / Clear To sent mechanism,
  121
RFID, 344
risk, 201
  acceptance, 176, 180
  analysis, 172, 175, 177, 178
  assessment, 172, 175, 177, 202
  avoidance, 176, 180
  breakdown structure, 175, 178
  CRITERIA, 202
  EVALUATION, 202
  factor, 171, 172, 175-178, 183, 184
  identification, 172, 175, 177
  impact, 176, 179-181
  management, 172, 174-176, 202
  matrix method, 176, 178
  mitigation, 176, 180
  probability, 176, 180, 181
  source, 202
  transference, 176, 180
  treatment, 202
  value, 172, 177, 179-181, 183, 184
Road transport, 197
roadside unit (RSU), 343
round-robin, 42, 47-49, 51, 59, 61, 67-70

routing, 23, 25, 32, 34, 37, 40-43, 46-52, 56-63,
  65-74, 78, 80-83
Routing approach
  proactive, 116
  reactive, 116
RPN
  Risk Priority Number, 206

S

SaaS, 172
secondary users, 363, 385
Security, 28, 34, 35, 72, 73
Self-Organizing Network, 247
sensor, 23-28, 30-36, 38-43, 45, 46, 50-52, 54-
  56, 58, 61, 63, 69, 70, 72-74, 76, 77, 78-82
Sensor Networks (SN), 129
Signal to Interference plus Noise Ratio, 251
Sink, 32, 33
Small Cells Interference, 256
Soft
  Data Fusion, 373
  soft targets, 201
  targets the buildings and the events, 201
software-defined radio, 363
spatial distribution, 346-350, 352-356, 359
spatiotemporal, 346, 347
Spatio-Temporal Knowledge, 142
spectral efficiency, 247, 268
spectrum sensing, 363, 364, 368, 370, 372, 373,
  381, 384, 386, 387
speed, 343, 346, 34-350, 354, 359
store–carry–forward, 344
SUMO, 359
Sun SPOT, 27, 30
superstrate, 235
symbol error rate, 218

T

Tag
  Checker software, 163
  Generator, 163
talkspurt, 311-319
tamper-proof hardware, 157
temperature, 24-26, 32, 36-40, 53, 55, 69
TinyOS, 25-27, 31, 42, 58, 82
Tmote Sky, 26, 27, 28, 40, 45, 56
traffic stream, 343-345
transmission
  power, 102
  reliability, 217
Trust, 158
Trusted Third Parties (TTPs), 160

U

ubiquitous networks, 259
USRP, 363, 373, 379, 381, 384, 387

V

V2R, 343, 345
V2V, 343-345, 348
V2X, 344
VANETs, 343, 344, 359
VDSL, 218
vehicle-to-roadside (V2R), 343
vehicle-to-vehicle (V2V), 343
Vehicular ad hoc networks, 343
Veins, 359
velocity, 344-346, 348, 359
video broadcasting-cable, 218
VoIP, 311, 312, 319

W

warehouse, 37, 38, 40, 41, 43, 44, 46, 55, 69

WAVE, 343
WDNC, 326
Weibull fading channels, 218
weight of a
    CC-edge, 103, 104
    direct edge, 102
wireless communication, 218
    systems, 217
    network, 23, 27, 28, 32, 74, 82, 155
WSN
    wireless sensor network. *Voir*
WSNs, 344

X

XNCC, 326

Z

ZigBee, 26, 30, 32, 40, 45, 55, 56, 58
ZRP (Zone Routing Protocol), 117

www.ingramcontent.com/pod-product-compliance
Lightning Source LLC
Chambersburg PA
CBHW080705220326

41598CB00033B/5315